Lecture Notes on Data Engineering and Communications Technologies

Volume 195

Series Editor

Fatos Xhafa, Technical University of Catalonia, Barcelona, Spain

The aim of the book series is to present cutting edge engineering approaches to data technologies and communications. It will publish latest advances on the engineering task of building and deploying distributed, scalable and reliable data infrastructures and communication systems.

The series will have a prominent applied focus on data technologies and communications with aim to promote the bridging from fundamental research on data science and networking to data engineering and communications that lead to industry products, business knowledge and standardisation.

Indexed by SCOPUS, INSPEC, EI Compendex.

All books published in the series are submitted for consideration in Web of Science.

Andriy Semenov · Iryna Yepifanova ·
Jana Kajanová
Editors

Data-Centric Business
and Applications

Modern Trends in Financial and Innovation
Data Processes 2023. Volume 1

 Springer

Editors
Andriy Semenov ⓘ
Faculty of Information Electronic Systems
Vinnytsia National Technical University
Vinnytsia, Ukraine

Iryna Yepifanova ⓘ
Faculty of Management and Information
Security
Vinnytsia National Technical University
Vinnytsia, Ukraine

Jana Kajanová
Department of Economics and Finance
Comenius University Bratislava
Bratislava, Slovakia

ISSN 2367-4512   ISSN 2367-4520   (electronic)
Lecture Notes on Data Engineering and Communications Technologies
ISBN 978-3-031-54011-0   ISBN 978-3-031-54012-7   (eBook)
https://doi.org/10.1007/978-3-031-54012-7

This Springer imprint is published by the registered company Springer Nature Switzerland AG
The registered company address is: Gewerbestrasse 11, 6330 Cham, Switzerland

Paper in this product is recyclable.

Preface

Several universal problems are emerging in countries that have embarked on the path of developed economy forming. The source of these problems is the system of creating and using effective information communication technologies between innovative companies and funding sources. A large number of innovative companies are emerging in these countries, competing for funding sources. Although these innovative companies work in different areas of the economy, they all have to present their proposals in a form required by funding sources. As a result, there is a need to develop universal methods and communication technologies for passing from databases that characterize an innovative company to databases that provide/create an opportunity for financial structures to compare different companies with each other. This volume summarizes the range of possible development directions for communication technologies between innovative enterprises and financial organizations using the experience of Ukrainian economy as an example.

Consequently, in the preface we provide short outlines of every study included in the volume. Starting with the first chapter titled "Organizational, Financial, and Informational Restructuring in the System of Demonopolization of the Transport Complex of Ukraine in the Conditions of the Knowledge Economy", this study examines and analyzes 25 subjects of natural monopolies of Ukraine, in particular 9 railway transport enterprises, 9 aviation transport enterprises, 5 road transport enterprises. The knowledge economy is considered as an environment for processes of restructuring and demonopolization of the transport complex. Forms, methods, and components of the restructuring of the transport complex are described. Demonopolization of the transport complex is aimed at reducing the negative impact of government monopolies and natural monopolies on the national economy. The priority areas of restructuring in railway transport during the martial law and post-war reconstruction are substantiated. The obtained results can be applied to increase the efficiency of financing utilization in the process of managing innovative processes in Ukraine.

In the next study titled "Modeling of the Key Threats and Identification of the Prospects for Social Security Support in the Sphere of Employment and Labor Payment in Ukraine," the authors extend their prior investigation on employment and labor payment. Using factor models, the study analyzes key threats and prospects for neutralization within Ukraine's labor market. Such factors as informal employment, youth unemployment, average wages, real wage indices, arrears in salary payments, and wage funds were examined. The deliberate exclusion of data from 2019 onward allows a focused analysis of pre-existing threats, acknowledging the challenges posed by the pandemic and the ongoing conflict in Ukraine. The authors emphasize the need for legislative, socio-economic, and managerial measures to address the complexities of the labor market and ensure the protection of social interests in employment and labor payment matters. Continuous analysis is highlighted as essential for timely adjustments in state policies to adapt to evolving social security landscapes.

In the next study titled "Formation of Marketing Competencies in Case of Startups Integration into the Intellectualized Market Space", attention is drawn to the importance of considering the marketing characteristics at all stages of a startup life cycle. The requirements for the formation of databases, which are necessary for effective communication between the startup team and its environment, have been identified. There are four key stages where these databases will be critical to success. The first stage is the generation of an idea, where it is necessary to lay down as many opportunities as possible for the further development of communication technologies. The second stage is the transition from the idea to a team creation and to a real business. This is where limitations of communication databases come into play. The third stage is a startup creation and the start of its activities, in which marketing constraints begin to come to the fore. The fourth stage is business scaling, in which marketing constraints become decisive for communication databases. The article describes in detail the key barriers to communication during integration of startups into the market space of Ukraine and possible marketing measures to eliminate them.

The subsequent chapter "Information Systems for Vehicles Technical Condition Monitoring" explores directions for modification of the existing system of monitoring the technical condition of road transport in Ukraine, which no longer meets the modern requirements for maintaining the vehicle operability. An information system and corresponding databases for monitoring operating conditions of vehicles were formed, a general information support of the system was proposed, and the following issues were described: investigation processes and evaluation of the operating conditions of vehicles; information model of a vehicle position; speed model of vehicle traffic modes. In addition, evaluation tools are presented: the type and condition of the road surface; state of transport infrastructure facilities; condition of an adjacent area of the road; fuel economy of vehicles in operating conditions, as well as an algorithm for adjusting the speed of vehicles. A mathematical model of the information system for evaluating parameters of the vehicle's technical condition was developed. Results of testing the mathematical model demonstrated its high predictive ability.

In the next study "Model for New Innovation Knowledge Spreading in Society" the authors propose to introduce a method for forming a tuple of knowledge as a set of databases from the objective characteristics of the manifestations of human activity (both in its activity and communication with social group). A method is proposed for introducing a metric in this tuple, which allows setting the task of knowledge dissemination as a change in the number of people who own a certain knowledge tuple. The diffusion mechanism of knowledge dissemination and the problem of spreading knowledge from one source (a university) in society is described in detail. A method of identifying the tuple of knowledge possessed by an individual, social group, or society was developed. Application of this method to important processes that accompany the development of society is discussed. It is shown that the obtained results can increase the efficiency of migration process management. A method of determining a set of tuples, which is necessary for the functioning of a developed country, is described. The possibility of applying the obtained results to increase the efficiency of regional development management is demonstrated.

In the next chapter titled "Informational and Digital Business Security in Tourism as a Component of the Coastal Region Competitiveness", the authors proposed to consider the peculiarities of tourist activities in the coastal region in the Smart City concept. The organization of travel agency activities in conditions of the global pandemic is described in detail, which is manifested in changes in the travel company operation (transition to remote service, greater flexibility in changing booking conditions, tours, constant control of conditions for entry and exit from countries, new forms of documents for tourists). The study was conducted using the analysis of digital marketing of travel companies in Zaporizhzhia, Ukraine. Databases for 2020-2021 were used. The authors described directions for considering the specific circumstances of recent years in Ukraine, when the number of electronic documents for entering other countries had increased.

Further chapter "Economic Assessment of Outsourcing of Intellectual and Information Technologies" demonstrates how to form databases that will increase the efficiency of outsourcing in IT companies. Methodological approaches used in Ukraine to analyze the effectiveness of outsourcing in IT companies are analyzed. To modernize such approaches, a new graphic-analytical method was developed, which is to provide a graphical comparative analysis and analytical justification of the effectiveness of decisions on IT outsourcing and implementation of IT functions by the company's forces. This method utilizes a comparison of the economic efficiency of internal and external outsourcing for IT companies. Application of the developed method at many enterprises in an industrial district of the city of Kharkiv, Ukraine, demonstrated its high efficiency. The results of the chapter open up new opportunities for application of databases and information communication technologies for outsourcing development in the field of information technologies in developing countries.

In the next chapter "Digitalization of Consumers' Behavior Model in the Dairy Market," the authors delve into the impact of rapidly changing digital economy on consumer behavior in the dairy market. Recognizing challenges such as elastic demand and diverse pricing approaches, the study aims to devise a consumer behavior model using digital tools. Based on a survey of 450 consumers in Ukraine, findings reveal that taste quality and product naturalness are primary considerations. The study evaluates brand strength using Fichbein's model, highlighting key factors influencing consumer choices. Noteworthy findings include popular places for purchasing dairy products, with supermarkets being predominant. The conclusion stresses the need for leading brands to defend their positions by enhancing competitive advantages and fostering consumer loyalty in the evolving digital landscape of the dairy market. This investigation provides valuable insights for producers navigating the complexities of digital consumer behavior in the dairy industry.

In the next study in "Modelling and Information Support for Assessing the Potential for Increasing the Financial Stability of Enterprises" the authors analyze the potential for increasing the financial stability of an enterprise. The potential for increasing the financial stability of an enterprise characterizes its ability to change the internal environment due to the implementation of relevant organizational, economic, and technological actions, as a result of which an increase in the level of financial stability of the enterprise will be ensured. To carry out an analysis of increasing the financial stability of the enterprise, it is necessary to evaluate the following factors: 1) a potential for increasing the amount of available income that the enterprise can use to repay loans and pay interest, 2) a potential for reducing the amount of loan capital, and 3) a potential for increasing the degree of flexibility of the enterprise. Databases were identified, based on which the potential for increasing the financial stability of the enterprise is assessed and relationships between them are specified. The model was tested on a sample of one hundred enterprises in the western region of Ukraine. Based on the results of the analysis of these enterprises, it was established that the availability of operational databases is crucial for the effective assessment of conditions for increasing the financial stability of the enterprise. An additional value of this study is that it opens up new opportunities for decision-making in regarding the financing of innovative enterprises in developing countries.

The chapter "Transformation of a Regional IT Cluster into a Cross-Border IT Cluster as a Direction of It Business Development Under the Conditions of Negative Influence of External Factors" analyses of the problem of transforming the regional IT cluster (IT cluster Lviv, Ukraine) into the Ukraine-Poland cross-border IT cluster, taking into account the negative impact of external factors. The possible results of integration of the Lviv IT cluster into the EU environment and creation of a cross-border cluster association with the ICT Polska Centralna Klaster are analyzed. The results indicate prospects of creating the cross-border cluster and confirm the possibility of obtaining a positive synergistic effect. Within the framework of the illustrative model, the stability of functioning of combined structures was investigated to determine the level of the synergy effect in the formation of cross-border clusters. The computer simulation of the model showed that the processes of cross-border cluster functioning are very complex and require constant adjustment to maintain it on a stable development trajectory.

In the study titled "Formation of Organizational Change Management Strategies Based on Fuzzy Set Methods", the authors use a method of fuzzy sets to define and justify change management strategies considering the degree of development and financial status of an organization at any moment in time. The model takes into account the factors that influence the company's strategy, and an equation is proposed for identifying the influence of target indicators of the construction enterprise development on the effectiveness of the changes. The method of adjustment, which takes into account the level of financial and market stability of the enterprise and its ability to change, was also studied. This study offers only one vector that allows to structure and formalize the change management process in the direction of forming, justifying, and supporting the company's strategy. It was found that the following enterprise factors influence the change management process: "flexibility of the management system", "competitiveness", "operational efficiency", and "adaptability". Modeling was carried out based on data from a sample of construction enterprises in Ukraine during 2007-2020. The proposed approach can be used to analyze change management strategies of companies operating in turbulent economic conditions and instability of the country's legal field.

In the subsequent chapter "Modeling the Level of Implementation of BIM by Enterprises as a Means of Optimizing the Cost" the authors argue that today's construction companies need to get additional competitive advantages, which can become new methods of processing databases and information modeling, such as building information modeling (BIM). In Ukraine, BIM must take into account the high level of uncertainty during project implementation. Therefore, to determine the expediency of BIM implementation by subcontractors, the authors proposed a new method of using simulation modeling technology, which takes into account the main barriers to the implementation of BIM models in practical activities in Ukraine. In addition to the imperfection of national standards or their absence, there is a lack of both demand from customers and own experience in using these technologies. The need for training and its high cost, the high cost of BIM technologies, and the small size of projects are also a barrier. It is proposed to base the calculations on a criteria of profitability of the capital invested in the construction project - ROI (Return On Investment), the assessment of changes which as a result of the implementation of BIM is provided using simulation modeling. It is emphasized that the simulation model should use Markov chains at a certain stage of calculations.

In the forthcoming chapter, "Digital Promotion as Innovative Business Management Technologies of Retail Chains", the authors explore the pivotal role of comprehensive marketing knowledge in the competitive landscape of the Romanian retail market. Focused on grocery and consumer goods, the study aims to develop and evaluate a digital promotion complex using platforms like Facebook, Google, and Instagram. By analyzing statistical information on internet usage, the authors propose tailored strategies to engage a wider audience, enhance customer loyalty, and attract new customers. Discussions and conclusions highlight the effectiveness of digital promotion in increasing brand awareness and fostering a loyal customer base. The study emphasizes the importance of strategic and integrated approaches, considering the specifics of the Romanian market, internet users, and communication channels. Evaluation metrics include conversion rates, ROAS, CPA, CPC, CPM, and CTR. Overall, the study reveals that a well-structured digital promotional strategy is profitable and efficient for retail companies in Romania, providing valuable insights for businesses aiming to navigate and thrive in the dynamic Romanian market.

In the chapter "Support for the Development of Educational Programs with Graph Database Technology", the authors tackle challenges in higher education by advocating for modern approaches to align educational programs with standards. Proposing the use of graph database technology, specifically Neo4j, the authors detail its application in modeling connections between educational components and competencies. Discussions highlight the significance of well-organized educational programs and the flexibility of the developed database, allowing users to make informed decisions for program improvement. Despite acknowledging limitations like the need for user knowledge in query languages, the authors emphasize the broad applications of graph database technology in educational digitization. The article encourages further research, suggesting potential directions such as knowledge graph modeling and text mining for competency compliance. Additionally, the integration of the developed graph database with text processing systems and program formation support systems is identified as a key area for future exploration.

The study "Information Technology for Identifying Hate Speech in Online Communication Based on Machine Learning" presents a robust approach to combat hate speech using machine learning. It covers data collection, preprocessing, labeling, and the evaluation of various classifiers like KNN, Naive Bayes, Decision Tree, Logistic Regression, and Random Forest based on accuracy and ROC AUC metrics. The conclusions highlight the system's effectiveness, with Logistic Regression and Random Forest as standout classifiers. The article emphasizes practical applications, including enhancing online safety, aiding content moderation, supporting human rights policies, promoting digital well-being, and raising awareness through education. It underscores the novelty and practical value in understanding hate speech dynamics and suggests future research areas such as advanced machine learning models, multilingual hate speech detection, contextual analysis, real-time streaming data processing, user feedback integration, transfer learning, ethical considerations, and multimodal approaches to enhance accuracy and ethical considerations in hate speech detection systems.

In the chapter "Method for Counting Animals in Motion for the Milking Plant Information Systems", the authors provided basic information about radio frequency identification (RFID) systems. Systems of radio frequency identification of animals are widely used on modern dairy farms as part of information systems designed to determine parameters of the technological process of milk production. The basic operation principle for such information systems is described in detail. An important factor for obtaining reliable information about the parameters of the technological process of milk production in the information systems of group milking facilities is the accurate accounting of animals during their movement to stalls. The method proposed by the authors increases the reliability of animal passage identification in comparison with existing methods. This ensures an increase in the accuracy of animal counting. The authors took into account the impact of errors in radio frequency identification of animals on obtaining information about milking parameters when using information systems of group milking units. As a result of the conducted experiments, detection of unidentified animals at group milking facilities increased from 60–70% to 80–90%.

In the last chapter "Cryptocurrency as a Tool for Attracting Investment and Ensuring the Strategic Development of the Bioenergy Potential of Processing Enterprises in Ukraine", a profound investigation unfolds into the prospects and challenges of employing cryptocurrencies within the agricultural sector, particularly in processing enterprises. Cryptocurrency has emerged as a financial instrument and as a transformative innovation in different sectors around the world. Its potential to revolutionize traditional financing mechanisms and stimulate economic growth is under intense investigation. A comprehensive study of the prospects and challenges of using cryptocurrencies in the agricultural sector, particularly in processing enterprises, is performed. The chapter ends with a detailed analysis using regression models and correlation studies, revealing the complex relationships between investment, green tariffs, cryptocurrency production, electricity generation, and computer power in biomineralization. These statistical analyses support the potential benefits and economic impact of incorporating biomine technology into processing enterprises. Finally, this chapter is an invaluable contribution to the discussion on integration of cryptocurrencies into agriculture. It is a beacon of innovative financial mechanisms that unlock the hidden potential of the bioenergy sector within agricultural processing companies and ultimately pave the way for sustainable and technologically advanced economic growth. This chapter is a testimony to the authors' dedication, rigorous research, and innovative insights, providing a compass for traversing the unknown waters of cryptocurrency integration in Ukraine's agricultural landscape.

Vinnytsia, Ukraine   Andriy Semenov
                                                    semenov.a.o@vntu.edu.ua

Vinnytsia, Ukraine   Iryna Yepifanova
                                                    yepifanova@vntu.edu.ua

Bratislava, Slovakia   Jana Kajanová
                                                jana.kajanova@fm.uniba.sk

Contents

Organizational, Financial, and Informational Restructuring in the System of Demonopolization of the Transport Complex of Ukraine in the Conditions of the Knowledge Economy

Tetiana Kosova⬤, Serhii Smerichevskyi⬤, Oksana Yaroshevska⬤, Oleksii Mykhalchenko⬤, and Larisa Raicheva⬤

Abstract The restructuring of the transport complex is defined as a change in its structure due to the optimization and rationalization of its organizational management scheme, the movement of material, financial, and information flows in order to ensure the competitiveness of transport enterprises and the demonopolization of the market of transport services based on the knowledge economy. The forms, methods, and components of the restructuring of the transport complex have been highlighted. A description of the content and purpose of organizational, financial, and informational restructuring has been given. Demonopolization of the transport complex is defined as a strategic goal of its development, aimed at reducing the negative impact of monopoly power and natural monopoly segments on the national economy. The distribution of spheres of functioning of the transport complex according to the level of development of competition has been substantiated in order to choose the appropriate methods of state regulation aimed at demonopolization. The knowledge economy is considered as an environment of the processes of restructuring and demonopolization of the transport complex; it has been proved that innovation and investment activities and modernization should be their integral component, aimed at reducing operating costs, improving the quality of customer service and transport safety. Requirements for the development of an investment declaration and supporting documents have been proposed. The conceptual principles of carrying out organizational, financial, and informational restructuring in railway transport based on the integration of the functional object system and the vertical model, tariff differentiation, the spread of two-part tariffs, the elimination of cross-subsidization,

T. Kosova (✉) · S. Smerichevskyi · O. Yaroshevska · O. Mykhalchenko
National Aviation University, 1 Liubomyra Huzara Ave, Kyiv 03058, Ukraine
e-mail: t_d_kosova@ukr.net

L. Raicheva
International Humanitarian University, 33 Fontans'ka Rd, Odesa 65000, Ukraine

1

the creation of new market segments, the expansion of the range of transport and related services have been substantiated. The priority areas of restructuring in railway transport in the conditions of martial law and post-war reconstruction have been substantiated, including within the framework of the concept of socially responsible restructuring.

Keywords Organizational · Financial · Informational · Restructuring · System · Demonopolization · Transport complex · Ukraine · Knowledge economy

1 Introduction

The features of modern national economies are the wave-like character of development, financial instability, significant influence of the knowledge economy, and innovative and informational development incentives. The processes of transformation of digital technologies from an integrative factor into a destructive one, the emergence of direct and indirect confrontation of national economies, the unity and struggle of the processes of economic concentration and deconcentration, social differentiation, and democratic equalization of population incomes are controversial [1]. The establishment of steady-state conditions of national economies and their separate industry complexes is determined by the nature of the interaction of demand and supply for goods, products, goods, services, the degree of their differentiation, the state of the competitive environment, etc. [2]. It is provided by the presence of mechanisms of healthy economic competition between many competing entities, represented by both small enterprises and large monopolies. The high level of monopolization of the national economy has a negative impact on the economic, social, and political life of the country, which needs to be eliminated by reducing the amount of monopoly power of individual subjects and demonopolization of the domestic economy in general.

The transport complex of Ukraine is the basis of the infrastructure of the national economy, a complex entity that provides both mass and differentiated services and is represented by segments that differ in the nature of the interaction between demand and supply: monopolistic, oligopoly, and competitive ones. The main components of the transport complex are railway, road, pipeline, sea and river, aviation transport, etc. A special segment of the monopolistic segment in transport is the natural monopoly inherent in railway transport, pipeline transport, and maintenance of transport terminals, ports, airports, etc.

The restructuring of any economic complex is a systemic transformation that changes the conditions of the internal and external environment of the functioning of its subjects. Each means of transport has its own specific areas of restructuring and general ones in those spheres that are naturally monopolistic or compete with each other. Being a complex economic phenomenon, restructuring has various aspects (namely, organizational, financial, and information aspects); it is oriented towards demonopolization and is based on innovative mechanisms in the system of the transport complex of Ukraine; that is, it is implemented in the environment of the knowledge economy. All of the above determines the relevance of the research topic devoted to organizational, financial, and informational restructuring in the system of demonopolization of Ukraine's transport complex in the conditions of the knowledge economy.

2 Literature Review

Issues of organizational, financial, and informational restructuring, and demonopolization in the conditions of the knowledge economy are in the center of attention of many scientists. Gubíniová K., Štarchoň P., Vilčeková L., Bartáková G.P., Brtková J. consider marketing communications as a tool of information restructuring [3]. Panevski D., Peráček T., Rentková K. analyze the implementation of various mechanisms of the law and models of information systems, which, on the one hand, ensure efficiency and ease of access to the necessary information, and, on the other hand, guarantee financial security and protection against financial intelligence [4]. Kiß H., Sulíková R. investigate the impact of the knowledge economy on the development of personnel and the training of qualified leaders who determine the success or failure of the organization [5]. Busse V., Strauss C. develop crowdfunding mechanisms as an alternative to financing startups and projects based on the collective cooperation of people (donors) who voluntarily pool their money or other resources [6]. Vančová M., Ivanochko, I. investigate the mechanisms of organizational restructuring of business processes of multinational IT companies based on long-term innovativeness [7]. In the article of Poplavska Z., Komarynets S. a model of the influence of environmental instability on the flexibility and performance of the enterprise based on organizational restructuring is proposed [8]. Kuzmin O., Ovcharuk V., Zhezhukha V. consider restructuring as a tool to help achieve the established goals of the company ("ends/means" principle) in the context of transposed projections of the system of balance indicators and administration systems in enterprise management [9]. Kuzmin O., Zhezhukha V., Gorodyska N. investigate the implementation of engineering projects as a tool for organizational restructuring of the enterprise under three potential conditions: uncertainty, certainty and risk [10].

Investigating modern aspects of the restructuring of transport enterprises, Vlasenko D. O. offers an adaptive approach to the development of management decisions in the conditions of an unstable, fast-moving market environment, to adapting transport enterprises to the conditions of market competition [11]. Evaluating possible options for the organizational restructuring of railway transport at the industry level, Eitutis H. D., Bozhok A. R. offer a German version of successive improvement of the territorial-functional management structure based on the integration of the functional-object system and the vertical model [12].

Natural monopolies are a key factor for structural changes and economic development in infrastructure industries in various countries around the world. Their regulation should be flexible and correspond to the stage of development of the industry, the technological and institutional changes taking place in it. Lysytsia K. S. has summarized the model of regulation of natural monopolies within the framework of the European Union, which provides for the separation of natural monopoly rights from potentially competitive types of activities and the privatization of the latter, accounting for the complexity of the transition to a competitive environment and the synthesis of two regulatory policies (the policy of introducing competitive mechanisms and state regulation) [13]. The work by Pozniakova O. examines the problem of the creation of hidden monopolies as a result of the concentration of ownership through the institution of trust management, which complicates the information base about the ultimate beneficial owners since the existing property registers focus on the right to own property and change the owner [14]. To solve the problem, it is proposed to create a General Property Register, which will contain information about property relations in their dynamics, not only taking into account the right of ownership but also use and disposal, that is, about the final beneficiaries.

In his article, Nikolayev Y. analyzes the advantages of using a cluster approach to the development of the transport complex, and proposed a strategy for the reconstruction of the Illichivsk Sea Trade Port as a full-fledged element of the Odesa transport cluster [15]. Zorina O. I. summarizes a variety of quantitative indicators for assessing the degree of monopolization of markets within the established product and geographic boundaries in global and domestic practice [16], which are used to assess the degree of monopolization of railway transport in the field of passenger transportation.

Shemaiev V. V. emphasizes that the investment policy of natural monopoly entities in the infrastructure sector has specific features compared to other economic entities, since the objective function, namely, maximizing the shareholder's profit, has certain limitations and obligations: in railway transport, there are social commitments to subsidize the cost of passenger and suburban transportation or transportation of privileged categories of citizens; on sea transport, these are the requirement to provide capital investments taking into account the observance of equal business conditions for port operators; air transport should meet the requirements for ensuring safe air navigation service or access to airport infrastructure [17]. We agree with Shyrokova O. M. [18] that the key factor in making a decision to carry out the restructuring of a transport and infrastructure enterprise is the growth of the economic efficiency of its operation.

Appreciating the scientific output of the mentioned authors, it should be recognized that they focus on certain aspects of the investigated problems of the functioning of the transport complex of Ukraine, however, further development is needed to clarify the interrelationships of various forms of restructuring, demonopolization, knowledge economy, etc.

3 Materials and Methods

The basis of the conducted research is the dialectical method and a systemic approach. The transport complex of Ukraine is considered as a system, the sub-systems of which are separate types of transport and related types of activities, as well as spheres differing in the degree of development of competition. In turn, the transport complex is interpreted as a subsystem of the national economy, which has close cooperative ties with business entities of other types of economic activity. From the standpoint of the dialectical method, the knowledge economy is the environment for the development of the researched processes, demonopolization is the target of the development of the transport complex, and restructuring is the tool for its achievement. The information base of the research is the regulatory and legal base of Ukraine, current scientific publications of leading foreign and domestic scientists, and official data of the Antimonopoly Committee of Ukraine.

4 Results

In Ukraine, the vast majority of transport and infrastructure enterprises are formed according to the holding (network) principle, when the parent company manages subsidiaries with separate balance sheets or there is a single corporate center and a network of branches in the form of separate structural units in different regions. A significant segment of the transport complex of Ukraine is represented by natural monopolies, the existence of which is due to the following factors: the natural rights of the monopolist, the interests of the economic security of the state, and greater efficiency of satisfaction in the absence of competition. Unlike artificial monopolies created on the basis of legislative acts, a natural monopoly exists objectively, indifferent to the efforts of the state and legal norms. Despite their important role in the economy of any type, natural monopolies distort the effect of market laws of pricing, and it is advisable to limit them in the interest of increasing the efficiency of social reproduction. If the elimination of the natural monopoly segment is impossible or difficult, then it should become an object of effective state management.

As of June 30, 2022, there were 398 enterprises in the consolidated list of subjects of natural monopolies of Ukraine, including 25 of them are in the field of transport infrastructure [19], including the following business entities (including state enterprises):

- 9 railway transport enterprises: JSC "Ukrainian Railways" (Kyiv), State territorial and sectoral association "South-Western Railway" (Kyiv), State territorial and sectoral association "Lviv Railways" (Lviv), Production unit "Poltava Locomotive Depot" of the Regional branch office "Southern Railway" of JSC "Ukrainian Railways" (Poltava), Production unit "Poltava Construction-Installation Operation Department" of the Regional branch office "Southern Railway" of JSC "Ukrainian Railways" (Poltava), Separated "Hrebinka Train Depot" of the Regional branch office "Southern Railway" of JSC "Ukrainian Railways" (Hrebinka), the Production Structural Subdivision "Kremenchutsk Territorial Administration" of the branch "Center for construction and installation works and operation of buildings and structures" of JSC "Ukrainian Railways" (Hrebinka), SE "Ukrainian State Center for Railway Refrigerated Transport Ukrreftrans" (Kyiv Region, Fastiv), SE "Management of Industrial Enterprises of the State Administration of Railway Transport of Ukraine" (Kyiv);

- 9 aviation transport enterprises: Regional structural branch of the Kyiv District Center "Kyivcentraero" of the Air Traffic Service of Ukraine (Kyiv Region, Boryspil District, Gora Village); ME "Kyiv International Airport" (Zhuliany), ME "Boryspil International Airport" (Boryspil), LLC "Odesa International Airport" (Odesa), SE "Konotop Aircraft Repair Plant (AVIAKON)" (Konotop), LLC "Master-Avia" (Kyiv), SE "Lviv State Aviation Repair Plant" (Lviv), SE Ministry of Defense, Odesa Aircraft Repair Company "Odesaviaremservice" (Odesa), Special Aviation Detachment of the Operational Rescue Service of the Civil Defense of the Ministry of Emergency Situations (Nizhyn);
- 5 road transport enterprise: SE "Kyiv Regional Road Administration" of the State Joint Stock Company "Automobile Roads of Ukraine" (Kyiv), Branch "Holoprystan road operation section" of SE "Kherson Avtodor" of JSC "Automobile Roads of Ukraine" (Gola Prystan city), SE "Zhytomyr Oblavtodor" of JSC "Automobile Roads of Ukraine" (Zhytomyr), PJSC "Uzhhorod Automobile Transport Enterprise 12107" (Uzhhorod), Public JSC "Odesavtotrans" (Odesa);
- 1 pipeline transport enterprise: LLC "Operator of the gas transport system of Ukraine" (Kyiv);
- 1 sea transport enterprise: SE "Ukrainian Sea Ports Authority", Illichiv Branch (Kyiv).

As a natural monopoly, the railway industry of Ukraine was formed on the basis of a centralized and concentrated form of state ownership in the bowels of the administrative command system of the Soviet Union. The monopolist is the national carrier of cargo and passengers Joint-Stock Company "Ukrainian Railways", established at the end of 2015 as the legal successor of all rights and obligations of the State Administration of Railway Transport of Ukraine, as well as subordinate enterprises and institutions that had the status of separate legal entities. Thus, the state structure was transformed into a joint stock company, 100% of whose shares belong to the state. The share of "Ukrainian Railways" in transportation by all modes of transport is: more than 75% in terms of cargo turnover, almost half of the total volume of passenger turnover. A significant share of the transport services market defines the important role of "Ukrainian Railways" as a large taxpayer. The legal basis for the functioning of JSC "Ukrainian Railways" is the Laws of Ukraine "On Railway Transport" [20]; "On the peculiarities of the formation of a joint-stock company of public railway transport" [21].

SE "Air Traffic Services" performs the functions of ensuring high-quality and safe air navigation services in the airspace on the market of air transport services in

Ukraine and in Europe. The legal basis of activity is the Chicago Convention on International Civil Aviation, the ICAO Manual on Air Navigation Services Economics, and the Air Code of Ukraine [22].

SE "Administration of Sea Ports of Ukraine" is designed to promote the development of the maritime transport infrastructure of Ukraine and increase the competitiveness of Ukrainian sea ports in the Azov and Black Sea basins. The sphere of management is the economic activity of marine terminals and maintenance of loading and unloading works in the port for export-import operations of enterprises. The legal basis of activity is the Law of Ukraine "On Sea Ports of Ukraine" [23].

LLC "Operator GTS of Ukraine" is a natural monopoly, founded by JSC "The Main Gas Pipelines of Ukraine", which provides natural gas transportation to consumers of Ukraine and the countries of the European Union. Created in 2019, since January 1, 2020, it is a certified operator of the gas transportation system of Ukraine and is completely independent of vertically integrated enterprises. The mission of LLC "Operator GTS of Ukraine" is to ensure the development of a competitive, transparent, and non-discriminatory gas market, a reliable gas supply for Ukrainian and European consumers. The legal basis is the Law of Ukraine "On Pipeline Transport" [24].

State transport enterprises have special relations with the budget system, which are regulated by the Procedure for deducting part of the net profit (income) to the state budget by state unitary enterprises and their associations [25]. The part of the net profit (income), which is allocated to the state budget based on the results of financial and economic activity in the first half of 2022, is determined in the amount of 30% for the following state enterprises in the field of transport and related industries:

- river and sea transport: Administration of River Ports, Administration of Sea Ports of Ukraine, "Ukrvodshlyakh", Reni Commercial Seaport, Mykolaiv Sea Trade Port, Odesa Sea Trade Port, Chornomorsk Sea Trade Port, Izmail Commercial Seaport, Yuzhny Sea Trade Port. Unfortunately, the Mariupol Sea Trade Port and the Berdyan Sea Trade Port, which are included in this list, are located in the temporarily occupied territories;
- air transport: Lviv Danylo Halytskyi International Airport, Boryspil International Airport;
- electric transport: Kyiv state regional technical inspection of city electric transport;
- project and service enterprises: "Ukrservice of the Ministry of Transport", "State Motor Vehicle Research Institute", Ukrainian State Enterprise for Maintenance of Foreign and Domestic Vehicles "Ukrinteravtoservice".

A common feature of state regulation of natural monopolies in the CEE countries is the creation of its combined mechanism, which uses both traditional and modern methods. The traditional ones include: the creation of a special institutional structure of regulatory bodies, a regulatory framework, the use of special state rights ("golden share"), and various methods of building the organizational structure of an enterprise (auctioning, holding, corporatization). Modern methods are: the use of the stock market as a regulator of the company's behavior, the creation of a special institutional structure of shareholders ("solid cores", strategic investors), a partnership of the state and private sectors; foreign investment, mergers, and acquisitions.

Demonopolization and limitation of the negative impact of natural monopolies on the national economy is a strategic goal of the development of the transport complex of Ukraine, the tool for achieving which is restructuring. The main reasons and motivations for the restructuring of the transport complex of Ukraine are the ineffectiveness of the existing organizational, financial, and informational structure, the external symptoms of which are:

- at the micro-level, inconsistency of the quality of transport services with growing consumer demands and international standards; high physical and moral depreciation of fixed assets, insufficient pace of their renewal and modernization; low technological conditions of the material and technical base and transport infrastructure;
- at the meso-level, low investment attractiveness of the industry; slow improvement of transport technologies and their insufficient connection with production, trade, warehouse, and customs technologies; insufficient level of employee motivation and low wages;
- at the macro-level, an insufficient level of using the advantages of Ukraine's geopolitical position and the possibilities of its high international transit potential; low efficiency of state management and regulation in the transport complex, weakness of state support for its innovative and investment development.

The processes of demonopolization of transport enterprises are based on the transition from uniform to differentiated tariffs without changing the traffic volumes, which will ensure the improvement of financial results in the case of lower individual costs compared to industry averages. Under the conditions of tariff differentiation, the total profit of the enterprises of the transport complex may decrease compared to the single tariff, which is based on the costs of the enterprises exceeding the industry average. One of the forms of a differentiated tariff is a two-part tariff, which consists of a permanent part, which is independent of the volume of cargo and passenger transportation, and a variable part, dependent on the specified volume. State antimonopoly regulation should ensure symmetrical distribution of information between providers of transport services and their consumers in terms of setting competitive transport tariffs.

The components of the restructuring of the transport complex are:

- organizational restructuring: changing the forms of ownership of business entities and their organizational and legal status; unbundling of enterprises and reduction of their concentration; changing the forms and methods of production management, the production and technological structure of the transport complex, including due to its modernization; reengineering of business processes of the provision of transport services and the format of relations with counterparties and the state; development of entrepreneurial management methods; market-oriented personnel training; implementation of modern methods of personnel management and schemes for increasing work motivation; formation of corporate culture and improvement of the quality of corporate management;

- financial restructuring: the corporatization of enterprises, development of prospectuses for the issue of stocks and bonds; use of fund financing mechanisms; attraction of foreign capital through the creation of joint ventures; development of forms and methods of interaction of business entities with banking and non-banking financial and credit institutions; definition of the procedure for repayment of problematic receivables and payables that arose due to the crisis of non-payments in the industry; registration of indebtedness by promissory notes; use of factoring schemes; capital structure optimization; streamlining the allocation of budget grants and subsidies by establishing clear criteria; control of the legality and effectiveness of their use; growth of financial autonomy of industry enterprises;
- information restructuring: ensuring rapid and relevant exchange of information between state administration bodies and business entities, between buyers and sellers of transport services, which contribute to the transparency and openness of the use of budget funds, formation of transport tariffs, elimination of consumer discrimination in terms of pricing; transparency of information about the results of operational, financial, investment activities, about the state of corporate management; timely publication of the company's financial and non-financial statements, management report, etc.

In the market of transport services, there is an exchange of information between buyers and sellers, which should be relevant and contribute to the growth of general well-being. Unreliable disclosure or distortion of information can lead to a reduction in the welfare of consumers and their price discrimination due to the abuse of the monopoly position of sellers, the processes of mergers and acquisitions, vertical or horizontal integration, high spending, outdated technologies for the provision of transport services and their sales. Transport companies should set differentiated tariffs for their services in conditions of their limited number, ensuring the effectiveness of their pricing policy not by the criterion of increasing the company's profit, but by the impact of this policy on changing the welfare of both consumers and producers, that is, public welfare in general.

The features of a natural monopoly on railway transport of Ukraine are: the network nature of the organization of the infrastructure and the provision of transport services (the system of transport lines as a single integrated complex, a single control center in on-line mode); the presence of barriers for new business entities to enter the rail transportation market, including due to the presence of numerous regulatory and control bodies; the need for significant investments to create an alternative transport business; low investment attractiveness for private capital due to a long payback period, high risks of implementing investment projects, a long period of depreciation of fixed assets; a significant volume of fixed costs, which determines efficiency under the condition of growing effect of scale; specificity of material assets (railway tracks, stations, rolling stock) and their narrow specialization; low elastic demand from consumers for transport services that satisfy the primary needs for the transportation of passengers and cargo in space and cannot be limited. The services of subjects of natural monopolies have no substitutes in the transport market, so their demand is less dependent on price and tariff fluctuations compared to other services. The most monopolized segment of railway transport is the long-distance transportation of raw materials, when the consumer has no acceptable alternatives for their replacement and is forced to pay monopolistically established tariffs for transport services, the demand for which is inelastic at the price of transportation [26].

The systematization of the features of a natural monopoly allows to highlight its advantages as a form of railway transport: large-scale investments for the development of the infrastructure network, reduction of operational costs due to the scale effect, increased opportunities for concentration and redistribution of financial resources for the implementation of significant capital investments, the introduction of innovations, risk allocation, etc. The specified advantages and high social significance require the effectiveness of state regulation of the natural monopoly segment of railway transport, both in the sphere of the public and private sectors. Methods of state regulation are divided into direct and indirect ones. Direct methods include: creation of special authorized regulatory bodies; direct management of enterprises by the state (municipality) as the owner; competitive access to the market (or regulation through contractual relations); establishment of regulated tariffs; control of the quality of services and the level of prices by state and public bodies for the protection of consumer rights, regional authorities and trade unions. Direct methods are state support in the form of grants and subsidies, budget financing, budget lending for the development and modernization of transport and communication infrastructure, etc. The purpose of using direct methods is to balance the economy of railway transport, and indirect methods are to maintain its equilibrium state.

The objects of state regulation of natural monopolies in railway transport are: existence of a development strategy; technical and technological licensing; level and structure of tariffs; freight volume; rate of return; market entry and exit; mergers and acquisitions; quality of transport and related services; compliance with the safety and environmental requirements of railway transportation; ownership structure of subjects of natural monopolies; methods of accounting and formation of financial reporting; ensuring commercial and financial transparency [27].

The guarantee of the effective state regulation of natural monopolies in the interest of stimulating the country's economic development is: ensuring the balance of interests of all interested parties, including the state, minimizing the rent of business entities from their own monopoly position, including by improving the efficiency of transport management in the electronic logistics market (Vincze et al. [28]). The main tasks of reducing the negative impact of natural monopolies on the development of railway transport [29] are:

- control of their monopoly position and protection of the interests of consumers of railway transport services;
- provision of full coverage of economically justified costs;
- creation of investment resources for expanded reproduction and modernization of the material and technical base;
- regulation of prices and tariff policy regarding transport and related services;

- state control over compliance with established tariffs for transport and compound services;
- quality control of transport and related services at economically reasonable prices;
- monitoring of the transport and related services market conditions;
- creation of conditions for the development of competition in order to remove the commodity market from the state of natural monopoly;
- limiting the influence of monopolists on state policy;
- creation of conditions for the sustainable development of state monopolies.

A retrospective analysis of the development of railway transport in the vast majority of countries of the world, except for the USA and Canada, shows the growing tendencies towards concentration, merger, and absorption of corporations, and nationalization of private property, which led to the emergence of natural monopolies and the spread of state management in the industry [30]. The main disadvantages of the development of railway transport are [31]: considerable state control while limiting budget funding, reduction in the volume of cargo and passenger transportation, lack of budget compensation for the transportation of preferential categories of citizens and cargo, inefficient management due to high operating costs, a high level of monopoly power and its abuse, loss of competitive positions in favor of road transport in the market of suburban passenger and small cargo transportation [32].

In Western Europe and the countries of Central and Eastern Europe (CEE), the goal of reforms in railway transport is to separate national railway companies from the ministries of transport and to ensure open access to the infrastructure for foreign railway companies. In the CEE countries, a "transitional" model has been created, which ensures vertical integration with the participation of the dominant company and allows competition to be created through horizontal branches, which are regional railways. JSC "Ukrainian Railways" has a functional-object organizational structure of management, while the objects are: transportation, finance and economy, tariffs, equipment and technologies, social sphere, resources, investments, and foreign economic activity. Demonopolization of its activities will be facilitated by the following measures: division of branches that belong to the competitive sector of the economy from its composition; creation of passenger transportation operating companies, including their entry into the market with their own fleet of locomotives; creation of regional infrastructure maintenance enterprises (RIME) and regional transportation management centers (RTMC), which should be vertically subordinated to JSC "Ukrainian Railways". The sphere of activity of RIME is: tracking distances, power supply, signaling and communication, construction-installation and regulatory works, information and computing centers, and scheduled preventive and emergency repairs of infrastructure facilities. Management of the specified objects is carried out through sectoral and functional services. RTMC manages the technological process of train movement organization through the dispatching apparatus of all railway stations. Organizational restructuring of JSC "Ukrainian Railways" will allow for a harmonious combination of branch and regional forms of management; to eliminate duplicate functions of railways and their directorates, to create regional enterprises based on them, to increase the safety of train traffic and the quality of passenger service due to effective management.

The general directions of restructuring of railway transport are: renewal and modernization of transport infrastructure and rolling stock; introduction of the latest technologies for the provision and sale of transport services, development of new types of services of transport enterprises, capture of new sectors of the transport market, redistribution of spheres of influence in the existing market, increase in the volume of cargo and passenger transportation, their profitability, improvement of existing international transport corridors and creation of new ones, integration of domestic transport enterprises to the European transport system.

A specific direction of the restructuring of railway transport is the separation of potentially competitive segments from natural monopolies through the limitation of the conditions for raising tariffs, the need to join the single European transport space and the fulfillment of the requirements of the Directive 2008/57/EC of the European Parliament on the interoperability of the railway system within the Community [33]. In European countries, there are three main models of the functioning of railway transport: the first is the separation of operators of infrastructure and railway transport as independent legal entities with separate accounting and reporting; the second is the creation of a single business entity, which is a legal entity subordinating independent subsidiary operating companies; the third is a holding model which is a synthesis of the two previous models. The choice of outlined models of railway transport management by European countries is free despite institutional factors [34]: technical and technological, political, and economic ones. However, a mandatory condition for their application is organizational and financial restructuring in the following forms: corporatization through transformation into a joint-stock company or a holding joint-stock company; separation of functions of state administration and economic activity; establishment of separate accounting of income and expenses for freight and passenger transportation; creation of regional railway transport companies based on one of the types of transportation; commercialization of their activities and the development of a competitive segment in the market of railway transport services. State-private partnership is one of the tools for financing infrastructure projects in the fields of natural monopolies, its common types are: sale of assets (sale of a unit, subsidiary, package of securities or other asset for accumulation of financial resources or reinvestment in new projects), concession (granting the right to use some of its functions to a private legal entity by the state to achieve public goals), a project "from scratch", contracts for management or leasing.

To ensure the legality and efficiency of the use of funds in investment activities, subjects of natural monopolies and state enterprises of the transport and infrastructure sector should develop an investment policy. We suggest that the following documents should be drafted for it: main document (investment declaration) and auxiliary documents (firstly, methodological recommendations for assessing the financial and economic efficiency of investment projects, which contain the procedure for calculating indicators and their evaluation criteria; secondly, the regulation of investment management, which should take into account requirements of the Law of

Ukraine "On Management of State Property Objects" [35]). The investment declaration should become a program document for the development of the transport and infrastructure enterprise, which should reveal the following basic provisions: the types and purpose of investment activity, its role in achieving medium-term and long-term goals, the main objects of real and financial investments, the content of the investment management process for capital and financial investments, the principles of forming an investment portfolio of projects, fixing investment projects by corporate center (parent company) and business units (subsidiaries), communication between them and the distribution of responsibilities. The purpose of investment activities of transport and infrastructure enterprises in the field of demonopolization should be to ensure the competitiveness of infrastructure facilities (river and sea ports, airports, railway stations, etc.), to achieve the payback of investment projects taking into account the factors of time, risks and inflation, to carry out organizational and financial restructuring with in order to reduce operating costs [36]. The principles of formation and implementation of the investment policy of transport and infrastructure enterprises in the system of their demonopolization are: obtaining funds for budget financing on a competitive basis; targeted use of the investment component of tariffs and fees; prohibition of discrimination in the access to financial credit and stock market resources; achievement of medium and long-term goals of the enterprise recorded in the investment declaration; application of a unified approach to the investment management process; an alternative approach to the development of investment projects in order to select the most effective ones; priority of started investment projects before starting new ones; application of state-private partnership; making capital investments in accordance with the needs of national security and defense, as well as the requirements of the Law of Ukraine "On Priority Areas of Innovative Activity in Ukraine" [37]; transparency and publicity of investment activities. The main types of investment activities of transport and infrastructure enterprises in the process of organizational and financial restructuring are: capital investments; lease or concession of assets; joint activity agreements; portfolio investment in shares, bonds, other financial instruments; investment and other types of contracts with accepted investment obligations; state-private partnership.

The restructuring of a transport enterprise can be defined as a process of directed evolutionary change of the current state in order to achieve the new qualitative state, which is determined by its financial and economic state and the availability of resources for the implementation of relevant measures and will ensure flexibility, mobility and reliability of functioning, the necessary preservation of the quality and competitiveness of transport logistics services. Restructuring should have an economic justification and a plan that is developed according to the following stages:

1. Clarification of adopted operational and financial strategies and goals on the basis of determining the reasons for restructuring and the consequences of its failure. New strategies, goals, sub-goals, and criteria for the degree of their achievement should be formulated.

2. A SWOT analysis, in which the strengths and weaknesses of the transport enterprise, favorable opportunities and threats in terms of assessing the market potential, transportation opportunities, and human, material, and financial resources are analyzed.
3. Formation of the "problem field" of the transport enterprise, identification of key problems, development of alternative options for restructuring, assessment of their consequences, selection of the most rational option according to the criteria characterizing the degree of achievement of the set goal, clarification of the desired terms of achievement of the goal.
4. Development of a program for the implementation of the selected option of the restructuring of the transport enterprise, which details the selected priority areas of activity, a list of specific works and responsible performers, expected results, deadlines, necessary human, material, and financial resources; systems for managing the progress of work implementation and motivation of performers.

Along with other natural monopolies, railway transport has a weighty economic, social, political, and defense significance for the national economy. The economic condition of a large number of business entities, their competitiveness in the domestic and foreign markets, and the costs of transporting passengers directly affect the financial condition of households, as well as social stability in the state, depending on the costs of transporting goods. Given the monopoly position of railway transport, this branch of the national economy is subject to state regulation.

Let us consider the conceptual foundations of the organization of the process of financial restructuring of railway transport enterprises as the basis of their strategic development [38]. At the preparatory stage, the purpose of financial restructuring, the optimal timing of its implementation, the sequence of implementation of individual measures, ways of adapting to the external environment, the required volume of investments, and the optimal source of financing are determined. The need for financial restructuring of railway transport enterprises is determined by the following factors: cross-subsidization of passenger transportation at the expense of cargo transportation; debt to the state for budget compensation for the transportation of preferential categories of passengers; artificially raising tariffs for business entities for freight transportation, which leads to an increase in the full cost of finished products and goods and distorts indicators of economic efficiency; the opacity of financial flows, which complicates the implementation of state financial control over the legal and effective use of budget funds and reduces investment attractiveness. The main goal of the financial restructuring of railway transport enterprises is to improve financial results based on the optimization of operating costs while qualitatively meeting the solvent demand for transportation and maintaining infrastructure facilities and rolling stock in good condition.

The information source for substantiating the plan of measures for the financial restructuring of railway transport enterprises is: internal information (data from business plans, investment memorandum, accounting (in terms of analytical accounting of income and expenses by types of transportation, distribution of general production costs), management accounting, financial and management reporting), external information (data on the amount of financing from external sources (the budget, external investors, banking institutions, the stock market), the state of the passenger and freight transport market, supply and demand data, the volume of passenger traffic on preferential transportation). Financial, economic, and SWOT analyses are used to justify measures for the financial restructuring of railway transport enterprises [39]. The analysis of strengths, weaknesses, threats, and opportunities made it possible to identify them (Table 1).

Table 1 Analysis of strengths, weaknesses, threats, and development opportunities of railway transport enterprises

Strengths	Weaknesses
– A sufficient level of development and high capacity of the transport services market; – Significant experience of functioning in market conditions; – High level of profitability of freight transportation, – Geographical spread over the entire territory of Ukraine and neighboring countries; – Significant transit potential; – Provision of transport services in a complex with related ones (including advertising, hotel and restaurant business); – Sufficient provision of land, material and human resources; – Stability of cargo turnover and passenger turnover according to the seasons	– Lack of a clear strategy and legislative framework that would regulate the processes of organizational and financial restructuring; – Lack of a special body for regulating natural monopolies; – Lack of staff awareness about international standards for the provision of transport services; – Limited use of innovative technologies; – Obsolescence of the technological structure of the railway industry, rolling stock, other specific assets; – Low level of transport security; – Imperfect management of financial results; – Low staff motivation, aging, staff turnover; – Lack of corporate culture and social responsibility
Opportunities	Threats
– Significant potential for quantitative and qualitative development, – Investment attractiveness due to the possibility of providing additional volumes of transportation; – Possibility of meeting the growing demand for cargo and passenger transportation in the conditions of the post-war recovery of the economy; – Existence of prerequisites for the creation of new market segments; – Close cooperative ties with enterprises of other types of economic activity	– Location of a significant part of the material and technical base in the temporarily occupied territories; – Risks of loss and damage to infrastructure and rolling stock due to military operations and shelling; – Limited amount of state support; – Disruption of traditional logistics connections for the movement of passengers and goods under martial law; – Decrease in demand for cargo and passenger transportation in conditions of reduction in GDP and external population migration; – Uncertainty regarding changing the organizational and legal form and owners; – An emerging gap between the fair and accounting value of assets; – A moratorium on the implementation of investment projects under martial law

Source Own development

Taking into account the nature of weaknesses and the degree of threats, it can be concluded that the greatest risks of the activity of railway transport enterprises in modern conditions are related to their activities in the conditions of martial law. Therefore, not economic tasks, but the increasing role of railway transport in ensuring the country's defense capability, the prompt restoration of damaged infrastructure and rolling stock due to military operations and shelling come to the fore among the anti-crisis measures. As the territories captured by the enemy are de-occupied, the issue of rebuilding territorial railway complexes and completing the started processes of organizational restructuring will be back on the agenda. In the conditions of the post-war recovery of the economy, we should expect an increase in the demand for freight and passenger transportation and the creation of conditions for the completion of the financial restructuring of railway transport.

Streamlining of the organizational structure of management and strengthening of the financial condition of railway enterprises will allow to eliminate their weaknesses and spread the use of innovative technologies; to improve the state of railway management, rolling stock, and other specific assets; to increase the level of safety of transportation; to increase staff motivation, attract highly qualified specialists, progressive youth; to complete the formation of corporate culture and increase social responsibility.

Particular emphasis should be placed on defining the dialectic of the relationship between restructuring on the one hand, and social responsibility and corporate governance on the other. Unfortunately, railway transport companies lag far behind business entities of other types of economic activity in terms of adopting corporate governance codes, forming and disclosing non-financial reports, increasing the environmental friendliness of business processes, occupational health and safety issues, implementing social programs, information openness etc. The large-scale restructuring of railway transport is of a commercial nature and is intended to solve the financial and economic problems of the industry. However, the concept of sustainable development is based not only on the requirements of profit growth and maximization of the market value of the enterprise, but also on moral principles, care for the environment, employees, social sphere, etc. Therefore, the restructuring of railway transport as a system of organizational, economic, financial, legal, and technical measures should be based on the principles of sustainable development, i.e. responsibility for environmental, social, and economic consequences for society [40]. In modern conditions, the concept of sustainable development is embodied in the responsibility of enterprises for their decisions, for the impact of their activities on the future, for the ability to harmonize the interests of stakeholders and meet the needs of employees, for participation in the defense of our state against the military invasion of the Russian Federation and the post-war recovery of its economy. Thus, the close dialectical relationship between restructuring and social responsibility, as well as their interdependence, allows drawing a conclusion about the emergence of the concept of socially responsible restructuring of railway transport. It is based on the voluntarily accepted responsibility of the enterprise for any possible risks during and as a result of changes in its structure, form of ownership, organizational and legal form, and sources of financing [41, 42].

5 Conclusions

1. The restructuring of the transport complex is understood as the change in its structure through the optimization and rationalization of its organizational management scheme, the movement of material, financial, and information flows in order to ensure the competitiveness of transport enterprises and the demonopolization of the market of transport services based on the knowledge economy. According to the forms and methods of implementation, restructuring can be divided into external restructuring (measures are implemented at the macroeconomic level within the framework of a sectoral reform, improvement of management methods) and internal restructuring (measures are implemented at the microeconomic level by a separate business entity).

2. The components of the restructuring of the transport complex are: organizational component as a change in the forms of ownership of business entities and their organizational and legal status; unbundling of enterprises and reduction of their concentration; change of forms and methods of production management; financial component includes increasing the efficiency of financial resource management, strengthening the financial situation due to the increase in budget allocations, funds of banking institutions and the stock market; informational component includes the provision of rapid and relevant information exchange between state administration bodies and business entities, between buyers and sellers of transport services, which contribute to transparency and openness in the industry.

3. Demonopolization of the transport complex is considered as a strategic goal of its development, aimed at reducing the negative impact of monopoly power and natural monopoly segments on the national economy. For the demonopolization of the transport complex, it is necessary to: make a rational division of the spheres of its functioning into three groups and take appropriate measures: (1) naturally monopolistic (development of the institutional mechanism of state regulation, improvement of antimonopoly policy, approval of justified methods of calculation of tariffs); (2) areas where competition is potentially possible (stimulation of directions of demonopolization of the railway transport market, support for new players' free entry into it, attraction of private investments as a result of structural reforms); (3) spheres where competition is natural (antimonopoly regulation and support of a competitive environment).

4. The knowledge economy is considered an environment for the processes of restructuring and demonopolization of the transport complex, it has been proved that innovative investment activity and modernization should be their integral component, aimed at reducing operational costs, improving the quality of customer service and transport safety. Requirements for the development of an investment declaration and accompanying documents have been proposed.

5. The conceptual principles of carrying out organizational, financial, and information restructuring in railway transport based on the integration of the functional-object system and the vertical model, tariff differentiation, the spread of two-part tariffs, the elimination of cross-subsidization, the creation of new market

segments, the expansion of the range of transport and related services have been substantiated. The priority areas of restructuring in railway transport in the conditions of martial law and post-war reconstruction have been substantiated, including within the framework of the concept of socially responsible restructuring.

Prospects for further research are the development of proposals for the restructuring of Ukrainian seaports in the conditions of martial law and post-war recovery.

References

1. Rivera Rios, M.A., Lopez, J.B.L., Veiga, J.G.: The fifth global Kondratiev. Low economic performance, instability and monopolization in the digital age. Marketynh i menedzhment innovatsii **2**, 270–291 (2018)
2. Kobets, V.M.: Vplyv taryfnoi polityky transportnoi kompanii-monopolista na suspilnyi dobrobut. Ekonomika ta derzhava **5**, 45–47 (2009)
3. Gubíniová, K., Štarchoň, P., Vilčeková, L., Bartáková, G.P., Brtková, J.: Marketing communication and its role in the process of creating rational awareness of generation Z representatives. In: Kryvinska, N., Poniszewska-Marańda, A. (eds) Developments in Information & Knowledge Management for Business Applications. Studies in Systems, Decision and Control, vol. 376, pp. 203–221. Springer, Cham (2021)
4. Panevski, D., Peráček, T., Rentková, K.: Analysis of the practices of financial intelligence units (FIUs) and other anti-money laundering agencies within EU. In: Kryvinska, N., Poniszewska-Marańda, A. (eds) Developments in Information & Knowledge Management for Business Applications. Studies in Systems, Decision and Control, vol. 376, pp. 241–269. Springer, Cham (2021)
5. Kiß, H., Sulíková, R.: Modern approaches to leadership development—an overview. In: Kryvinska, N., Poniszewska-Marańda, A. (eds) Developments in Information & Knowledge Management for Business Applications. Studies in Systems, Decision and Control, vol. 376, pp. 271–287. Springer, Cham (2021)
6. Busse, V., Strauss, C.: Crowdfunding and uncertain decision problems—applying Shannon entropy to support entrepreneurs. In: Kryvinska, N., Poniszewska-Marańda, A. (eds) Developments in Information & Knowledge Management for Business Applications. Studies in Systems, Decision and Control, vol. 376, pp. 289–304. Springer, Cham (2021)
7. Vančová, M.H., Ivanochko, I.: Factors behind the long-term success in innovation—in focus multinational IT companies. In: Kryvinska, N., Poniszewska-Marańda, A. (eds) Developments in Information & Knowledge Management for Business Applications. Studies in Systems, Decision and Control, vol. 376, pp. 441–483. Springer, Cham (2021)
8. Poplavska, Z., Komarynets, S.: Modelling the external economic environment instability impact on the organizational flexibility of the enterprise. In: Kryvinska, N., Greguš, M. (eds) Developments in Information & Knowledge Management for Business Applications. Studies in Systems, Decision and Control, vol. 330, pp. 1–20. Springer, Cham (2021)
9. Kuzmin, O., Ovcharuk, V., Zhezhukha, V.: Administration systems in enterprises management: assessment and development. In: Kryvinska, N., Greguš, M. (eds) Developments in Information & Knowledge Management for Business Applications. Studies in Systems, Decision and Control, vol. 330, pp. 201–229. Springer, Cham (2021)
10. Kuzmin, O., Zhezhukha, V., Gorodyska, N.: Simulating the processes of engineering payments establishment during the engineering projects implementation at enterprises. In: Kryvinska, N., Greguš, M. (eds) Developments in Information & Knowledge Management for Business

Applications. Studies in Systems, Decision and Control, vol. 330, pp. 231–271. Springer, Cham (2021)

11. Vlasenko D.O.: Suchasni aspekty restrukturyzatsii transportnykh pidpryiemstv. Naukovyi visnyk Uzhhorodskoho natsionalnoho universytetu. Seriia: Mizhnarodni ekonomichni vidnosyny ta svitove hospodarstvo **23**(1), 34–38 (2019)

12. Eitutis, H.D., Bozhok, A.R.: Restrukturyzatsiia zaliznychnoho transportu Ukrainy na osnovi rehionalno-haluzevoi modeli upravlinnia. Visnyk ekonomiky transportu i promyslovosti **68**, 85–93 (2019)

13. Lysytsia K.S.: Rynochnye uslovyia razvytyia estestvennыkh monopolyi v stranakh Tsentralnoi y Vostochnoi Evropы (na prymere zheleznodorozhnoho transporta). Visnyk Pryazovskoho derzhavnoho tekhnichnoho universytetu. Ser.: Ekonomichni nauky **22**, 36–44 (2011)

14. Pozniakova, O.: Improvement of the methodology for the property registry formation as a tool preventing the development of hidden monopolies. Tekhnolohycheskyi audyt i rezervy proyzvodstva **3**(4), 45–48 (2020)

15. Nikolayev, Y.: Development strategy and restructuring of Illichivsk commercial sea port as an integral part of Odessa transport cluster. Ekonomichni innovatsii **50**, 237–247 (2012)

16. Zorina, O.I.: Otsinka realnoho stupeniu monopolizatsii zaliznychnoho transportu v sferi pasazhyrskykh perevezen. Visnyk ekonomiky transportu i promyslovosti **26**, 25–28 (2009)

17. Shemaiev, V.V.: Investytsiina polityka subiektiv pryrodnoi monopolii transportno- infrastruk-turnoho sektoru. Dorohy i mosty **21**, 47–58 (2020)

18. Shyrokova O.M.: Model otsinky ekonomichnoi efektyvnosti funktsionuvannia zaliznych-noho transportu v protsesi restrukturyzatsii. Zbirnyk naukovykh prats Dnipropetrovskoho natsionalnoho universytetu zaliznychnoho transportu imeni akademika V. Lazariana. Problemy ekonomiky transportu **1**, 122–125 (2011)

19. Antimonopoly Committee of Ukraine. Summary list of subjects of natural monopolies. https://amcu.gov.ua/storage/app/sites/1/%D0%B7%D0%B2%20%D0%BC%D0%BE%D0%BD D0%BE%D0%BF%D0%BE%D0%BB%D1%96%D1%97/zvedeniy-perelik-subektiv-prirod nikh-monopoliy-za-cherven.pdf (2022)

20. On railway transport: Law of Ukraine. https://zakon.rada.gov.ua/laws/show/273/96-%D0% B2%D1%80#Text (1996)

21. On the peculiarities of the formation of a joint-stock company of public railway transport: Law of Ukraine. https://zakon.rada.gov.ua/laws/show/4442-17#Text (2012)

22. Air Code of Ukraine: Code of Ukraine. https://zakon.rada.gov.ua/laws/show/3393-17#Text (2011)

23. About seaports of Ukraine: Law of Ukraine. https://zakon.rada.gov.ua/laws/show/4709-17# Text (2012)

24. On pipeline transport: Law of Ukraine. https://zakon.rada.gov.ua/laws/show/192/96-%D0% B2%D1%80#Text (1996)

25. On the approval of the Procedure for deducting part of the net profit (income) to the state budget by state unitary enterprises and their associations: Resolution of the Cabinet of Ministers of Ukraine. https://zakon.rada.gov.ua/laws/show/138-2011-%D0%BF#Text (2011)

26. Melnikov, S.V.: Dynamic monopoly pricing under the purchase price effect. Aktualni problemy ekonomiky **12**, 417–422 (2014)

27. Smerichevskyi, S., Kosova, T., Tryfonova, O., Bezgina, O., Solomina, G.: Financial—accounting support of marketing strategies of energy efficiency of coal mines. Sci. Bull. Natl. Mining Univ. **1**, 163–169 (2022)

28. Vincze, R., Karovič, V., Kavalets, I.: The efficiency of transport-management in the E-logistics marketplace. In: Kryvinska, N., Greguš, M. (eds.) Developments in Information & Knowledge Management for Business Applications. Studies in Systems, Decision and Control, vol. 420, pp. 239–255. Springer, Cham (2022)

29. Rudiaha, I.M.: Zaliznychnyi transport Ukrainy yak subiekt pryrodnoi monopolii: hospodarsko-pravovyi aspekt. Ekonomichna teoriia ta pravo **2**, 216–229 (2016)

30. Diehtiar A. O. Derzhavne rehuliuvannia zaliznychnoho transportu yak pryrodnoi monopolii. Mizhnarodnyi naukovyi zhurnal "Internauka" **2**(1), 26–30 (2017)

31. Makovoz, O.V., Borysov, B.I.: Osnovni aspekty restrukturyzatsii zaliznychnoho transportu Ukrainy. Visnyk ekonomiky transportu i promyslovosti **49**, 69–73 (2015)
32. Kosova, T., Voronkova, O., Kliuchka, O., Kostynets, I.: Financial control in the sys-tem of budgetary security of the state and regions under decentralization conditions in Ukraine. Financ. Credit Activ. Probl. Theory Pract. **1**, 140–148 (2021)
33. Directive 2008/57/EC of the European Parliament and of the Council of on the interoperability of the rail system within the Community (Recast). https://eur-lex.europa.eu/legal-content/EN/ALL/?uri=CELEX:32008L0057 (2008)
34. Petrenko, E.: Formation of institutional background logistizatcija transport during restructuring passenger complex railways of Ukraine. Zbirnyk naukovykh prats Derzhavnoho ekonomiko-tekhnolohichnoho universytetu transportu. Ser.: Ekonomika i upravlinnia **32**, 124–133 (2015)
35. On the management of state-owned objects: Law of Ukraine. https://zakon.rada.gov.ua/laws/show/185-16?find=1&text=%D1%96%D0%BD%D0%B2%D0%B5%D1%81%D1%82#w1_1 (2006)
36. Kosova, T., Smerichevskyi, S., Ivashchenko, A., Radchenko, H.: Theoretical aspects of risk management models in economics, marketing, finance and accounting. Financ. Credit Activ. Probl. Theory Pract. **3**(38), 409–418 (2021)
37. On priority areas of innovative activity in Ukraine: Law of Ukraine. https://zakon.rada.gov.ua/laws/show/3715-17#Text (2011)
38. Oliinyk, HYu.: Finansovo-ekonomichni pokaznyky ta analiz plato-spromozhnosti, konkuren-tospromozhnosti ta finansovoi stiikosti pidpryyemstv zaliznychnoho transportu na stadii pidhotovky do restrukturyzatsii. Vodnyi transport **2**, 106–114 (2012)
39. Shkulipa, L.V.: Zastosuvannia SWOT-analizu dlia otsinky dotsilnosti provedennia restruktu-ryzatsii na zaliznychnomu transporti Ukrainy. Visnyk sotsialno-ekonomichnykh doslidzhen **1**, 168–175 (2013)
40. Shkulipa, L.V.: Sotsialno-vidpovidalna restrukturyzatsiia zaliznychnoho transportu v Ukraini yak providna determinanta v SSVZT. Zbirnyk naukovykh prats Derzhavnoho ekonomiko-tekhnolohichnoho universytetu transportu. Ser.: Ekonomika i upravlinnia **21–22**(1), 66–73 (2012)
41. Kryvinska, N., Auer, L., Strauss, C.: Managing an increased service heterogeneity in a converged enterprise infrastructure with SOA. IJWGS **4**, 440 (2008). https://doi.org/10.1504/IJWGS.2008.022546
42. Kryvinska, N., Auer, L., Strauss, C.: The place and value of SOA in building 2.0-generation enterprise unified vs. ubiquitous communication and collaboration platform. In: 2009 Third International Conference on Mobile Ubiquitous Computing, Systems, Services and Technologies (2009). https://doi.org/10.1109/UBICOMM.2009.52

Modeling of the Key Threats and Identification of the Prospects for Social Security Support in the Sphere of Employment and Labor Payment in Ukraine

Kseniia Bondarevska ⓘ

Abstract Following the study, the conceptual basis for a mechanism for assessing the main threats to social security in Ukraine's employment and labour wage sectors and the prospects for their neutralization was developed. In the framework of the factorial model of the main threats to social security in the Ukrainian labour market, we studied the dynamics of the parameters of employment and wages and investigated the impact of each threat on the coefficients of the labour supply–demand ratio. In the process of analyzing threats to employment and wage security, it assessed the impact of parameters such as informal employment, youth unemployment, average monthly wages, real wage index, wage arrears rate and wage fund. The distribution of the general factors affecting social security in the labour market demonstrates the instability of the dynamics of changes in the indicators analysed between 2006 and 2018, as well as the significant negative severity of the escalation of the crisis phenomenon between 2008–2009 and 2014–2015, which, due to their low control and spontaneity, reflect the high level of indicators for employment and salaries since 2019, without including the assessment of labour market parameters in the context of the worsening risks to social security during the Pandemic (2020–2021) and the war in Ukraine (2022), which was deliberately not taken into account in the modelling process.

Keywords Social security · Social risks · Social threats · Labor market · Employment · Social security in the labor market · Modeling · Labor payment

K. Bondarevska (✉)
University of Customs and Finance, Street Vladimir Vernadsky, 2/4, Dnipro 49000, Ukraine
e-mail: Kseny-8888@i.ua

A. Semenov et al. (eds.), *Data-Centric Business and Applications*, Lecture Notes on Data Engineering and Communications Technologies 195, https://doi.org/10.1007/978-3-031-54012-7_2

1 Introduction

The process of economic transformation reflects the transition to the information society and knowledge economy, which is characterized by the current phase of social development, particularly in terms of changes in production methods, the nature of production relationships and ownership relations. They are responsible for significant changes in the labour market, resulting in the emergence of new changes in traditional occupations, the emergence of non-standard forms of employment, and the growth of structural unemployment. These processes have a major impact on the development of the entire economic system, forming a fundamentally new quality of social and labour relations focusing on innovation and establishing priority for information factors.

At the same time, changes in both the global and domestic economies pose some risks in the area of national labour and employment and contribute to the emergence of threats to social security in the labour market. The threat of growth in informal employment, youth unemployment, the exit of the elderly from the labour market, and the decline in national income levels are particularly important. This diagnosis must not only include understanding the current situation of threats, but also include its predictive characteristics and the possibility of adjusting the data received.

The results of research on social security and the neutralization of threats in the fields of employment and wage compensation are reflected in the scientific works of foreign and national authors. Among the publications of foreign researchers, T. Vilhagen, K. Crouch, B. Menihert and G. Standing are noteworthy in highlighting the conceptual foundations of safety in the context of the development of human potential and the functioning of the labour market [1–3]. Works of national researchers O. Gishnova, O. Doronina, L. Lisohor, U. Sadova, O. Sydorchuk, and L. Shaulska identified important aspects of labour market development and social security provision [4–8]. Scientists S have applied scientific and practical approaches to economic and mathematical modeling and the use of tools to assess the well-being of the population and the migrant component of socio-economic relations. Kozlovskyi, D. Bilenko, M. Kuzheliev, L. Nikolenko, and O. Peresada were used [9, 10].

At the same time, it is particularly urgent to update the existing threats in the fields of employment and labour wages and to establish a basis for neutralizing them under modern conditions. In view of the important scientific achievements of national and international scientists in the field of labour market development and social security, and the scientific theoretical and practical importance of the proposed problems, taking into account the main threats to its development in the phase of economic environment transformation under high dynamism and instability of economic conditions, the authors are objectively obliged to justify the strategic orientation to ensure social security.

2　The Mechanism of Assessment of the Directions for Overcoming Threats to Social Security in the Labor Market in Ukraine

In order to evaluate the main threats and their prospects, the author proposes an approach based on expert research methods, economic and mathematical modeling, and dynamic projections of problems and threats in the labour market affecting the provision of social security (Fig. 1).

Recognizing the importance of evaluating the prospects for neutralization of threats in the area of social security in the labour market, the formation of a comprehensive mechanism is a combination of efforts from all the above-mentioned subjects that are aimed at the objective of the evaluation using procedures including resources, methods, stages and principles. The aim of the assessment is not only to assess the importance of the major threats to social security on the labour market, but also to assess their prospects for neutralization. In the process of practical application of experts' surveys and economic-mathematical modeling, it is necessary to take into account the availability of scientific-methodological, human, financial, material-technical, time and information resources, which determine the success of assessments. At the same time, the assessment of the prospects for neutralizing the key threats to social security in the labour market should be based on the principles of objectivity (reliability and accuracy of information as a basis for further justification of the directions to solve the problem); exhaustiveness (taking into account information data and indicators that fully reflect the threats to social security in the labour market and how to overcome them); Self-sufficiency of indicators and criteria (the absence of homogeneity in the process of determining the list of experts' questions and the absence of duplicate indicators during the construction of the economic-mathematical model); priority (focusing only on key, priority issues and indicators); priority (concentrating only on key, priority issues and indicators).

The assessment procedure involves the passage of its standard stages, including:

I.　The preparatory phase (setting of assessment objectives, determining assessment tools, gathering the necessary information);

II.　Evaluation (Expert survey methods include the process of expert answers to questions; in the case of economic and mathematical modeling, the second stage of evaluation involves the processing of the information sequence with the help of a computer and the construction of appropriate models);

III.　The processing of assessment results is the processing of data received based on expert assessment results and economic-mathematical modeling.

IV.　The final stage involves the formulation of conclusions and the confirmation of promising directions for ensuring social security on the labour market, provided that its main threats are neutralized.

The basic component of assessing key threats to social security in the labour market is to determine its severity and direction of neutralisation, taking into account the opinions of experts—leading scientists and practitioners in the field of labour and

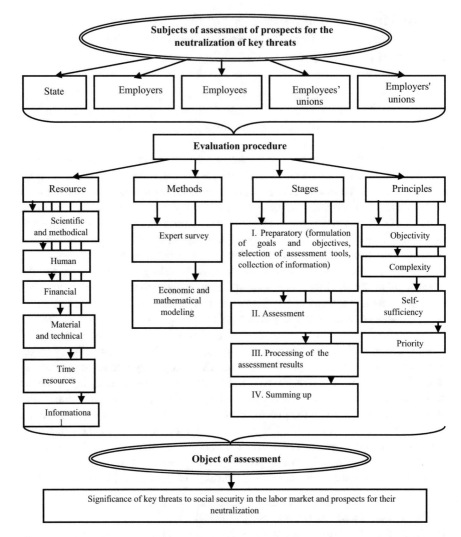

Fig. 1 Conceptual vision of the mechanism of assessment of the directions for overcoming threats to social security in employment and labor payment in Ukraine. *Source* Development of the author [11]

population employment. Thus, expert assessments offer the opportunity to obtain a formal description of complex problems, whose solution is impossible without the participation of a statistical research method based on a mathematical model [12, p. 24]. The expert survey therefore becomes a basis for further application of economic and mathematical modeling methods and provides important information for drafting strategies for ensuring social security in the Ukrainian labour market. Therefore, it is necessary to conduct an expert survey involving only well-known specialists who have significant theoretical and applied professional experience and can provide comprehensive answers to important questions about the prospects for the development of the labour market, as well as quantitative assessment of the seriousness of the main threats and the importance of factors affecting the formation of social security conditions in the labour market. By applying the above-mentioned methodology for diagnosing threats and prospects for the development of the labour market, it is possible to obtain independent and reliable information, which can be further used in the next phase of the assessment mechanism.

The objective of the expert survey "The threats to social security in Ukraine's labour market and methods of its neutralization" was to identify the threats and justify the direction of the further development of the labour market in the context of the guarantee of social security. Respondents gave answers to open and closed types using a scale of 1–10 (the minimum score being 1 point, the maximum score being 10 points). The results of the expert survey are reflected in the scientific publication "The main threats and prospects for supporting social security in the Ukrainian labour market" [13].

3 The Modeling of the Threats and Its Changes in the Context of Social Security in the Labor Market of Ukraine

An important component of the mechanism for assessing the main threats and the prospects for their overcoming in the Ukrainian labour market is the determination of strategic directions for solving the above-mentioned problems using economic modeling tools. The special quality of modeling is that it allows to reproduce certain characteristics of the object studied with its help, with the help of equations and algorithms, and to study changes in the behavior of the object under certain conditions as close as possible to the actual ones [14, 15].

Within the model phase, it is appropriate to identify the main threats to social security in the labour market:

   I. Setting model objectives;
   II. The determination of the assessment parameters and their numerical values;
  III. The collection of the necessary information and its processing by software; and
  IV. Implementation of a defined model using a selected software product;
   V. Validation of the model and Evaluation of the accuracy of the results;
  VI. Identification of the influence of different factors and assessment of the significance of each of them;
 VII. The characterization of the results of the modeling;
VIII. Determining strategic priorities for the provision of social security in the labour market, taking into account the results obtained.

The objective of economic and mathematical modeling is to assess the main threats to social security in the labour market under conditions of dynamic change in key parameters. Consider the major threats posed in the Ukrainian labour market and the eradication of social security in the labour market, and create an appropriate model based on expert research showing that such threats and negative phenomena such as informal employment growth, youth unemployment, low wage compensation, etc. We believe that the parameters should be selected.

Consequently, the main modeling parameters are defined as follows:

- Level of informal employment (%);
- Unemployment rate for 15–24 years of age (%);
- Unemployment rate for 25–29 years of age (%);
- Unemployment rate for 30–34 years of age (%);
- Average Monthly Lung (UAH);
- Real Lung Index (%);
- Unpaid Lung and Lung Fund (times).

In particular, the above-mentioned list of indicators assesses phenomena both in terms of employment stability (formal employment levels, youth unemployment rates) and in terms of salary stability (average monthly wages, real wage index, ratio of unpaid salaries to wage funds).

The labour supply–demand ratio (the ratio of vacancies to registered unemployed) is chosen as a result indicator reflecting the nature of the strategy for ensuring social security in the labour market and determining the degree to which the labour market has reached equilibrium and the corresponding situation at the time of analysis (Kds). Thus, if this coefficient reaches level 1, a perfect correlation between labour demand and supply and a balanced labor market situation would be achieved, which unfortunately is not possible in the real economic situation. If the number of open jobs exceeds the number of registered unemployed, the labour market situation will show a shortage of labour, indicating that demand for labour exceeds supply. At the same time, the number of unemployed exceeding the number of vacancies indicates a labour surplus in the labour market, determining that labour supply exceeds demand. Thus, the adjustment of Kds values to the maximum possible is an important condition for achieving balance in the labour market and ensuring the social security of the population in terms of employment.

The study period was 2006–2018. This will enable the identification of existing trends and their changing factors, as well as the formulation of new projections for future events, as provided for by a mechanism for assessing threats to social security in the labour market and the direction of their neutralization. Threat modeling provides opportunities to identify trends in the labour market and the drivers of these changes in terms of the economic cycle of the labour market (before and after the 2008 economic crisis and the events in Ukraine in 2014). In this study, the employment and wage indicators of 2019 were intentionally not included in the modeling strategy, as labour market parameters were not assessed in the context of an aggravated social security risk during the pandemic (2020–2021) and the Ukrainian war (2022) due to their low capacity for management and spontaneity.

Figure 2 shows trends in labour supply and demand that influence the outcome measurement over the research period [16–19].

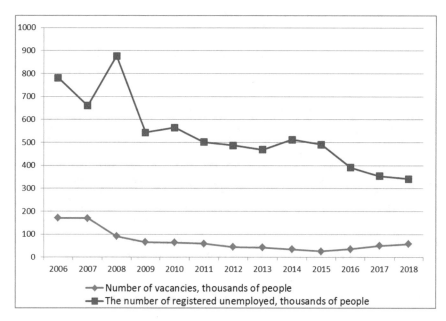

Fig. 2 Labor demand and labor supply in Ukraine, thousands of people. *Source* Defined by the author based on the information: [16–19]

The Ukrainian labour market therefore currently has a significant imbalance between supply and demand, as confirmed by statistical data. During the 2008 crisis period, labor demand declined significantly and the number of unemployed increased, resulting in the largest gap between the number of vacancies and the number of employees seeking employment. In general, the volume of labour demand decreased by 65.7% and the volume of supply by 56.2% during the study period. As a result, the decline in the number of job openings is much higher than the decline in the number of registered unemployed, as evidenced by the decline in the supply demand ratio (Fig. 3).

The lowest supply–demand ratio was recorded during and after the crisis (2008 and 2015), while the number of jobs opened declined with the increase in the number of unemployed. At the same time, it should be noted that since 2016, there has been a certain improvement in the dynamics of the indicators investigated, confirming a positive growth trend, as opposed to the unstable dynamics of previous years.

In order to preserve positive trends and prevent the spread of threats in the labour and employment sectors, it is appropriate to define several models for the implementation of social security strategies in the labour market, which will be reflected in changes in the values of Kds. In our opinion, the most appropriate modeling method is factor analysis, which involves a type of information compression, i.e. the transition from a set of values of elementary properties m with a quantity of information $n * m$ to a limited set of elements of the matrix of the factor mapping $m * r$, why $r < m$. Therefore, factor analysis methods for assessing threats enable visualization

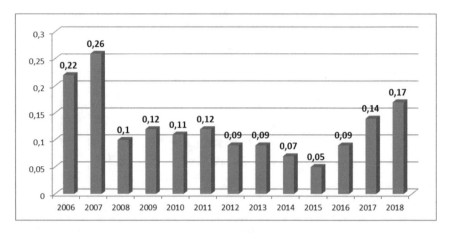

Fig. 3 The coefficient of labor supply and labor demand correlation in Ukraine (Kds) during 2006–2018. *Source* Defined by the author based on the information: [14–17]

of the structure of dynamic phenomena and processes, thus predicting their future development [20].

Furthermore, factor analysis has the advantage of reducing a large number of variables X_j to small quantities, called factors. At the same time, strongly correlated variables are combined with the first factor and less correlated variables with other factors. In these conditions, the main goal of factor analysis is to find factors that best explain the relationship between variables. Another advantage of this method is that it not only allows you to determine the existence of a relationship between individual aspects of the object being investigated, but also the extent of this relationship.

The condition of the effective using of this method is the identification of the complex of elementary features X_j ($j = \overline{1, n}$, where n is the number of initial features under consideration), the interplay of which implies the existence of hidden factors F_r ($r = \overline{1, m}$, m—the number of hidden factors), which are installed as a result of the generalization of the initial sings [20].

Unlike the main component methodology, the factor analysis methodology is based on a real assessment of the possibility of examining the main threats to social security in the labour market. This takes into account the fact that the variation in basic characteristics with general factors is impossible to explain, as part of the variation remains a hidden characteristic of the threat to social security in the labour market being studied. The standardized value of the characteristic value of the i-th object observed x^*_{ij} simultaneously is represented by a linear dependency of the form [20]:

$$x^*_{ij} = a_{j1} f_{1i} + a_{j2} f_{2i} + \cdots + a_{jm} f_{mi} + a_j d_{ij}, \tag{1}$$

where d_{ij}—the latent feature that summarizes the characteristic variation of an elementary feature; a_j—coefficient of the wage with the characteristic factor.

The dispersion of standardized value x_{ij}^* is decomposed according to the estimates of the variance of the common factors and characteristics from Eq. (1) as follows:

$$D(x_{ij}^*) = a_{j1}^2 + a_{j2}^2 + \cdots + a_{jm}^2 + d_j^2 = 1. \tag{2}$$

The method of principal (main) components may be considered as the special case (2), when $d_j^2 = 0$ and $D(x_{ij}^*) = a_{j1}^2 + a_{j2}^2 + \cdots + a_{jm}^2 = 1$.

At the same time, it is worth noting that $a_{j1}^2 + a_{j2}^2 + \cdots + a_{jm}^2$ is generality, and it is the proportion of total dispersion that can be explained by general factors. It can be denoted as h_j^2. The characteristic is the share of dispersion that is not explained by general factors, or the contribution to the total dispersion of the trait x_j^* of some characteristic hidden factor.

In the factor analysis technique, the factors are distinguished in sequence: the first factor explains the largest part of the prime variation, the second factor explains the small part of the variance, and the second factor after the first latent factor is the third factor. These factors are independent of each other, i.e. they are not associated, because each subsequent factor is determined to maximize the remaining variation of the previous factor.

To study the main threats to social security in the labour market, we selected one of the methods of factor analysis, the main factor method. This is a method of identifying major factors that have a significant impact on the level of social security in the labour market.

According to the features of using of the method of main factors in the reduced correlation matrix $\boldsymbol{R}_h$, the main diagonal is no longer units, but the characteristics of the totality h_j^2. When we use the method of main factors, the generality h_j^2 was determined using the square of the multiple coefficient of correlation. According to Iberl rules, it is known that with the growth in the number of variables at the constant number of factors, the lower limit of the generality estimate h_j^2 is approaching to the true value of generality. This method provides the opportunity to calculate the generality h_j^2 with the reverse matrix of correlation $\boldsymbol{R}^{-1}$ [20].

According to the classic model of factor analysis, the equations in the process of definition the coefficients with general factors F_r are written as follows [20]:

$$Z_j = a_{j1} F_1 + a_{j2} F_2 + \cdots + a_{jm} F_m + a_j D_i \tag{3}$$

where D_j is the characteristic factor.

The solution of the equation is as follows under the condition of the maximization of sums [20]:

$\sum_{j=1}^{m} a_{j1}^2 = \lambda_1$ is the first maximum of the dispersion of elementary features $(D(Z_j))$;

$\sum_{j=1}^{m} a_{j2}^2 = \lambda_2$—the second maximum that remained after λ_1 dispersion, etc., is reduced to the definition of eigenvalues λ_r and eigenvectors l_r symmetric matrix $\boldsymbol{R}$ from equation $(\boldsymbol{R} - \lambda_r \boldsymbol{I}_m) l_r^T = 0$, where $\boldsymbol{I}_m$—unit dimension matrix m. If the values λ_r and l_r are known, the coefficients $a_{jr}\, a_{jr}$ can be calculated using the formula

$A = L^T \lambda^{\frac{1}{2}}$, where $\lambda^{\frac{1}{2}}$—diagonal matrix, r–th element of the main diagonal of which is $\sqrt{\lambda_r}$, L^T—matrix of normalized vectors l_r.

In practice, there are various techniques and methods for finding the parameters of the model of the main factors λ_r l_r, a_{jr}, they can, conditionally, be easily divided into two large groups.

The first group focuses on the algorithm of the main component method and, in essence, repeats it, the only difference is that the calculation is done on the basis of data from the reduced correlation matrix, not the usual pair correlation matrix. In this case, everyone gets the m values of eigenvalues λ_r λ_r and eigenvectors l_r. The second group combines techniques that allow the values of the individual and eigenvectors to be set sequentially. The following steps are performed after a preliminary check of the sufficiency of the informational sufficiency of the main factors already selected. This approach can be called classical in factor analysis. In order to improve the interpretation of the main factors, the orthogonal return of the general factors can be performed using the "varimax" method. The Varimax method calculates the quality criteria V_r for the structure of each factor [20]:

$$V_r = \frac{m \sum_{j=1}^{m} a_{jr}^4 - \left(\sum_{j=1}^{m} a_{jr}^2 \right)^2}{m^2} \tag{4}$$

The "varimax" method provides the maximum simplification for the description of columns in the matrix of factor mapping. It is possible to improve the factor structure individually and obtain factors with high loads for some variables and low loads for others. This return of common factors allows you to clearly define the name of each common factor.

Taking into account the significant differences in the measuring units of the selected indicators, in particular the level of informal employment (X1), the unemployment rate of 15–24 years (X2), the unemployment rate of 25–29 years (X3), the unemployment rate of 30–34 years (X4), the average monthly wage (X5), the real wage index (X6), the ratio of arrears from wage payments and the wage fund (X7), the standardisation was carried out to allow an analysis in a single population.

The creation of generalized factors reflected by certain behavioural strategies was carried out using the SPSS software complex, resulting in the determination of three main factor models. Thus, Table 1 shows data on the total variance of the structure model of the strategy by the method of the main components.

Based on the above data obtained when determining the total variation of the components, the loading matrix of the estimated factors of threats to social security in the labour market for 2006–2018 can be calculated using the main components method (Table 2).

In order to investigate the coherence of results obtained by various methods of factor analysis, we used the correlation coefficient calculated in accordance with the formula (5):

Table 1 The results of calculating the full dispersion according to the method of main components

Components	Initial eigenvalues			Sums of squares of extraction loads		
	Total	% variance	Cumulative %	Total	% variance	Cumulative %
X1	3.900	65.713	65.713	3.900	65.713	65.713
X2	1.752	15.027	70.740	1.752	15.027	70.740
X3	0.732	10.454	91.195	0.732	10.454	91.195
X4	0.316	4.512	95.706	–	–	–
X5	0.176	2.520	98.226	–	–	–
X6	0.105	1.504	99.730	–	–	–
X7	0.019	0.270	100.00	–	–	–

Source Defined by the author based on using the SPSS software package

Table 2 The results of calculating of the load matrix of the estimated factors of threats to the social security in the context of labor market sphere in 2006–2018 years

Initial signs	The main factors		
	F_1	F_2	F_3
X_1	0.944	0.906	−0.654
X_2	1.391	0.479	0.456
X_3	1.312	0.815	−0.739
X_4	−0.107	0.224	−1.892
X_5	0.104	0.097	−0.040
X_6	0.398	−0.352	0.851
X_7	0.298	0.024	0.892

Source Defined by the author based on using the SPSS software package

$$\phi_{pq} = \sum_{j=1}^{m} a_{j(p)} a_{j(q)} \left/ \sqrt{\sum_{j=1}^{m} a_{j(p)}^2 \sum_{j=1}^{m} a_{j(q)}^2} \right. \tag{5}$$

where p and q are compared general factors in the first and second solutions of the factor analysis, $a_{j(p)}$, $a_{j(q)}$—the coefficients of factor loads on the j-th feature in the generalized factors, respectively, in the first and second solutions of the factor analysis.

For all three general factors, the coherence coefficient was greater than 0.95. As a result, all considered options give well-adapted results. Table 3 shows the values of the matrix of components of three selected models of threats to social security in the labour market by the method of the main factors.

Based on the obtained general factors, the model equations characterized three options for modifying the main threats to social security in the labour market based on seven factors (X1–X7) for 2006–2018 were determined:

Table 3 The results of the calculating of the component matrix in the contex of the factor models

Initial signs	Components		
	F_1	F_2	F_3
X_1	0.222	0.939	0.004
X_2	−0.688	−0.679	−0.071
X_3	−0.820	−0.491	−0.040
X_4	−0.886	−0.354	−0.173
X_5	0.885	0.139	0.294
X_6	0.053	0.731	0.519
X_7	−0.046	0.079	−0.978

Source Defined by the author based on using the SPSS software package

$$F1 = 0.22X1 - 0.69X2 - 0.82X3 - 0.89X4 + 0.89X5 + 0.05X6 - 0.05X7 \tag{6}$$

$$F2 = 0.94X1 - 0.68X2 - 0.49X3 - 0.35X4 + 0.14X5 + 0.73X6 + 0.08X7 \tag{7}$$

$$F3 = 0.004X1 - 0.07X2 - 0.04X3 - 0.17X4 + 0.29X5 + 0.52X6 - 0.98X7 \tag{8}$$

Accordingly, the following conclusions can be drawn from the obtained factor model. According to the first general factor F1 (factor model No. 1), positive correlations can be traced to early indicators such as the level of informal employment, monthly average wages, and real wage indexes. The negative values of the initial indicators of youth unemployment in three age groups (X2, X3, X4) and the ratio of wage payments and wage funds reflect positive changes in the resulting indicators—the ratio of demand and supply—when reaching the corresponding values. The level of informal employment (X1) is positively correlated with performance indicators, as it somewhat eases the burden on the labour market. At the same time, it should be noted that the risks of social nature associated with an increase in the scope of informal employment, which are not taken into account in the model, are expressed in the lack of social protection for informally employed persons, the reduction in tax revenues to the budget, and the general shadow of the economy, etc.

According to factor model No. 2 (after the second generalized factor F2), trends are reflected in the positive correlation of all initial characteristics, with the exception of youth unemployment (X2, X3, X4). At the same time, it should be noted that the highest level of negative impact on the indicators of results is precisely the level of unemployment among young people aged 15–24. The positive value of the ratio of acquittal from the pay of wages and the wage fund reflects a disproportionate influence within the framework of this factor model, which proves that even a slight increase in that ratio will not lead to a significant reduction in the performance indicator. Other indicators of wage security (real wage index and average monthly wage size) are important positive factors in the labour market, especially when the real labour income of the population increases.

Factor Model No. 3 (F3) reflects this distribution of initial characteristics and is negatively correlated with indicators such as the unemployment rate for all ages of young people and the ratio of wages to wage funds. At the same time, the most serious adverse effects are observed in relation to the payment of wages' arrear indicator, and the positive effects are observed in relation to the real wage index. This reflects the importance of wage security in the implementation of national labour market development policies. In the framework of this model, all other indicators have not such a significant impact on the resulting characteristics.

The details of the first factor model (F1) of dynamics 2006–2018 are shown in Fig. 4.

On the basis of the model's formula, it is possible to observe the need to reduce the main threats to social security in the labour market, reflecting the general threat X indicator, and to increase the productive characteristics of Y, determined by the coefficient of demand-demand ratio (Kds). At the same time, the determination coefficient confirms the accuracy and reliability of the results of the simulation and confirms the relatively high dependence of the variation of the general threat indicator on the variation of independent variables ($R^2 = 0.8097$).

According to the obtained data, two crises had a major impact on the development of the labour market in 2009 and 2014. During these periods, significant deterioration was observed in the analyzed indicators, reflecting the transition from positive to negative values on the distribution graph. In particular, this is reflected in threats such as the growth of youth unemployment rates for all youth groups and the reduction

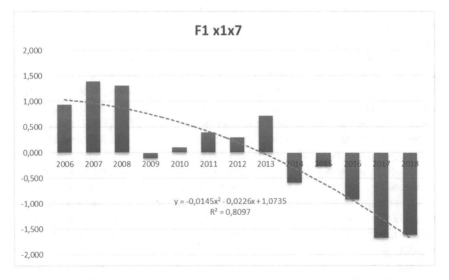

Fig. 4 The distribution results of the factor model of threats to social security in the labor market No. 1(F1) in 2006–2018 years. *Source* Defined by the author

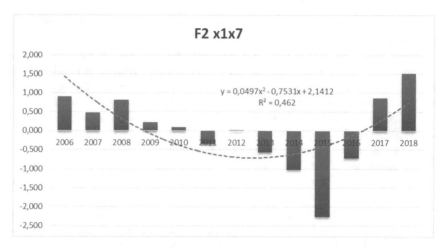

Fig. 5 The distribution results of the factor model of threats to social security in the labor market No. 2 (F2) in 2006–2018 years. *Source* Defined by the author

in the real wage index. In addition, in the crisis period 2009 and 2014, there was a significant increase in informal employment, unemployment and wage arrears in relation to the wage fund, as well as a decline in real wages reflecting the real income situation of the population.

Factor model No. 2 (F2) clearly characterized the good improvement of the labour market in recent years, in particular due to the improvement of public policy in the field of population employment (Fig. 5).

However, given the low determination coefficient ($R^2 = 0.462$), we consider that the distribution of the general threat indicator based on the second factor model is not completely reliable.

In determining the obtained data according to factor model No. 3 (F3), it is worthwhile to note that the first model and its similarity are clearly reflected in the post-crisis and post-crisis periods of labour market development: 2008, 2009, 2014, 2015. Therefore, the first deterioration of the labour market and the country's macroeconomic situation was caused by the impact of the 2008–2009 global financial and economic crisis; the second crisis was caused by military conflicts and events in eastern Ukraine, which had a negative impact on labour demand, taking into account the reduction in the number of jobs due to the liquidation of enterprises and thus increased the number of unemployed people. Based on the average reliability of the obtained results, as demonstrated by the determination coefficient ($R^2 = 0.597$), it is desirable to use the obtained results to justify further directions of state labour policy in order to ensure the neutralization of threats to social security. Details of the results of factor model No. 3 are shown in Fig. 6.

Since social security in the labour market is expressed in its two components—employment and wage security—we believe it is beneficial to carry out economic and mathematical modelling of its main threats separately in these areas. Therefore,

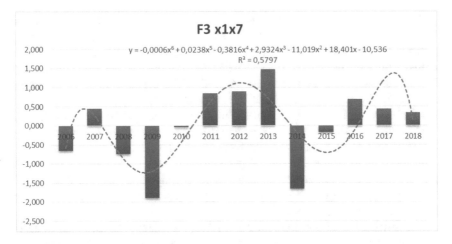

Fig. 6 The distribution results of the factor model of threats to the social security in the labor market No. 3 (F3) in 2006–2018 years. *Source* Defined by the author

in order to construct a factor model of the main threats to employment security, four threats indicators have been chosen, including: informal employment (X1), youth unemployment rates between 15 and 24 years of age (X2), youth unemployment rates between 25 and 29 years of age (X3), youth unemployment rates between 30 and 34 years of age (X4). According to the selected indicators, a component matrix of two generalized factors was calculated using the main component method (Table 4).

Under the condition that the specified results of the calculation of the component matrix are used, the equations obtained from the F1(ES) and F2(ES) models of employment security (ES) threats are as follows:

$$F1(ES) = -0.79X1 - 0.98X2 - 0.95X3 - 0.91X4 \tag{9}$$

$$F2(ES) = -0.6X1 - 0.06X2 - 0.24X3 - 0.34X4 \tag{10}$$

Table 4 The results of calculating the matrix of components of the context of factor models of the employment security threats

Initial signs	Components	
	F_1	F_2
X_1	−0.794	−0.597
X_2	−0.975	−0.063
X_3	−0.951	−0.241
X_4	−0.907	−0.339

Source Defined by the author based on using the SPSS software package

According to the above equations, a threat to employment security such as an increase in youth unemployment, such as an increase in youth unemployment, is particularly significant and has a negative impact on the resulting indicators, if data is used within the framework of the F1(ES) factor model. At the same time, it has the highest weight in the age group of 15–24. The negative value of the initial sign coefficient testifies to the negative role of strengthening the threats identified to social security in the labour market. In terms of the F2(ES) factor model, the most important value of the X1 coefficient in the investigated indicators reflects the significant negative impact of the increase in informal employment among the population. Furthermore, the second generalization factor is negatively correlated with the Youth Unemployment indicator, thus reducing the value of the factor itself. At the same time, with the rise in youth unemployment, the effectiveness coefficient of the studied factors decreases.

The distribution of the first generalized factor of threats to employment security within the F1(ES) factor model characterized the instability of the dynamics of changes in the analyzed indicators during 2006–2018, which identified periods of positive effects of the reduction in the level of representation of threats to employment security and negative effects of their increase (Fig. 7).

The resulting equation of the factor model F1 (ES) reflects the calculation of performance characteristics Y (demand and supply coefficients) when a general indicator of employment security (X) is present, and the determination coefficient reflects the close connection between the studied characteristics, thus forming the appropriate reliability level of the results obtained ($R^2 = 0.762$).

The results of the distribution of the second generalized factor of threats to employment security according to the F2(ES) factor model show the significant negative gravity of the deterioration of the economic crisis in 2014–2015, reflecting the high negative value of the factor (Fig. 8). The value of the determination coefficient is

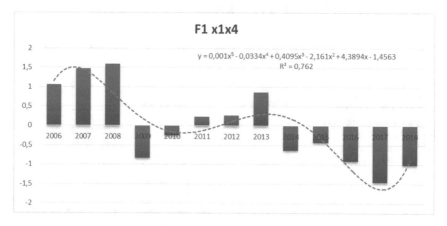

Fig. 7 The distribution results in the context of factor model of employment security threats F1(ES) in 2006–2018 years. *Source* Defined by the author

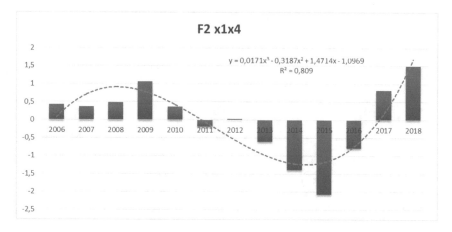

Fig. 8 The distribution results in the context of factor model of employment security threats F2(ES) in 2006–2018 years. *Source* Defined by the author

slightly higher than the previous model ($R^2 = 0.809$), allowing the possibility to further treat the results obtained in order to determine the impact of employment security threats on performance indicators.

Given the importance of threats to wage security (SS), we consider it appropriate to evaluate the impact of the relevant threats and model the generalized factor in the form of two variants of the factor model. The following indicators are selected for the purpose of the study: average monthly salary (X5), real wage index (X6), ratio of wages paid to wage fund (X7). Table 5 shows the value of the component matrix for two factor models identified by the main factor method.

Using the defined results of the component matrix calculation, the wage-security threat models F1(SS) and F2(SS) can be obtained in the form of equations:

$$F1(SS) = 0.93X5 + 0.69X6 - 0.21X7 \tag{11}$$

$$F2(SS) = 0.781X5 + 0.521X6 - 0.069X7 \tag{12}$$

Table 5 The results of the calculating of the matrix of the components in the contex of factor models of salary security threats

Initial signs	Components	
	F_1	F_2
X_5	0.93	0.781
X_6	0.69	0.521
X_7	−0.21	−0.069

Source Defined by the author based on using the SPSS software package

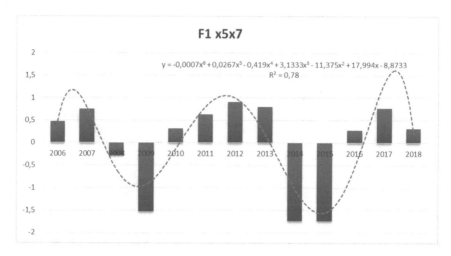

Fig. 9 The distribution results in the context of factor model of the salary security threats F1(SS) in 2006–2018 years. *Source* Defined by the author

According to the obtained equations, the increase in the ratio of wages to the wage fund (X7) has a negative impact on the effective sign of the two factors models that have been formed. At the same time, increases in average monthly wages and real wage indicators have a positive impact on labour supply and demand ratios, which are, in our view, a kind of indicator of social security levels in the labour market. The distribution of the first generalized factor for salary security threats formed in the factor model F1(SS) is shown in Fig. 9.

Thus, the crisis phenomena in the area of wage security are evidenced by the distribution of the general threat index in the years 2008, 2009, 2014 and 2015, as evidenced by the negative values of the initial signs studied. At the same time, the 2014–2015 crisis period is of greater severity, reflecting the increasing threat of increased wage payments due to the deterioration of the financial situation of economic entities and the decrease in the level of the real income of the population due to the intensification of inflationary processes. The reliability of the factor model is confirmed by the determination coefficient R^2, which is 0.78.

The distribution of the second generalized factor on the threat of salary security within the framework of the factor model F2(SS) shows positive changes in the labour market, taking into account some improvement in the country's monetary policy and in recent years in the social sphere, which, in turn, has a positive impact on the direction of salary organisation (Fig. 10).

The determination coefficients of this factor model correspond to a high level of reliability of the obtained results ($R^2 = 0.97$) and provide the possibility to use the obtained equations to further calculate effective indicators reflecting the situation of social security in the Ukrainian labour market.

It is particularly important to identify and show the main threats to social security in the labour market and their prospects for change in order to formulate strategic

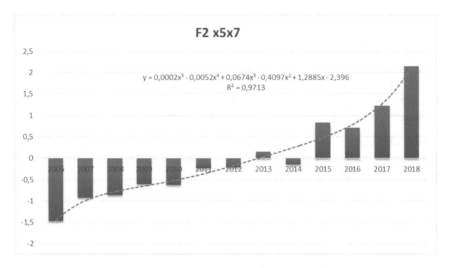

Fig. 10 The distribution results in the context of factor model of the salary security threats F2(SS) in 2006–2018 years. *Source* Defined by the author

directions to improve the situation by predicting relevant trends for each simulated threat. The results of the analysis of threat factors showed that the dynamics of threat change were unstable during the period 2006–2018. This clearly reflects the crisis periods 2008–2009 and 2014–2015, during which both the threats to job security and the threats to salary security increased (in particular the factors F1 and F2 of social security threats in the labour market). Consequently, at the current stage of the development of Ukrainian socio-economic relations, through the development of effective national policies based on strategic priorities to ensure social security in the labour market, the spread of youth unemployment is the main factor increasing the threat to employment security and the trend towards the reduction of real wages increases the threat to employment security The spread of youth unemployment is the main factor increasing the threat to employment security.

The important step in taking into account the results of modelling and its subsequent approval is the application of forecasting tools, which include expectations for significant qualitative changes in the future. Given the characteristics of the author's research on the key threats to social security in the labour market and the reflection of its impact on the ratio of labour supply and demand, it is desirable to use a prediction method based on a system of equations of interconnected dynamics that enable to obtain estimates not only of effective and factor signs.

In order to improve the quality of forecasting research, it is desirable to take into account such forecasting principles as:

– the aims determined to achieve the expected results of forecasting calculations;
– the systematic, which includes considering social security in the labour market as a system of interconnected phenomena, which is affected by major threats in the

form of informal employment, the growth of youth unemployment and problems related to wages;
- the accuracy that reflects the possibility of applying the maximum approximation of predictive calculations to the actual social security system on the labour market through accurate descriptions of the existing situation;
- the effectiveness of forecasting methods and indicators consisting of the application of modern forecasting methods and the use of reasonable indicators.

However, given the complexity of the Ukrainian socio-economic situation linked to Russian military aggression, it is not sufficiently relevant to predict the level of major threats in the fields of employment and salary.

4 Conclusion

Given the current trends in the functioning of the Ukrainian labour market, it is remarkable that, among the promising trends in the development of the labour and employment sector, negative factors causing the amplification of the crises and the amplification of existing problems are expected to dominate, determining the existence of a serious threat to social security. Following the study of the dynamics of employment and salary security parameters within the framework of the main threat factors model, the impact of each threat on the coefficient of labour supply demand ratio has been determined. For this purpose, the results of assessments of the impact of parameters such as informal employment levels, youth unemployment levels, average monthly wages, real wage indexes, wage arrears rates and wage fund were taken into account. The distribution of the general factors threatening social security in the labour market proves the instability of the dynamics of changes in the indicators analysed for the period 2006–2018 and the significant negative gravity of the deepening of the crisis phenomenon in 2008–2009 and 2014–2015, which reflect the high value of the factors.

All of these measures require the establishment of a set of measures aimed at ensuring the protection of the social interests of the population in the areas of employment and labour compensation that are fully based on legislative, socio-economic, organizational and managerial influence and are to be taken into account in the process of establishing effective state and regional policies, strategic priorities at the level of the economic unit. It should be noted that a timely adjustment of state policy and the adaptation of the relevant measures to existing changes is only possible if a thorough analysis of the situation of social security is carried out.

References

1. Hijzen, A., Menyhert, B.: Measuring Labour Market Security and Assessing Its Implications for Individual Well-Being. OECD: Social, Employment and Migration Working Papers. No. 175. OECD Publishing, Paris (2016). https://www.oecd-ilibrary.org/docserver/5jm58qvzd 6s4-en.pdf?expires=1582033692&id=id&accname=guest&checksum=6A7E7DA3E7903FA 4B447A117CF3E5CA7
2. Crouch, C.: Flexibility and security in the labour market: an analysis of the governance of inequality. ZAF **43**, 17–38 (2010). https://doi.org/10.1007/s12651-010-0031-9. https://link.spr inger.com/article/10.1007/s12651-010-0031-9#citeas
3. Wilthagen, T.: Managing social risks with transitional labour markets. In: Mosley, H., O'Reilly, J., Schömann, K. (eds.) Labour Markets, Gender and Institutional Change: Essays in Honour of Günther Schmid, pp. 264–289 (2002)
4. Grishnova, O.A., Kharazishvili, Yu.M.: Quality of life in the social security system of Ukraine: indicators, level, threats. Ukr. Econ. **11–12**, 157–180 (2018)
5. Shaulska, L., Doronina, O., Naumova, M., Honcharuk, N., Bondarevska, K., Tomchuk, O.: Cross-country clustering of labor and education markets in the system of strategic economic management. REICE: Revista Electrónica De Investigación En Ciencias Económicas **8**(16), 166–196 (2020). https://doi.org/10.5377/reice.v8i16.10681
6. Lisohor, L.S.: Forecasting the development of the labor market in Ukraine: problems and prospects. Labor Mark. Popul. Employ. **1**, 54–56 (2012)
7. Sadova, U.Ya., Stepura, T.M.: Jobs as the basis of a new employment policy and reproduction of the quality of human potential in the country. Labor Mark. Popul. Employ. **2**(55), 15–22 (2018)
8. Sydorchuk, O.G.: Social Security: State Regulation and Organizational and Economic Support: Monograph, 492 pp. LRIDU NADU, Lviv (2018)
9. Kozlovskyi, S., Bilenko, D., Kuzheliev, M., Lavrov, R., Kozlovskyi, V., Mazur, H., Taranych, A.: The system dynamic model of the labor migrant policy in economic growth affected by COVID-19. Glob. J. Environ. Sci. Manag. **6**(SI), 95–106 (2020). https://www.gjesm.net/art icle_40322_9d782b96c1e7cd7c4bef2547b0393905.pdf
10. Kozlovskyi, S., Nikolenko, L., Peresada, O., Pokhyliuk, O., Yatchuk, O., Bolgarova, N., Kulhanik, O.: Estimation level of public welfare on the basis of methods of intellectual analysis. Glob. J. Environ. Sci. Manag. **6**(3), 355–372 (2020). https://www.gjesm.net/article_38259_b 5e58d0c075bd7a9717cd6ab455f76b9.pdf
11. Bondarevska, K.V.: Strategy of Implementation of Social Security in the Labor Market of Ukraine. Dissertation on the receipt of the scientific degree of doctor of economic sciences. Vasyl' Stus Donetsk National University, Vinnytsia (2021). https://abstracts.donnu.edu.ua/art icle/view/9734
12. Petyak, Y.F.: Methodology of surveying experts to identify factors of information security of mobile devices. Proceedings **1**(50), 23–29 (2015)
13. Bondarevska, K.V.: Key threats and prospects of social security support in the labor market of Ukraine. Econ. Manag. Org. (2021). https://jeou.donnu.edu.ua/article/view/11122/11021
14. Kryvinska, N., Kaczor, S., Strauss, C., Gregus, M.: Servitization strategies and product-service-systems. In: 2014 IEEE World Congress on Services (2014). https://doi.org/10.1109/SER VICES.2014.52
15. Stoshikj, M., Kryvinska, N., Strauss, C.: Service systems and service innovation: two pillars of service science. Procedia Comput. Sci. **83**, 212–220 (2016). https://doi.org/10.1016/j.procs.2016.04.118
16. Economic Activity of the Population of Ukraine 2008: Statistical Collection, 232 pp. Derzhkomstat of Ukraine, Kyiv (2009)
17. Economic Activity of the Population of Ukraine 2018: Statistical Collection, 205 pp. State Statistics Service of Ukraine, Kyiv (2019)
18. Economic Activity of the Population of Ukraine 2009: Statistical Collection, 205 pp. Derzhkomstat of Ukraine, Kyiv (2010)

19. Economic Activity of the Population of Ukraine 2014: Statistical Collection, 207 pp. State Statistics Service of Ukraine, Kyiv (2015)
20. Business Analytics of Multidimensional Processes. Factor Analysis (2020). http://ebooks.git-elt.hneu.edu.ua/babap/8-1-id8-1.html

Formation of Marketing Competencies in Case of Startups Integration into the Intellectualized Market Space

Serhii Smerichevskyi⬤, Olha Polous⬤, Inna Mykhalchenko⬤, and Larysa Raicheva⬤

Abstract Due to the significant interest in the creation and operation of startups in existing mature markets this study takes place. Attention is paid on the marketing competencies of startup as driving force of this type of companies throughout their life cycle. To achieve the goals of the study, process, system, structural and situational approaches were used. The paper presents the model of startup life cycle with the allocation of opportunities for the formation of marketing competencies, which contains four main stages. The first stage is the generation of an idea, which highlights the competencies of entrepreneurial culture, creativity, development of ideas, i.e. the ability to analyze, to master modern knowledge and apply them in practical situations. The second stage is the intention, i.e. the transformation of the idea into the real business, which includes the ability to multiply scientific values, work in team, use marketing tools in innovations, analyze consumer behavior and identify market characteristics. The third stage is the launch of new business (startup), which includes the ability to measure, quantify the probability of business idea success, determine the necessary of tangible and intangible resources, business planning, search for funding, risk-taking and leadership skills. The fourth stage is the expansion with the ability to demonstrate multifaceted and complex competencies, leadership, strategic orientation and coordination, looking for funding opportunities, the ability to operate on international markets. The key barriers to the integration of startups into the Ukrainian market space are presented, with possible marketing measures to

S. Smerichevskyi · O. Polous (✉) · I. Mykhalchenko
National Aviation University, Liubomyra Huzara Ave, 1, Kyiv 03058, Ukraine
e-mail: olha.polous@npp.nau.edu.ua

S. Smerichevskyi
e-mail: serhii.smerichevskyi@npp.nau.edu.ua

I. Mykhalchenko
e-mail: inna.mykhalchenko@npp.nau.edu.ua

L. Raicheva
International Humanitarian University, Fontans'ka Rd, 33, Odesa 65000, Ukraine
e-mail: larisa_1991@ukr.net

eliminate them. We consider the analysis of effective tools and technologies of high-tech marketing, which would take into account the specifics of attracting venture financing at different stages of the startup life cycle, to be a promising area of further research.

Keywords Life cycle of the organization · Innovations · Opportunities · Development · Human capital · Financing · Capitalization

1 Introduction

Companies are classified according to different characteristics, distinguish between different types and kinds. In modern business conditions, attention is paid to startup companies as financial projects, the goals of which are the rapid return on financial investment and profit. Important aspects of startups development, their integration into the market space and the formation of key competencies of their founders are explored in the works of Choi S.-K., Han S., Kwak K.-T. [1], Del Bosco B., Chierici R., Mazzucchelli A. [2], Duckworth V., Farrell F., Rigby P., Smith R. M., Sardesh-mukh S. R., Combs G. M. [3], Harms R., Schwery M. [4], Kopera S., Wszendybyl-Skulska E., Cebulak J., Grabowski S. [5], Linton G. [6], Loufrani-Fedida S., Hauch V., Elidrissi D. [7], Passaro R., Quinto I., Thomas A. [8], West G. P., Taplin I. M. [9], Yang X., Sun S. L., Zhao X. [10]. The success of these companies is to respond quickly to market challenges, create new products and promote them. In order to work effectively, a startuper must have certain skills and develop competence in specific areas.

The purpose of the study is to determine the marketing competencies that startup founders must have in order to ensure targeted management of production processes, commercialization of innovative developments and ideas and successful integration of a new company into an intellectualized market space.

Modern economics has fairly wide range of tools that have formed variety of views on the understanding of the concept of "startup". It should be noted that this concept has many synonyms or close in meaning terms. For example, a spin-off company is a company that has got separated and been created on the basis of another business structure or academic structure; "start-up company" is a small innovative company, usually a science-intensive, venture company pre-sowing (start-up) cycle. Also, this concept is traditionally associated with the sector of high-tech and science-intensive products (services), which requires investment of knowledge and possession of relevant creative competencies, introducing into economic practice terms such as "high-tech startup", "successful expert" or "growth company". Of course, this type of company did not appear by itself, it was preceded by certain historical and economic factors and events.

In our opinion, intensifying the formation of marketing competencies is impossible without ensuring the effective integration of startups into the market space of the state, because they are an important stage in the life cycle of any company, which can further ensure its growth economy with new talents. In turn, the existing human capital is the basis for the formation of viable elements of a startup as an innovative form of business, ensuring the formation of a unique business concept, developing the necessary competencies and finding investment, forming a positive image and dedicated, active team. At the same time, marketing competencies may differ depending on the form of organization of startups, which necessitates their comprehensive research, understanding of the main features and synergistic combination of existing approaches to their understanding. Startup life cycle analysis is necessary to identify opportunities for the formation of marketing competencies. It's very often than new projects and ideas die on the stage of implementation, not to mention the stage of profitability and development. The effective functioning of the startup guarantees the success of business and further development.

Certain stages of creating a startup as well as the types of its financing and legal design have been appeared in the economic development of Ukraine. But it should be analyzed current stage and trends in startup development. It's important to find measures that will promote the further development of domestic startups, to recognize key components on which financial structures pay attention deciding to invest their funds in the development of the startup. Disclosure of these aspects will provide an opportunity to better understand the nature, operation and development of startups with the aim of formation of marketing competencies.

2 Materials and Methods

The study is based on the main provisions of the organization's theory, management and strategic planning, as well as the concept of stakeholders and the life cycle of the organization, the theory of decision making and the theory of motivation. To achieve the goals of the study, process, system, structural and situational approaches were used. The developed model is based on the decomposition method, which involves consideration of existing problems by groups and a deterministic comparison of each group with factors in the development of a startup at a certain stage of its life cycle.

Today, an extremely complex scientific problem is the calculation of the economic efficiency of a particular startup on which directly depend future management decisions, and, accordingly, marketing competencies that can be formed by participants of the startup at future stages of its development. Scientific practice has not developed a single methodological approach to determining the economic efficiency of a startup, which complicates the comparative evaluation of two or more projects to determine their competitiveness in an intellectualized market. Each individual approach considers the object of evaluation quite subjectively, based on specific sources of internal and external information on the functioning of the startup. Thus, none of the existing approaches can be defined as universal, as it has its advantages and limitations and cannot guarantee a single approach to the evaluation of startups, even within one sector of the economy. Therefore, an extremely important step in the context of economic support for the formation of marketing competencies is the choice of method that is best able to reflect the characteristics and conditions of the startup.

As a rule, the initial stage of a startup is associated with research aimed at obtaining a working prototype (a product that is the result of realizing the intellectual potential of startup founders), which requires appropriate financial costs, which correlate with the complexity of the developed product and often require the founders of relevant marketing competencies, which would attract the necessary resources for its creation and to finance certain stages of the process. At the same time, already at the initial stage of its development, the startup faces significant barriers to market entry, which delay the start of commercialization of development results.

Comparative analysis and identification of potential competitors that are typical for traditional enterprises is significantly complicated for startups due to lack of information about their own activities and the activities of potential competitors, due to the desire of founders to hide trade secrets and the form of development they use (for example, interaction with a specific accelerator). As a result, it is impossible to even estimate the future value of the company based on its comparison with other startups due to the high degree of uncertainty in the external environment of its operation.

An analysis of studies related to the problems of innovative and technological development suggests that the issues of creating and developing start-ups are currently not sufficiently developed, especially in terms of methodological support for the development of marketing competencies of their founders, without providing science and practice with proper methodological tools. The problem of low motivation of the authors of business ideas for the practical implementation of the research results is due to the lack of simple tools that characterize the startup development technology and the possibility of its successful implementation. As a result of the incompleteness of the competencies of the authors of business ideas and the lack of the possibility of improving the project through interaction with other participants in the system, the problem of the quality of projects also arises. Since each startup is unique and common tools for working with them are not enough, an individual approach to the modernization of each startup is necessary. As a rule, managers develop technology and the product does not understand how to bring it to the market and make money on it, as a result of which the problem of commercializing the results of startups arises, which consists in the lack of vision of the business component of startup managers.

3   Results and Discussion

By formulating the definition of "startup", approaches to understanding its nature and characterization of this type of business, it is important to take into account the dominant trends in world markets, which have been actively transformed in recent times. Table 1 shows the prerequisites that contributed to the formation of traditional definitions of startups. This approach to the study of the concept of "startup" allows not only to define its essence, but also to form key features of modern startups (international nature, openness, going beyond one structure, independence, mostly virtual format, creativity, etc.) and to some extent to predict possible transformations of this form of doing business under the influence of the latest trends in socio-economic development.

However, despite the above systematization of the prerequisites for the formation of startups, today there is no single approach to defining the term "startup". Basic existing approaches are shown in Table 2.

Consideration of the approaches listed in Table 2 allows us to conclude on the market orientation of startups (according to the approaches of E. Rice, S. Blanc, L. Rayner, P. Graham) [11–14], which emphasizes the importance of further marketing classification of startups to establish their role in the process of human capital intellectualization. It should be noted that the basic approach that is most common and used in economic practice in defining the concept of "startup" is the approach proposed by S. Blank and B. Dorf in "Startup: the founder's desktop book" [12].

Despite the fact that a startup is a relatively young form of doing business, there are many approaches to the classification of startups. Thus, B. Claris uses such classification features of small innovative companies as market demand/technological

Table 1 Prerequisites for the formation of startups as an innovative form of business

Time period	1980s–1990s	1990s–2000s	2000s–nowadays
Key global trends	The development of the computer revolution and the boom in venture capital	The development of the Internet and the emergence of large number of Internet companies and the boom of "dotcoms"	The development of a knowledge-based economy based on accelerated STP, globalization of world economic relations and deregulation
Start-up forms	"Garage" firms and internal corporate offices, which were the initial phase of development of high-tech companies	Fast-growing innovative firms in the field of IT, small firms-innovators and imitators	Independent small structures—the result the of the dissemination of knowledge from large corporations, universities in the open innovation market (internationalization nature, going beyond one structure, openness)

Source Systematized by the authors based on the study of literature sources

Table 2 Approaches to the definition of "startup"

No.	Source/author	Definition	Feature of definition, approach
1	D. Ponomarev [15]	"Startup is an organizational form of innovation. This is a company with a short history of operations, which is in the process of developing or researching promising markets"	This definition represents a startup as a company, but a startup is primarily an innovative project for which a company can be created
2	achupryna.com	"A startup is a web project or site that aims to generate revenue quickly"	The definition reflects the initial concept of a startup when it was identified with an internet project
3	pashigrev.com	"Startup is a company that is at the initial stage of its activity; organized on the money of the founders and seeks to increase capitalization as the product develops in the hope that it will be in demand"	The definition characterizes the company's presence at the initial stage of existence and its desire to increase capitalization as the product
4	E. Reese [11]	"A startup company is any newly created organization designed to develop new product or service in conditions of extreme uncertainty"	The definition reflects the presence of a high degree of risk associated with uncertainty and rapid changes in the external environment of the company
5	S. Blank and B. Dorf [12]	"Startup is a temporary structure that seeks a renewable, profitable and scalable business model"	Startup is not considered a full-fledged company, only as a tool for development
6	L. Rayner with co-authors [14]	"Startup is a company that is usually engaged in the design and implementation of innovative processes of development, validation and research of key markets"	This definition almost identifies a startup project with a marketing company
7	US Small Business Administration	"Startup is a technology that is usually technology-oriented and has a high potential for development"	The definition considers only the innovative side of the startup project
8	P. Graham [15]	"Startup is a company designed for rapid growth"	The definition is not specified, there is no clear explanation of the ways of development
9	A. Yevseychev [7]	"Startup is a process of implementing an idea in a short time and, as a rule, with limited resources of an atypical project, which is new"	Generalized definition, which combines a number of features of a startup project and emphasizes limited resources

supply and technological certainty/uncertainty [1]. This classification is character-ized by the fact that within its practice, the most common spin-in companies were organized by specialists who are from recognized and successful companies.

Traditionally, such companies work to develop products and technologies that are focused on the goals of the "parent company" (to some extent, the so-called "subsidiaries"). In this case, all economic activities, including venture capital, are reflected in a separate balance sheet of the company. In the process of growth and consistent overcoming of technological stages spin-in-company is absorbed by the "parent" company, which can use it as a lever to attract further rounds of funding to achieve maximum stages of profitability [16].

On the other hand, S. Koster in his work [5] identifies four possible forms of orga-nization of startups. Two dimensions are used for classification: "resource sharing" and "parent company support". Individual startups are based on resources, which are mostly received from an entrepreneur, an individual. Spin-out companies, in turn, use resources that have been accumulated in other companies, but the formation or matu-ration of which is not supported by these companies. At the same time, traditional spin-off companies, like spin-outs, are built on existing resources, but are supported by the "parent" company during the maturation phase. And support is a multi-level and continuous process. In addition, 100% of organizational spin-off companies are launched by the "parent" company by separating its branches or divisions and transforming them into a new independent business.

Most Western experts in the field of startup research (S. Koster, B. Clarice, N. Morey and A. Hirman, S. Blank and E. Rees, etc.) argue that academic spin-off companies are quite stable structures and have a higher survival rate compared to other forms of startups within one sector of the economy. They associate this phenomenon with a specific organizational scheme and "ecosystem" of their existence, which allow even inefficient organizations to continue their existence. Also, according to the authors considered above, academic spin-off companies benefit from relationships with the "parent" research organization or university (in terms of transferring their image, reputation and other positive externalities) and focus on R&D (availability of technological base and accordingly trained personnel). The main purpose of these startups is to commercialize their own ideas and promising developments in order to capitalize on the university's share in the venture. In contrast, corporate spin-off companies are focused on the commercialization of their own ideas and promising developments in order to occupy profile/non-profile promising markets and make a profit. According to J. Moore, the reasons that allow large corporations to benefit by transferring their own developments to external legal entities are as follows:

- product development in a particular niche within a large-scale corporation is too costly;
- the promotion of new technologies requires greater entrepreneurial initiative and motivation of key players (in a large corporation it is difficult to achieve this level of return);
- such projects are associated with high risks, so it is more profitable for corporations to share them with venture investors;

- it is simply impossible to engage in various activities at once, so corporations seek to focus on key business processes, transferring secondary to external partners;
- the "parent" company always has the opportunity to buy a spin-off company [17].

S. Koster expanded the classification of startups by including a new type of startups that exist in practice—"cluster spin-off companies", the basis for which is to support the innovation cluster. The author also adds a number of subtypes of academic and corporate spin-off companies, using criteria such as availability or attraction/acquisition of resources (technology, ideas, scientists, companies, etc.), type of founder. In light of the active spread of clustering processes in the world, accompanied by the development and expansion of relevant startups network based on innovation clusters, S. Koster emphasizes the need to acquire knowledge about the peculiarities of their creation and cluster support. Knowledge of the specifics of such startups helps their founders to develop an effective plan for their integration into innovation clusters, the ecosystem of which is different from the usual fragmented markets. S. Koster also notes the existence of significant differences between corporate and academic startups in the middle of their group, which leads to the presence of corresponding differences in their creation and further management.

Based on the concept of technological systems proposed by S. Glazyev, it is possible to conduct another classification of startups based on their assignment to a certain technological system. Accordingly, startups that work on the principles of the existing (newest) way are fundamentally new and will strengthen the transition to the next technological order. In other words, in the Ukrainian reality, the transition to the seventh technological order is possible by increasing the number of high-tech startups with their potential to intellectualize human capital. Knowledge of innovative methods of creating such startups and the specifics of their management, which distinguishes them from the management of startups that existed under previous technological systems, can increase the number of high-tech projects and accelerate the transition to the latest technology, thereby enhancing human capitalization. This approach finds its support in the classification of startups presented by S. Blank, which is based on market types [12].

In early 2000, researchers began to actively analyze the life cycle of startups, highlighting the characteristics of each stage and identifying management practices and marketing competencies that will promote their active growth. We propose to consider a life cycle model aimed at identifying the stages that startups go through in their lifetime (identifying startup ecosystem entities to operate for obtaining the necessary knowledge and resources) [7, 9]. For each stage, the life cycle model determines the specific of functioning, organizational resources and features of the entrepreneur, external actors involved in support activities, through which startups can gain the necessary knowledge (experience, human capital) and resources to develop entrepreneurial features, achieve concrete results and overcome possible crises and problems or continue to develop (Fig. 1).

According to Fig. 1, the model is based on the division of the startup life cycle into four stages, which can be described as follows:

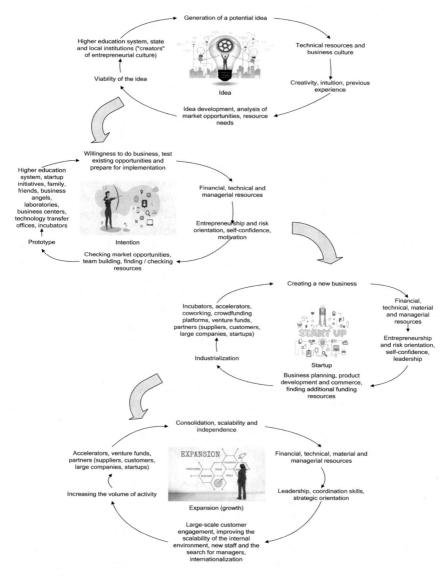

Fig. 1 Startup life cycle model with opportunities for the formation of marketing competencies

1. The idea. A potential startuper (or founding team) focuses on developing and generating a possible idea. In particular, he or she is trying to understand whether an idea has the potential (whether it is innovative) that can solve a major problem or meet the needs of the consumer [18]. In other words, he or she seeks to understand whether there is a market opportunity to implement a startup [3]. A potential startuper also begins to identify the necessary resources and their availability, without even considering starting a new business. At this stage, the

most important features of startuper are usually creativity, intuition and previous experience, which support (and provide) the generation of potential business ideas [19].

In addition, the startuper needs technical resources aimed at analyzing and adequately determining the viability of the idea. Thus, the startuper must be provided with adequate technical resources and set of entrepreneurial skills, culture, as well as develop the necessary personal abilities and competencies. The key external actors in this phase are the higher education system, startup competitions and local government agencies as the main providers of knowledge and support for business idea development [8]. This initial stage is characterized by a significant share of uncertainty and possible risks [20].

2. Intention. At this stage, the potential startuper focuses on the ability to turn a business idea into a real business. It is crucial that market opportunity testing can be done using the so-called "truth moment" method, when potential customers need to show interest in a business idea [6]. At the same time, the startuper begins to determine the amount of financial resources needed and look for start-up funds and investors (usually family, friends, business angels, venture funds, etc.). In addition, in order to obtain technical and managerial resources, a potential startuper must cooperate with some subjects of the business ecosystem (incubators, technology transfer offices, business centers, universities, etc.) [21].

These resources are crucial to make intent more stable and viable in highly competitive environment, as well as to develop prototype of future product/service. Therefore, at this stage, the startuper should develop formal and informal relationships with various participants in the ecosystem, conducive to the startup, to obtain the necessary resources, capital and support [10]. The key features of a startuper at this stage should be motivation, risk assessment and propensity to do business in the selected market segment, as well as self-confidence. This stage can be considered complete when product/service prototype (or standard/protocol/beta, etc.) has been developed.

3. Startup. The startuper starts new business [15]. At this stage, start-up entrepreneurs must be able to measure, quantify the probability of business idea success and determine the necessary tangible and intangible resources for its successful development. Thus, the startuper can be considered as start-up entrepreneur who devotes his time and effort to founding a new viable, competitive and independent company. The launch phase includes technological and commercial development and formal business planning, as well as the search for additional and more fundamental resources to fund future activities [4]. To this aim, startupers can turn to traditional sources (financial institutions, venture funds) [22], as well as crowdfunding platforms [23]. In addition, incubators, accelerators and coworking spaces play an important role at this stage, which can help to obtain additional management, technical and physical resources. The startuper must build a system of agreements with reliable customers and suppliers, as well as develop relationships with other external partners. The key features of a startuper at this stage should be entrepreneurial and risky (risk-oriented) focus,

self-confidence and leadership. Also at this stage, a new product/service enters the market, and the startup should be able to issue the first invoice (be legalized).

4. Expansion (growth). According to S. Blank and B. Dorf, at this stage of the life cycle the startup should become independent [12, 24]. Startuper needs to develop new skills, such as processing more information, motivating and coordinating more employees, communicating with new customers and suppliers, finding international markets and partners, and developing the ability to delegate and distribute activities. The startuper must also show multifaceted and complex competencies and abilities in terms of leadership, strategic orientation and coordination, finding funding opportunities, etc. The key activities at this stage are mass involvement of customers, improving the scalability of the internal environment, new staff and hiring qualified managers (if necessary), internationalization of activities, selection of suppliers.

The start-up industry in Ukraine often faces a number of challenges related to bureaucratic difficulties in doing business, as well as contradictions in legislation that require a deep understanding of this aspect of the generated business idea, require significant time and financial investment. However, international experts in the field of venture financing note some improvement in this situation in Ukraine. Thus, the amount of venture capital investments in Ukrainian technology companies in 2020 reached a record figure of $571 million, according to AVentures Capital.

The largest investments in Ukrainian startups came from global companies that consider Ukraine as a country for R&D offices. The reports state that out of the total amount of investments Ukrainian startup-beginners were received $42 million, and at the stage of round A—$119 million.

It should also be noted that 62% of all investments in companies with Ukrainian roots in 2020 attracted four companies: GitLab, Creatio, Restream and Airslate. Reface.io was recognized as the fastest growing Ukrainian startup in 2020 with 70 million downloaded applications since the beginning of 2020 (AVentures Capital, 2021) [25].

Thus, today most startups should strive to improve marketing competencies to better understand the business interests of their main customers, interact with big business and build a mechanism of interaction and cooperation with other participants in the innovation ecosystem to develop and further intellectualize their activities. In order to highlight measures that will promote the further development of domestic startups, we have identified key barriers to the integration of startups into Ukrainian market space (Fig. 2).

Thus, based on the results of research and study of the peculiarities of startups in today's realities, we can identify a number of measures that will further integrate startups into the domestic market space and ensure the formation of marketing competencies at all stages of their life cycle:

1. Shifting priorities to conducting activities in one target segment and its preliminary in-depth analysis. According to J. Moore [26], it is necessary to choose only one segment in which the company will build the foundation to meet the needs of its customers. If there are several such segments, it will be more difficult

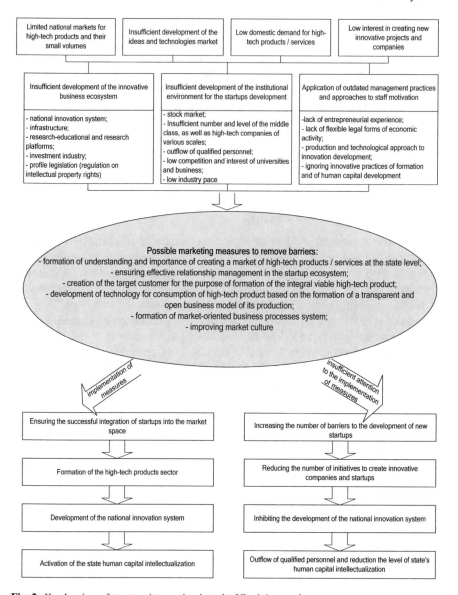

Fig. 2 Key barriers of startups integration into the Ukrainian market space

to scale innovative business, when it will be necessary to modify the product to different customer requirements and choose specific marketing tools to boost sales. Thus, focusing on one segment maximizes the benefits of segmentation. According to B. Cooper and P. Vlaskovic [27], most startups fail not because of a lack of product development, but because of a lack of market development. In addition, the search for the target segment should be carried out in stages, constantly testing and checking their hypotheses in relation to consumers, the satisfaction of which requires the development of appropriate innovative solutions. The importance of this idea is also confirmed by S. Blank in his proposed concept of consumer development.

2. Rational approach to creating a minimally viable product at the appropriate stage of the startup life cycle and its further development. According to E. Rice, the Minimum Viable Product (MVP) is a version of the product that allows the team to collect the maximum amount of verified customer information at minimal cost. However, S. Blank interprets it as a minimal set of characteristics that make up the product, stimulating feedback from consumers. In both cases, MVP illustrates the key problem (need) of consumers and demonstrates a solution that can offer a developed innovative product [28]. Accordingly, within the framework of this idea it is necessary to determine the narrowest set of features of such product, for which the customer is willing to pay when purchasing the first version. At the same time, the final MVPs test the business model itself, and their intermediate versions—its components that can carry high risks. It should be noted that the MVP in a startup must evolve in stages, going from the concept of an innovative product to full-scale marketing. In addition, the MVP needs to be tested and verified taking into account customer feedback (partner networks), determining how well it solves the identified problem, which will allow it to be corrected in the future. This approach will further expand the MVP within the key business idea after the emergence of larger number of its users.

3. Focusing on the use of open business models. The key startup strategy includes a search engine capable of maintaining the viability of the business and testing the hypothesis using its basic elements based on feedback from potential users, buyers and partners (the required ecosystem). Thus, in the early stages of startup development it is important to form a business model project, i.e. to determine how the company plans to work in the interests of users and in its own interests, as well as check and confirm its correctness: product characteristics, pricing, distribution, customer engagement strategy. To do this, it is very important to adhere to the principles of continuity and speed—it is necessary to quickly develop key elements of the business model and promptly receive feedback from the consumer ecosystem in order to improve it. In addition, the type of developed business model must be open, which will promote the development of the idea of joint creation of market value with all participants in the startup ecosystems.

   According to Blanc, in most developed and promising sectors, user feedback and openness are more important than secrecy, and the best results are obtained through constant communication with consumers. In our opinion, it is advisable to use the structure developed by A. Osterwalder and I. Pinier in their work [27] as templates for developing a startup business model.

4. Clear definition of the startup leader and his human capital (team). According to some modern venture investors, the success or failure of a startup largely depends on the quality of the team (as well as the personality of the founder): its efficiency, coherence, experience and potential (the idea is responsible for only 5% of successes, another 25% is luck, 70%—depends on team). Thus, in 90%

of cases, the reasons for the startup to fall into the "valley of death" is the low level of business training of the team (including internal conflicts between key members) and only 10%—errors from R&D [29].

It should also be noted that when making an investment decision, venture investors primarily evaluate the team—the "intellectual basis" of the startup. The team is also important for business angels in the early stages of startup development, as well as for seed investors and funds that make small investments in startups in the early stages of their development. These can be explained by the fact that investors have less information about the startup that they intend to finance. In this situation, they have to focus primarily on the experience of the team, its business behavior and participants.

5. Understanding of definition features of necessary infrastructural functioning frameworks. Startups are constantly operating in conditions of limited resources, which is often defined as a source of hidden opportunities for them, as it forces to rationally dispose of those resources that are available. The main asset of modern startups is usually intellectual property, so their key business functions are related to its ownership and development. As usual, startups do not need laboratories, complex equipment, but they need specialists who can manage innovation processes. It is important to note that most startups have no experience with intellectual property—there is no understanding of the role of this intangible resource in the company's development strategy, and to the protection of know-how is given very little attention [30]. In addition, the key technological stages, as well as all the processes that form the cost, must also be implemented independently. In addition to the material infrastructure for the successful operation of the startup it's necessary to have intangible components. Key attention should be paid to the openness of the communication platform of the startup's interaction with other ecosystem participants for the joint creation of market value. Accordingly, startup participants should have access to all the resources needed to organize such technology platform.

6. Orientation of startup business processes on the market. According to S. Blanc, the main principle of building successful startup is to build processes based on the consumer (Customer Development principle), not on the product (Product Development principle). It should be noted the importance of active application of marketing principles as the basis of the idea of "open innovation", which will form an effective ecosystem around the customer and focus on the market from the beginning, helping to intellectualize human capital of all participants [31].

Thus, in order of effective startup, team of professionals, market presence (sufficient size) and market prospects (consumer needs), as well as technological potential (embodied in an innovative product in demand on the market) are necessary. These key components help the startup functioning effectively and developing into a full-fledged large-scale business. Financial structures (venture investors, business angels, etc.) pay attention to them when deciding to invest their funds in the development of a startup [32].

The implementation of the above conditions will allow the startup to integrate into the market space, taking into account the following priorities:

- focus on service (application of customization and personalization technologies);
- joint creation of value (network approach involving all participants of the ecosystem);
- the presence of relationships based on trust and personal contacts both inside the company and outside;
- strategic segmentation;
- lack of rigid hierarchy (small integrated teams, networks) and adhocracy (in which all decisions are made together);
- intellectualization of consumers, communication with them (personalized), dialogue with them;
- application of innovative marketing tools (co-creation, crowdsourcing).

7. Understanding the peculiarities of attracting venture financing. In the initial stages (risky for the startup), the company (as usually) owns only one intangible asset—the idea of future business and a team of entrepreneurs (their human capital). Therefore, traditional sources of capitalization are not available at this stage. The main source of financing is venture capital, which, unlike credit, is provided on non-repayable, interest-free and unsecured basis. The venture investor, by providing his funds for using, exchanges them for a share in the initial capital, hoping to receive significant "premium" for the risk due to the multiple increase in the value of the startup during its development. It is important to note that a venture capitalist not only invests his own funds, but also acts as a professional manager of investment pool funds provided by third parties and organizations that trust his business competencies and expect to receive dividends.

Therefore, in relation to the level of risk at each stage of startup development, venture investors should be divided into appropriate categories of investment specialization and strategy. At the initial stage of creating and developing a startup, the risks of the project are quite significant and currently requires significant investment. However, an investor can expect a significant share of the business in exchange for relatively small amounts of financing. In the process of successful growth of the startup, the risks are reduced, the company requires more and more "heavy" investments to start production and gain market share, but at this stage the investor can count on a smaller share of "premium" on investment. Thus, startups are financed in several stages, each of which involves a "specialized" investor who is interested in investing in specific business processes.

4 Conclusions

For the effective formation of marketing competencies of startups during entering intellectualized markets, the interest in cooperation of all participants of the national innovation system of the state is fundamentally important. At the same time, each of the stakeholders must clearly understand their own goals, focus their efforts on solving specific tasks and have appropriate motives for further activities. The state can increase the volume and quality of innovation projects, promote the introduction of innovative business on the world market and create a competitive innovation sector of the economy. Innovative accelerators will receive new innovative projects of high quality, able to compete with their global counterparts and bring profits. Startup founders will have the opportunity to grow into a full-fledged company capable of generating higher profits. The above-mentioned cooperation of all participants of the innovation process will help to accelerate the receipt of the desired result on the basis of a comprehensive approach to solving problems in the formation of marketing competencies.

The unity of process, system and structural approaches to startup and marketing competencies management will maximize the benefits of its implementation. Thus, consideration of indicators of the level of development of the startup and the compliance of marketing competencies as an element of internal business processes and further adjustment of negative elements will ensure stable development of the startup. The system approach allows to look at the startup as an open system that actively interacts with other components of the innovation system of the state. At the same time, the internal structure of the startup combines the resources and components it needs to achieve its goals, and the external—ensures the receipt of the necessary resources and has a positive or negative impact on its operation. The structural approach allows grouping of all problems and factors influencing the intensification of startup development in order to analyze them in detail and ensure the formation of conditions for further operation and development of the startup. In some cases, in order to develop marketing competencies aimed at providing flexibility to respond to changes in the external environment of startups and develop new management practices, a situational approach can be used to bring the existing management system in line with market challenges caused by economic, political, social and technological shifts.

Thus, in order to integrate startups into the market space and intensify the intellectualization of human capital, it is necessary to address to the concept of consumer development—to the market ("customer development") and gradually move away from the priority of the concept of product development, due to the realities of modern business processes that dominate on the high-tech Ukrainian and world markets.

Further integration of startups into the market space and ensuring the formation of marketing competencies is possible through activities in one target segment, creating a minimally viable product, using an open business model, focus on team implementation of the startup, defining the necessary infrastructure framework, market orientation, involvement venture financing. It should be taken into account the key barriers to integration into the market space, focusing on the stage of expansion of the startup. Building the startup should initially be based on the application of marketing principles adapted to the sector of its operation, which is quite a difficult task, because at the moment there are no effective tools and technologies of high-tech marketing that would take into account the specifics of attracting venture funding at different stages of the startup life cycle, which is a promising area for further research.

References

1. Mohapatra, S.K., Prasad, S., Bebarta, D.K., Das, T.K., Srinivasan, K., Hu, Y.C.: Automatic hate speech detection in English-Odia code mixed social media data using machine learning techniques. Appl. Sci. **11**(18), 8575 (2021)
2. Alshalan, R., Al-Khalifa, H.: A deep learning approach for automatic hate speech detection in the Saudi Twittersphere. Appl. Sci. **10**(23), 8614 (2020)
3. Lingiardi, V., Carone, N., Semeraro, G., Musto, C., D'Amico, M., Brena, S.: Mapping Twitter hate speech towards social and sexual minorities: a lexicon-based approach to semantic content analysis. Behav. Inf. Technol. **39**(7), 711–721 (2020)
4. Watanabe, H., Bouazizi, M., Ohtsuki, T.: Hate speech on Twitter: a pragmatic approach to collect hateful and offensive expressions and perform hate speech detection. IEEE Access **6**, 13825–13835 (2018)
5. Bisht, A., Singh, A., Bhadauria, H.S., Virmani, J., Kriti: Detection of hate speech and offensive language in Twitter data using LSTM model. Recent trends image signal process. Comput. Vis. 243–264 (2020)
6. Al-Hassan, A., Al-Dossari, H.: Detection of hate speech in social networks: a survey on multilingual corpus. In: 6th International Conference on Computer Science and Information Technology, vol. 10, pp. 83–100 (2019)
7. Pawar, A.B., Gawali, P., Gite, M., Jawale, M.A., William, P.: Challenges for hate speech recognition system: approach based on solution. In: International Conference on Sustainable Computing and Data Communication Systems (ICSCDS), pp. 699–704. IEEE (2022)
8. Thiago, D.O., Marcelo, A.D., Gomes, A.: Fighting hate speech, silencing drag queens? Artificial intelligence in content moderation and risks to LGBTQ voices online. Sex. Cult. **25**(2), 700–732 (2021)
9. Fitria, T.N.: Artificial intelligence (AI) technology in OpenAI ChatGPT application: a review of ChatGPT in writing English essay. ELT Forum: J. Engl. Lang. Teach. **12**(1), 44–58 (2023)
10. Kwarteng, J., Perfumi, S.C., Farrell, T., Third, A., Fernandez, M.: Misogynoir: challenges in detecting intersectional hate. Soc. Netw. Anal. Min. **12**(1), 166 (2022)
11. Akuma, S., Lubem, T., Adom, I.T.: Comparing bag of words and TF-IDF with different models for hate speech detection from live tweets. Int. J. Inf. Technol. 1–7 (2022)
12. Chhabra, A., Vishwakarma, D.K.: A literature survey on multimodal and multilingual automatic hate speech identification. Multimed. Syst. 1–28 (2023)
13. Laaksonen, S.M., Haapoja, J., Kinnunen, T., Nelimarkka, M., Pöyhtäri, R.: The datafication of hate: expectations and challenges in automated hate speech monitoring. Front. Big Data **3**, 3 (2020)
14. Yadav, A.K., Kumar, M., Kumar, A., Shivani, Kusum, Yadav, D.: Hate speech recognition in multilingual text: Hinglish documents. Int. J. Inf. Technol. **15**(3), 1319–1331 (2023)
15. Duwairi, R., Hayajneh, A., Quwaider, M.: A deep learning framework for automatic detection of hate speech embedded in Arabic tweets. Arab. J. Sci. Eng. **46**, 4001–4014 (2021)
16. Velankar, A., Patil, H., Joshi, R.: A Review of Challenges in Machine Learning Based Automated Hate Speech Detection (2022). arXiv:2209.05294
17. Mossie, Z., Wang, J.H.: Social network hate speech detection for Amharic language. Comput. Sci. Inf. Technol. 41–55 (2018)
18. Ullmann, S., Tomalin, M.: Quarantining online hate speech: technical and ethical perspectives. Ethics Inf. Technol. **22**, 69–80 (2020)
19. Chiu, K.L., Collins, A., Alexander, R.: Detecting Hate Speech with GPT-3 (2021). arXiv:2103.12407
20. Prokipchuk, O., Vysotska, V., Pukach, P., Lytvyn, V., Uhryn, D., Ushenko, Y., Hu, Z.: Intelligent analysis of Ukrainian-language tweets for public opinion research based on NLP methods and machine learning technology. Int. J. Mod. Educ. Comput. Sci. (IJMECS) **15**(3), 70–93 (2023). https://doi.org/10.5815/ijmecs.2023.03.06
21. Roy, S.G., Narayan, U., Raha, T., Abid, Z., Varma, V.: Leveraging Multilingual Transformers for Hate Speech Detection (2021). arXiv:2101.03207

22. Defersha, N.B., Kekeba, K., Kaliyaperumal, K.: Tuning hyperparameters of machine learning methods for Afan Oromo hate speech text detection for social media. In: 4th International Conference on Computing and Communications Technologies, pp. 596–604. IEEE (2021)

23. Chhikara, M., Malik, S.K.: Classification of cyber hate speech from social networks using machine learning. In: 11th International Conference on System Modeling & Advancement in Research Trends (SMART), pp. 419–423. IEEE (2022)

24. William, P., Gade, R., Chaudhari, R., Pawar, A.B., Jawale, M.A.: Machine learning based automatic hate speech recognition system. In: International Conference on Sustainable Computing and Data Communication Systems (ICSCDS), pp. 315–318. IEEE (2022)

25. Fernando, W.S.S., Weerasinghe, R., Bandara, E.R.A.D.: Sinhala hate speech detection in social media using machine learning and deep learning. In: 22nd International Conference on Advances in ICT for Emerging Regions (ICTer), pp. 166–171. IEEE (2022)

26. Mozafari, M., Farahbakhsh, R., Crespi, N.: A BERT-based transfer learning approach for hate speech detection in online social media. In: Complex Networks and Their Applications VIII: Volume 1 Proceedings of the Eighth International Conference on Complex Networks and Their Applications Complex Networks, vol. 8, pp. 928–940. Springer (2020)

27. Sultan, D., et al.: Cyberbullying-related hate speech detection using shallow-to-deep learning. Comput. Mater. Continua **74**(1), 2115–2131 (2023)

28. Mykytiuk, A., Vysotska, V., Markiv, O., Chyrun, L., Pelekh, Y.: Technology of fake news recognition based on machine learning methods. In: CEUR Workshop Proceedings, vol. 3387, pp. 311–330 (2023)

29. Khanday, A.M.U.D., Rabani, S.T., Khan, Q.R., Malik, S.H.: Detecting Twitter hate speech in COVID-19 era using machine learning and ensemble learning techniques. Int. J. Inf. Manag. Data Insights **2**(2), 100–120 (2022)

30. Sandaruwan, H.M.S.T., Lorensuhewa, S.A.S., Kalyani, M.A.L.: Sinhala hate speech detection in social media using text mining and machine learning. In: 19th International Conference on Advances in ICT for Emerging Regions, vol. 250, pp. 1–8. IEEE (2019)

31. Stoshikj, M., Kryvinska, N., Strauss, C.: Service systems and service innovation: two pillars of service science. Procedia Comput. Sci. **83**, 212–220 (2016)

32. Kryvinska, N., Strauss, C.: Conceptual model of business services availability vs. interoperability on collaborative IoT-enabled eBusiness platforms. In: Internet of Things and Inter-cooperative Computational Technologies for Collective Intelligence, pp. 167–187 (2013)

Information Systems for Vehicles Technical Condition Monitoring

Volodymyr Volkov, Igor Gritsuk, Igor Taran, Tetiana Volkova, Volodymyr Kuzhel, Andriy Semenov(iD), and Oleksandr Voznyak(iD)

Abstract The current state analysis of road transport has shown that the existing system of monitoring the technical condition no longer meets modern requirements for maintaining the operational efficiency of cars. This can be solved by introducing into the cars' technical operation the principles of the "adaptive" system of managing cars' technical condition, the basis of which is the creation of information systems of organizational and functional control and support of the processes of their technical operation. A mathematical model of the information system for evaluating parameters of the vehicle's technical condition in operating conditions has been developed, and the "IdenMonDiaOperCon "HNADU-16"" information software complex has been created on its basis". As a result of experimental studies, with the help of an information software complex, the determination of the actual parameters of the technical condition of the car itself, the correction of the operating conditions, as well as the exact determination of the location and the exact time according to the parameters received from the navigation satellite systems performed by the GPS

V. Volkov · T. Volkova
Kharkiv National Automobile and Highway University, Kharkiv, Ukraine
e-mail: volf-949@ukr.net

T. Volkova
e-mail: wolf949@ukr.net

I. Gritsuk
Kherson State Maritime Academy, Kherson, Ukraine
e-mail: gritsuk_iv@ukr.net

I. Taran
Dnipro University of Technology, Dnipro, Ukraine

V. Kuzhel (✉) · A. Semenov
Vinnytsia National Technical University, Vinnytsia, Ukraine
e-mail: kuzhel_v@vntu.edu.ua

A. Semenov
e-mail: semenov.a.o@vntu.edu.ua

O. Voznyak
Vinnytsia National Agrarian University, 3 Soniachna St., Vinnitsia 21008, Ukraine

© The Author(s), under exclusive license to Springer Nature Switzerland AG 2024
A. Semenov et al. (eds.), *Data-Centric Business and Applications*, Lecture
Notes on Data Engineering and Communications Technologies 195,
https://doi.org/10.1007/978-3-031-54012-7_4

receiver, and the exchange of this information with workplace monitoring of vehicles and other participants in monitoring work processes of vehicles.

Keywords Road transport · Vehicle · Cars technical operation · Cars maintenance and repair adaptive system · Monitoring · Operating conditions · Information and software complexes

1 Introduction

Motor transport is the most important sector of the Ukrainian economy, which serves almost all sectors of the economy and population strata and contributes to the growth of mobility and the quality of the population.

Currently, Ukraine's car fleet includes more than 14 million units of cars, the structure of which is as follows [1]: trucks—15.5%, buses—2.6%, and cars—81.9%.

To date, the number of licenses for the right to carry out transportation received by legal entities and individuals of Ukraine is approximately 140,000, and the number of vehicles they use is up to 400,000 units [2]. According to the data of the Main State Inspection on Road Transport, the share of carriers with only one vehicle in operation is 61%; up to three vehicles—22.4%; up to five vehicles—7%; up to ten—5.4%, more than ten vehicles—4.3%.

The technical operation of cars, according to the definition [3–6], is one of the most important subsystems of road transport, which, in turn, is a subsystem of transport in the structure of a rather complex transport and communication program of the state. It is a complex of organizational and technical measures that ensure maintenance of the operational efficiency of vehicles. In the existing system of technical operation of cars, there was a planned and preventive system of maintenance and repair of cars, the essence of which is that maintenance is preventive and carried out according to the plan, and repairs are carried out as needed.

The importance of the technical operation of cars is confirmed by the fact that, for example, about 30 billion dollars are spent per year in the USA to maintain cars in working condition, and around the world, about 100 billion dollars are spent annually on the technical operation of cars. In the USA, the cost of operating one car per year is 1800–1900 dollars [6].

The main feature of the modern system of technical operation of cars on road transport for general use in post-Soviet countries [2, 3] is:

– the absence of a regulatory framework regarding the obligation of every owner of rolling stock to carry out a certain set of technical measures that ensure its efficiency and safety, which results in the loss of the mechanism for managing the level of technical condition of the car fleet through a flexible system of maintenance and repair on public road transport;
– the absence of the necessary information base of the industry in the form of a network of supporting enterprises, which previously allowed public road transport,

firstly, to control the implemented quality indicators and the reliability of vehicles in operation and, secondly, to make justified demands on car manufacturing plants;
– the ineffectiveness of the certification system for maintenance and repair services proposed by the state.

As a result, public road transport and, above all, small enterprises in the field of road transport found themselves in difficult conditions because they [2, 3]:

– are obliged to ensure the technical condition of rolling stock in accordance with state requirements for traffic safety and environmental safety of transport;
– they do not have the conditions (bases, equipment, personnel) to maintain the efficiency and necessary technical condition of the rolling stock;
– do not have a clearly legalized obligation to apply the maintenance and repair system and to perform the minimum amount of maintenance and repair work that can ensure the necessary performance and safety of vehicles.

Based on the results of the analysis of the modern state of road transport and its subsystem—technical operation of cars, it was found that the main part of cars in Ukraine is concentrated in small enterprises in terms of size and number. This led to an organizational and technological vacuum, the result of which is the practically uncontrolled operation of cars in most small enterprises of road transport, deterioration of the technical condition of vehicles, an increase in the number of traffic accidents caused by car malfunctions, and environmental pollution. The maintenance and repair system that exists in the technical operation of cars, which establishes for cars the average mileage and labor intensity of their technical effects and allows the application of a number of adjustment factors for a specific car, leads to a significant increase in costs for maintaining the operational efficiency of vehicles.

In connection with the use of built-in onboard diagnostics on cars, the development of satellite navigation systems and mobile communication, and modern information technologies, it became possible to carry out remote monitoring with an assessment of the level of the vehicle's technical condition. This, in turn, allows you to switch to an adaptive system of maintenance and repair of cars.

2 Innovative Systems of Vehicle Operation Processes, Organizational and Functional Support

The appearance in transport, for example, in aviation of "systems with full responsibility", such as FADEC (Full Authority Digital Electronic Control System) [7], allows for neutralization difficulties. The concept of FADEC is aimed at creating a single structure of on-board systems for managing the work processes of nodes and aggregates, control and diagnostic systems, systems of organizational and functional support for vehicle operation processes, which allows for the formation of information systems for organizational and functional support (collection, analysis and management of information flows) operation processes, that is, it allows to implement in practice the information support of products (ISP)/CALS/PLM technologies.

ISP/CALS/PLM technologies, i.e., information support of supplies and the life cycle of products—is a modern approach to the design, production, and operation of high-tech and knowledge-intensive products, which consists in the use of computer equipment and modern information technologies at all stages of the life cycle of products [8]. In the field of public road transport companies, the integrated information environment of ISP/CALS/PDM technologies is just being implemented. An example can be the Torque program, as the basis of the "automotive" FADEC concept, which is the first step towards the Failure Reporting Analysis and Corrective Action System (FRACAS) system and, accordingly, ISP/CALS/PLM technologies, which are designed to receive and display diagnostic information of the on-board self-diagnosis system. Today, it already "knows" how to display the current operating parameters of the engine, other systems, nodes, and aggregates, display and decipher "error codes", "erase errors" from the electronic control unit, automatically send the values of the parameter values monitored by the sensor to the integrated electronic information metaspace, where during six months you can see not only the current values of the controlled values at different times, but also see on the map the entire route of the vehicle during this period [8].

No less significant for ISP/CALS/PLM-technologies on motor vehicles of general use are the simplest (from the point of view of tasks solved on motor transport) electronic information systems such as [8]:

- GPS-Trace Orange, which provides based on the commercial transport monitoring system "Wialon" satellite monitoring and control services through the Web interface for vehicles equipped with a tracker or any other communicators with a GSM module [9, 10];
- M2M (Machine-to-Machine, Mobile-to-Machine, Machine-to-Mobile), which creates technologies that allow quite simply, reliably, and profitably to ensure the transfer of data between "smart" devices (smart devices), which are electronic machines capable of interacting with each other [11, 12];
- Fuel consumption control system, which is a set of modern rolling stock management "tools", based on satellite navigation of transport monitoring, which provides control of fuel consumption, axle load, the operating time of rolling stock, and their operating parameters [13, 14];
- Teletrack, which represents a specialized hardware and software complex for satellite monitoring, which consists of an on-board scanner—communicator (controller—communicator, various sensors that provide open architecture, scalability, flexibility of the monitoring system), software (server, dispatcher "Track Control") and which allows integrating this solution for monitoring transport into any management system of the enterprise, solving complex and non-standard tasks [15, 16];
- Dynafleet® is a Swedish transport information system or a single telematics product for cars (for example, Scania) that works throughout the European Union [17, 18].

To solve multifunctional tasks related to the commercial operation and technical operation of cars and their infrastructure, the authors developed a virtual transport and information system "HNADU TESA" (The Department "Technical Exploitation and Service of Cars" Kharkiv National Automobile and Highway University (KhNAHU))—a system of cellular monitoring of vehicles [19, 20].

3 Mathematical Model Development for Monitoring the Vehicles' Technical Condition in Operating Conditions

The following requirements have been formulated for the development of the monitoring system of "HNADU TESA":

- automatic determination of navigation parameters of vehicles (geographical coordinates, speed of movement, azimuth, altitude above sea level) [21];
- automatic determination of parameters of the technical condition of vehicles according to the indications of control devices of telematic navigators-receivers (availability and "quality" of power supply in the control system of engine work processes, etc.; state of locks (open/closed) of the doors of the passenger compartment (cabin); inclusion of sound systems, light signaling; the position of the tipper body; the operation of attached and additional equipment, the temperature regime, exceeding the permissible speed of movement; the level of liquids in containers (tanks, cisterns) and others) [22];
- automatic transmission of navigational and other information about the vehicle to small enterprises of road transport after a given time interval (frequency from 20s) [23];
- automatic transmission to small enterprises of road transport of out-of-order messages about changes in the parameters of the state of the vehicle when control devices or sensors are activated (the driver pressing the alarm button, changing the mode of operation of additional equipment, a long simple object, the object entering a certain zone or exiting her) [24];
- automatic recording of navigation information and information about the state of the vehicle in the non-volatile memory in case of loss of external communication channels, with subsequent automatic or on-demand sending of recorded data to small road transport enterprises [25];
- automatic monitoring of the vehicle's execution of the route or schedule of traffic with the provision of alarm communication in case of deviations [26];
- the possibility of selecting individual units of the vehicle to monitor their movement and condition in real-time [27];
- display in graphic form the location and parameters of the vehicle on vector electronic maps of the area [28];
- display of data on the location and state of objects in text form in the form of tables [29];

– display in small enterprises of road transport of emergency messages about changes in the state of the vehicle in the form of alarm windows with a warning signal [30].

Control functions of the "HNADU TESA" system:

– formation of electronic maps of the territory of control zones for tracking the movement of the vehicle [31];
– control and analysis of the actual mileage of the vehicle at certain time intervals;
– transmission of the dispatcher's commands to the executive devices of the vehicle (locking the starting system, turning off the engine, turning on the emergency signal system, calling the driver, controlling additional equipment) [32];
– voice communication between the dispatcher of the small enterprise of road transport and the drivers of the vehicle [33];
– automatic entry into the log of events of all actions taken by the dispatcher of a small road transport enterprise.

Functions of the "HNADU TESA" system regarding information storage and integration with external information systems:

– storing information in a single database (MS SQL or Interbase) [33];
– transformation of information into a format compatible with user information systems [34];
– data exchange with user information systems [35];
– creation of databases in the format of user archives [36].

According to the construction principle, the "HNADU TESA" system contains three main parts [37]:

– "Vehicles"—objects of monitoring;
– "information transmission networks"—GSM/GPRS, Internet;
– "information processing and storage system"—telematics server.

The vehicle—the object of monitoring in the "HNADU TESA" system must be equipped with navigation-communication and receiving-transmitting devices, made in the form of a compact module—a telematics receiver-navigator (hereinafter the scanner-communicator controller) with control and executive devices, as well as by means of text, voice and video communication.

In order to implement the adaptive system of maintenance and repair in the technical operation of cars, the following information and software complexes were developed and experimentally tested at the HNADU: "Virtual Mechanic "HADI-12", "Service Fuel Eco "NTU-HADI-12" (National Transport University—Kharkiv Automobile and Road Institute-12), "MonDiaFor" HADI-15", "IdenMonDiaOp-erCon "HNADU-16"", which are integrated into the "HNADU TESA" system for processing the received data. Information and software complexes work in the conditions of intelligent transport systems (ITS) and have organizational and functional capabilities for managing the work of road transport enterprises, assessing the impact of operating conditions on the technical condition and environmental safety of the vehicle, the current state and the possibility of predicting the vehicle's performance, as well as evaluate the technical and economic indicators of the vehicle.

The system for monitoring the technical condition of the vehicle in operating conditions is a complex dynamic system with a clearly ordered hierarchical structure and a branched network of relationships between its elements, which develops in space and time. It is determined by the technological processes implemented by its subjects in accordance with the target orientation within the subject area. In general, the model of the functioning of the entities monitoring the effectiveness of the functioning can be represented as a set of the following components [38]: models of objects participating in structural connections, models of parameters that determine the results of the functioning of subjects, modeling algorithms that establish the rules for the functioning of objects and changes in the values of their parameters, etc.

The development of an intelligent system for evaluating technical condition parameters is based on a general approach to the study of the "driver-car-road environment" (DCRE) system, which includes the system interaction of the components of the vehicle monitoring with the driver and on-board information complex; operating conditions of vehicles (road, transport, atmospheric and climatic conditions, and work culture) [4, 5, 22, 33–36, 39]; transport infrastructure and highway infrastructure (Fig. 1).

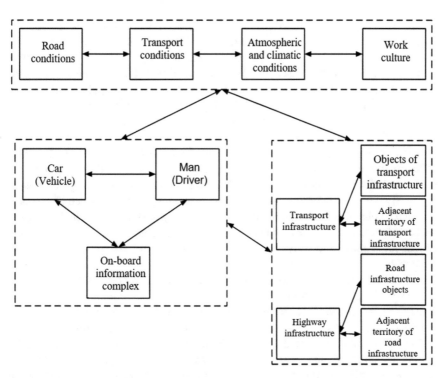

Fig. 1 The general diagram of system interaction of the system "driver—car—road environment" in ITS conditions

The process of monitoring the technical condition of the vehicle in operating conditions is the process of forming a single information function that describes the interaction of the vehicle in the form of parameters of the technical condition of the vehicle obtained with the help of the onboard information complex; of the driver, which is related to the process of transformation of information about the parameters of the technical condition and processes that depend on the physiological capabilities of the driver (person), the technical data of the vehicle and the degree of their resistance to the negative effects of the external environment; operating conditions of the vehicle [7] and the interaction of models of transport infrastructure parameters and highway infrastructure.

Formally, this mapping has the form (see Eq. 1) [9, 22]:

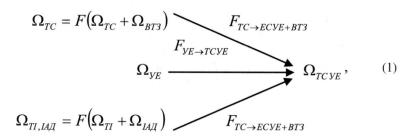

$$\Omega_{TC} = F\left(\Omega_{TC} + \Omega_{BT3}\right)$$

$$F_{TC \to ECVE + BT3}$$

$$F_{VE \to TCVE}$$

$$\Omega_{VE} \longrightarrow \Omega_{TCVE}, \qquad (1)$$

$$\Omega_{TI,IA\varPi} = F\left(\Omega_{TI} + \Omega_{IA\varPi}\right)$$

$$F_{TC \to ECVE + BT3}$$

where Ω_{TC} is a set of models of parameters of the technical condition of the vehicle, as $\Omega_{TC} = F(\Omega_{TC} + \Omega_{BT3})$ system interaction of parameters of the technical condition of the vehicle and the driver (person), which, in turn, is related to the process of transformation of information about the parameters of the technical condition of the vehicle means and processes that depend on the physiological capabilities of a person, the technical data of the vehicle and the degree of their resistance to the negative effects of the external environment; Ω_{BT3}—a set of models of the state of a person (driver) of a vehicle; Ω_{VE} is a set of models of parameters of operating conditions of vehicles; $\Omega_{TI, IA\varPi} = F(\Omega_{TI} + \Omega_{IA\varPi})$—set of models of parameters of transport infrastructure and highway infrastructure; $\Omega_{TC\ VE}$—a set of models of parameters of the technical condition of the vehicle under the relevant operating conditions; $F_{TC \to TCVE + BT3}$—functional display of models of parameters of the technical condition of the vehicle and the driver of the vehicle; $F_{TC \to TCVE}$ is the functional display of models of parameters of the technical condition of the vehicle; $F_{TI, IA\varPi \to TI + IA\varPi}$ is the functional display of models of parameters of transport infrastructure and highway infrastructure.

We consider it expedient to combine into a set of Ω_{TC} models of the parameters of the vehicle under the operating conditions of Ω_{TC} in interaction with Ω_{BT3}. At the same time, we assume that the functioning of the single system of the vehicle and the driver $F(\Omega_{TC} + \Omega_{BT3})$ changes in operating conditions in the form of technical and economic indicators of the vehicle. At the same time, we understand that the system adapts to different operating conditions, changing its operational properties [7]. Also, we consider it expedient to combine all the effects of the environment on the vehicle in the form of a change in models of operating conditions, models of parameters of transport infrastructure, and highway infrastructure in the form of a set of Ω_{YE} models of parameters of vehicle operating conditions.

On the basis of the above, in a generally unified form, the process of monitoring the technical condition of the vehicle under operating conditions is a process of transformation of information on the state and processes of functioning of the vehicle and operating conditions.

In the process of analysis and synthesis, the formation of possible variants of the schemes of the information system for monitoring cars in operating conditions in the parts of enforcement: identification of the vehicle, collection of data on the technical condition of the vehicle, monitoring and forecasting of the parameters of the technical condition of the vehicle, identification of the operating conditions of the vehicle, diagnosing the condition of the vehicle, checking the conformity of the condition of the vehicle, morphological analysis was used [10].

Formally, the indicated mapping has the form (see Eq. 2):

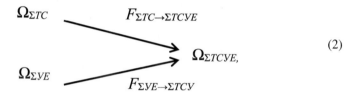

$$
\begin{array}{ll}
\Omega_{\Sigma TC} & F_{\Sigma TC \to \Sigma TCYE} \\
& \qquad\qquad\qquad\qquad \Omega_{\Sigma TCYE,} \\
\Omega_{\Sigma YE} & F_{\Sigma YE \to \Sigma TCY}
\end{array}
\tag{2}
$$

where $\Omega_{\Sigma TC}$ is a set of aggregate models of parameters of the vehicle's technical condition; $\Omega_{\Sigma YE}$ is a set of aggregate models of vehicle operating conditions parameters; $\Omega_{\Sigma TC\ YE}$ is a set of aggregate models of parameters of the technical condition of the vehicle in the relevant operating conditions; $F_{\Sigma TC \to \Sigma TCYE}$ is a functional display of aggregate models of parameters of the technical condition of the vehicle; $F_{\Sigma YE \to \Sigma TCYE}$ is a functional display of aggregate models of vehicle operating conditions parameters.

The peculiarity of the above analysis is that in the experimental system "Driver—car—road environment" for the formation of the basic morphological formula of the information system for monitoring cars in operating conditions, several main characteristics of the functional elements—morphological signs—characteristic for it were selected, each of which was previously a maximally complete list of various relevant variants of the technical expression of the given features was compiled [11]. For each morphological feature, the characteristic properties of the classifications, features of the car design, components of the monitoring system, operating conditions, etc., are given, which depend on the solution of the research problem and the achievement of the main goal of the functioning of the system "Driver—car—road environment" in operating conditions.

In the morphological matrix of the schemes of the vehicle monitoring information system in operating conditions, 12 features are allocated for the functional element "Car (Vehicle)", and for the classification element "passenger car", 4 features are additionally allocated for the classification element "bus"—1; for the classification element "truck"—2 signs. For the functional element "Car engine (Vehicle)" 4 signs are allocated. For the functional element "Equipping the vehicle with information and communication equipment", there are 3 signs. For the functional element "External networks", 1 sign is in 4 options. For the functional element "Monitoring the state of the vehicle and operating conditions", 3 signs are also allocated. For each of the 23 morphological features of the system, the main variants of their implementation were selected (from 2 to 10). A change in the constructive expression of a specific variant of any of the 23 features forms a new scheme for providing the vehicle monitoring information system in operating conditions.

In accordance with the provisions of [10, 12, 22], the research method is based on the morphology of the constituent objects of the monitoring system and allows systematic analysis of various structures of the object—the monitoring system, arising from the regularities of their structure.

The number of possible schemes of the information system for monitoring vehicles in operating conditions in the case of using the created morphological matrix is:

- for a passenger car (vehicle): $N = 8 \cdot 4 \cdot 7 \cdot 9 \cdot 4 \cdot 3 \cdot 6 \cdot 4 \cdot 10 \cdot 3 \cdot 4 \cdot 4 \cdot 4 \cdot 2 \cdot 2 \cdot 2 \cdot 2 \cdot 2 \cdot 2 \cdot 2 \cdot 2 \cdot 4 \cdot 4 \cdot 4 \cdot 6 = 4{,}749 \cdot 10^{13}$;
- for a bus: $N = 8 \cdot 4 \cdot 7 \cdot 5 \cdot 4 \cdot 10 \cdot 3 \cdot 4 \cdot 4 \cdot 4 \cdot 2 \cdot 2 \cdot 2 \cdot 2 \cdot 2 \cdot 2 \cdot 2 \cdot 2 \cdot 4 \cdot 4 \cdot 4 \cdot 6 = 1{,}691 \cdot 10^{12}$;
- for a truck (vehicle): $N = 8 \cdot 4 \cdot 7 \cdot 10 \cdot 7 \cdot 4 \cdot 10 \cdot 3 \cdot 4 \cdot 4 \cdot 4 \cdot 2 \cdot 2 \cdot 2 \cdot 2 \cdot 2 \cdot 2 \cdot 2 \cdot 2 \cdot 4 \cdot 4 \cdot 4 \cdot 6 = 2{,}368 \cdot 10^{13}$,

and for one variant of the vehicle when using the morphological matrix in the part of equipping the vehicle with information and communication equipment, external networks, monitoring the state of the vehicle and operating conditions: $N_1 = 768$. For a similar variant with the additional use of the morphological matrix in the part of the car engine (vehicle): $N_{11} = 12{,}288$.

Thus, the scheme of the monitoring information system for a C1 segment vehicle with an injection gasoline engine in operating conditions includes the following combinations of specified features (see Eq. 3):

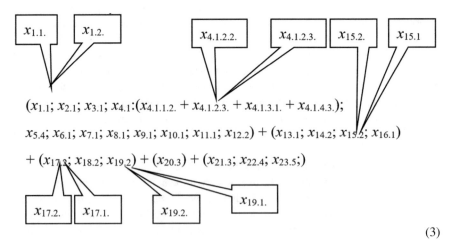

$(x_{1.1}; x_{2.1}; x_{3.1}; x_{4.1}{:}(x_{4.1.1.2.} + x_{4.1.2.3.} + x_{4.1.3.1.} + x_{4.1.4.3.});$

$x_{5.4}; x_{6.1}; x_{7.1}; x_{8.1}; x_{9.1}; x_{10.1}; x_{11.1}; x_{12.2}) + (x_{13.1}; x_{14.2}; x_{15.2}; x_{16.1})$

$+ (x_{17.2}; x_{18.2}; x_{19.2}) + (x_{20.3}) + (x_{21.3}; x_{22.4}; x_{23.5};)$

$$(3)$$

To conduct research in formula (3), the variable components of morphological formulas with different layouts for passenger vehicles of the *C1* segment are marked with arrows, and the arrows are green for passenger vehicles of the *C2* segment with a diesel engine. That is, formula (3) took the form (4) for the *C1* segment vehicle (see Eq. 4):

$$(x_{1.1}; x_{2.1}; x_{3.1}; x_{4.1} : (x_{4.1.2.2.} + x_{4.1.2.3.} + x_{4.1.3.1.} + x_{4.1.4.3.});$$
$$x_{5.4}; x_{6.1}; x_{7.1}; x_{8.1}; x_{9.1}; x_{10.1}; x_{11.1}; x_{12.2})$$
$$+ (x_{13.1}; x_{14.2}; x_{15.1}; x_{16.1}) + (x_{17.2}; x_{18.2}; x_{19.1}) + (x_{20.3}) + (x_{21.3}; x_{22.4}; x_{23.5}),$$

$$(4)$$

and for a *C2* segment vehicle (see Eq. 5):

$$(x_{1.2}; x_{2.1}; x_{3.1}; x_{4.1} : (x_{4.1.1.2.} + x_{4.1.2.3.} + x_{4.1.3.1.} + x_{4.1.4.3.});$$
$$x_{5.4}; x_{6.1}; x_{7.1}; x_{8.1}; x_{9.1}; x_{10.1}; x_{11.1}; x_{12.2})$$
$$+ (x_{13.1}; x_{14.2}; x_{15.2}; x_{16.1}) + (x_{17.1}; x_{18.2}; x_{19.2}) + (x_{20.3}) + (x_{21.3}; x_{22.4}; x_{23.5}).$$

$$(5)$$

Modern on-board systems for monitoring technical condition parameters in ITS conditions allow for vehicle identification, continuous automatic measurement of parameters characterizing the technical condition of the vehicle, diagnostics, namely control of the vehicle and its components, recognition, and prevention of failures in its operation, and ultimately—ensuring the functioning of the adaptive system of technical maintenance and repair of vehicles according to the technical condition [13, 14]. The work of the developed information software complex IdenMon-DiaOperCon "HNADU-16", state and operating conditions of the vehicle includes a set of stationary and mobile (on-board relative to the vehicle) information collection and transmission systems. Information exchange between ITS elements (on the example of IPC "IdenMonDiaOperCon "HNADU-16"" and the "HNADU TESA" system, namely the vehicle and transport infrastructure in the process of monitoring parameters of the technical condition in operating conditions is shown in Fig. 2.

The main principle of information exchange between ITS elements, namely the vehicle and the transport infrastructure in the process of monitoring the parameters of the technical condition in the conditions of operation and construction of the monitoring information system, is that the vehicle is not only an object of control and management but also a source of constantly updated information about the state of its operating conditions. That is, it is a modern control and measurement system that accumulates and stores information about the technical condition of the vehicle and the conditions of its operation within the traffic area, and also makes decisions when a dangerous emergency situation or a malfunction of the vehicle is detected.

The system's information support includes the following components:

- a system of collecting, accumulating, and distributing information about the technical condition of the vehicle in operating conditions;

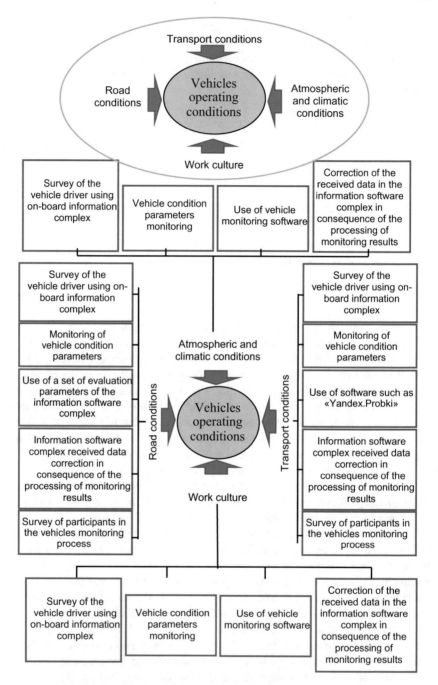

Fig. 2 The general scheme of methods of obtaining and forming information about the state and conditions of operation of the vehicle in ITS conditions

- automated tools for diagnosing the technical condition of vehicles and roads;
- database of geographic data on road conditions and highway infrastructure objects;
- data collection and transmission system;
- a set of tasks for monitoring the condition and planning the operating conditions of the vehicle;
- means of visualization of highway monitoring results and communication with the driver and other road users.

The general task of forming a methodology for the application of the classification of vehicle operating conditions in the information conditions of ITS, as a complex system, is based on obtaining information about the actual technical condition, methods, and means of its implementation when solving specific scientific and technical problems, assessment, verification of compliance with established restrictions, means for its provision, criteria for evaluating the obtained indicators and determining the relationship between them.

Information about the actual parameters of the technical condition of the vehicle is depicted in the study as a construction of a function:

- in the processes of monitoring and diagnosing the technical condition (see Eq. 6):

$$
\begin{cases}
F_{ts}\left(\overline{H}_t, t, \Delta t, \overline{X}_i(t), \overline{X}_i(t-\Delta t), \ldots, \overline{X}_i(t-n\Delta t), DTC_{s_i}, K_{t_i}\right) \Rightarrow S_{y.e.\text{T3}} \\[2mm]
\Omega_l^{m_i}\left(e_{y.e.\text{T3}}, r\right)^J = \Omega_l^{m_i}\left(\begin{cases} e_{y.e.\text{T3,тр}} \\ e_{y.e.\text{T3,дор}} \\ e_{y.e.\text{T3,а.к}} \\ e_{y.e.\text{T3,ке}} \end{cases}, r\right)^J = S_{y.e.\text{T3}}
\end{cases}
\tag{6}
$$

- in the processes of forecasting the technical condition (see Eq. 7):

$$
\begin{cases}
F_{(t+k\Delta t)}\left(\begin{array}{l} \overline{H}_{(t+k\Delta t)}, t, \Delta t, \overline{X}_i(t+k\Delta t), \overline{X}_i(t+(k-1)\Delta t), \ldots, \\ \overline{X}_i(t+(k-n)\Delta t), DTC_{s_i}, K_{t_{i(t+k\Delta t)}} \Rightarrow S_{y.e.\text{T3}}(t+k\Delta t) \end{array}\right) \\[2mm]
\Omega_l^{m_i}\left(e_{y.e.\text{T3}(t+k\Delta t)}, r\right)^J = \Omega_l^{m_i}\left(\begin{cases} e_{y.e.\text{T3,тр}(t+k\Delta t)} \\ e_{y.e.\text{T3,дор}(t+k\Delta t)} \\ e_{y.e.\text{T3,а.к}(t+k\Delta t)} \\ e_{y.e.\text{T3,ке}(t+k\Delta t)} \end{cases}, r\right)^J = S_{y.e.\text{T3}}(t+k\Delta t)
\end{cases}
\tag{7}
$$

where F_{ts} is the information about the parameters of the technical condition of the vehicle under the relevant operating conditions at the relevant time; $\overline{H}_t$ is the vector of the control bodies of the vehicle power plant (coordinate of control body sensors) at time t; t is the current time of the monitoring process; Δt is the time interval between measurements in monitoring processes; $\overline{X}_i(t)$ with $i = 1,\ldots, m$ is the characteristics of the technical condition of the vehicle in operating conditions, which are measured and included in the list of basic parameters of the technical condition of the vehicle in operating conditions); n is the number of intervals in past monitoring periods; t is the number of measured parameters of the technical condition of the vehicle; $DTC_{s_i} K_{t_i}$ is the results of vehicle DTCs (Diagnostic Trouble Codes) monitoring; Ω is the mapping operator; $S_{y.e.T3}$ is the system for determining the operating conditions of the vehicle (in the presented case, the system $S_{y.e.T3}$ is a display of the properties of the sub-objects for determining the operating conditions $e_{y.e.T3}$ of the vehicle and their relations r for m and for J in l); m_i is the number of means of obtaining information in vehicle; l is the connections between means of observation and sub-objects of determining the operating conditions of vehicle; $e_{y.e.T3}$ is a set of sub-objects of determining the conditions of operation of vehicles ($e_{y.e.T3,\text{тр}}$ is the transport; $e_{y.e.T3,\text{доp}}$ is the road; $e_{y.e.T3,\text{а.к}}$ is the atmospheric and climatic; $e_{y.e.T3,\text{ке}}$ is the culture of operation); r is a set of relations between the main operating conditions of the vehicle; J is the task of determining the operating conditions of the vehicle; $F_{(t+k\Delta t)}$ is the predicted information about the parameters of the technical state of the vehicle at the appropriate moment in time in the process of performing its functions (during the operation of the vehicle as intended) in the future on a bias interval of length ($t + k\Delta t$) depending on the known values in the past, in the given prediction interval δ with a given confidence probability p; k is the number of intervals of predicted values of the parameters of the technical condition in the future, determines the type of forecast—short-term, medium-term, etc. under the predicted operating conditions, respectively ($e_{y.e.\ T3(t+k\Delta t)}$).

Provision of the information system of the operating conditions of the vehicle $S_{y.e.\Sigma T3_i}(t)_i$ is built on the basis of server solutions $S_{y.e.T3_i}(t)_i$ according to provisions [4, 12, 22–26], a local source of information $S_{y.e.Vd_i}(t)_i$ and network databases $S_{y.e.Net_i}(t)_i$ (see Eq. 8):

$$S_{y.e.\Sigma T3_i}(t)_i = \left(S_{y.e.T3_i}(t)_i, S_{y.e.Vd_i}(t)_i, S_{y.e.Net_i}(t)_i\right) \tag{8}$$

This provides the possibility of creating a single centralized repository of information distributed in space, supporting a multi-user environment for obtaining information (editing), the possibility of remote user access, systematization of information, and its visual display in a single complex. In the process of developing information support processes for monitoring the parameters of the technical condition of the vehicle, taking into account the operating conditions, the available sources of information were collected in terms of the coordinates of the vehicle on the terrain in real-time, the model of the road, models of road infrastructure objects, territorial natural and man-made systems, obtained vehicle tracking results. Sources of information to ensure the functioning of the information system for monitoring the technical condition of the vehicle, according to the operating conditions, are presented in Table 1.

In the process of research and evaluation of the operating conditions of the vehicle, the geographic model of the road in Torque, Yandex.Maps were used, which were the basis of the analysis system and is a layer of linear objects with information interaction parameters (see Eq. 9):

$$F_{ts}(RV_{Road})_i = \begin{pmatrix} Ident_{RV_i}, Cat_{RV_i}, Cod_{Dil_i}, \text{Distanse}_{Descr_i}, \\ Type_{RVn_i}, Type_{Roadn_i}, Track_{V_i} \end{pmatrix}, \tag{9}$$

where $F_{ts}(RV_{Road})_i$ is the information is similar to the relevant parameters of the technical condition of the vehicle under the relevant operating conditions at the relevant time for the road information system; $Ident_{RV_i}$ is the identifier and traffic sections of the vehicle; Cat_{RV_i} is category and road; Cod_{Dil_i} is the code of the section of the road, $\text{Distanse}_{Descr_i}$ is the description of the section of the road, $Type_{RVn_i}$ is the number of traffic lanes, $Type_{Roadn_i}$ is the type of road surface, $Track_{V_i}$ is the width of the traffic lane.

The information model of the location of the vehicle on the highway was developed on the basis of the geographic model of the highway. Each part of the model is described by the following vector of parameters (see Eq. 10):

$$F_{ts}(RV_{Traffic})_i = (Ident_{RV_i}, Ident_{PRoute_i}), \tag{10}$$

Table 1 Sources of information for the information system monitoring of the technical condition of vehicles according to operating conditions

№	Parameter	Sources of information for the information system for monitoring the technical condition of the vehicle according to operating conditions
1	Parameters of the state and position of the vehicle on the map	http://view.torque-bhp.com/ http://ian-hawkins.com:8080/
2	Transport conditions for the operation of the vehicle, according to geolocation	http://yandex.ua/maps.ru/kharkov.htm
3	Atmospheric and climatic conditions of vehicle operation	http://meteoco.ru/ http://ready.arl.noaa.gov/READYcmet.php
4	Road conditions of vehicle operation	http://view.torque-bhp.com/ https://yandex.ua/maps/147/kharkiv/?lang=ru&ll=36.231202%2C49.990175&z=13)
5	Vehicle identification during operation in ITS conditions	http://view.torque-bhp.com/ http://carlife.in.ua/vin-kod

where $Ident_{RV_i}$ is the identifier and traffic section of the vehicle; $Ident_{P\,Route_i}$ is the identifier of sections of the vehicle traffic route.

The traffic route is formed from the route sections of the information model of the vehicle position on the road. It represents a certain path of movement of the vehicle, implemented in the form of a linear object and accompanied by the following vector of parameters (see Eq. 11):

$$F_{ts}(RV_{Marshrut})_i = \left(Ident_{RV_i}, Ident_{Route_i}\right), \tag{11}$$

where $Ident_{RV_i}$ is the identifier and traffic section of the vehicle; $Ident_{Route_i}$ is vehicle traffic route identifier.

The speed model of vehicle traffic modes is a table of linear events superimposed on the traffic route and has the following structure (see Eq. 12):

$$\begin{aligned}
F_{ts}\left(RV_{Route_{Properties}}\right)_i \\
= \big(Ident_{RV_i}, Ident_{Route_i}, Route_{Property_{yi}}, Ident_{SR_i}, \\
Coordinate_{First_i}, Coordinate_{End_i}, Value_{V_i}, Date_i, Base_{Speed_i}\big), \tag{12}
\end{aligned}$$

where $Ident_{RV_i}$ is the identifier and traffic section of the vehicle; $Ident_{Route_i}$ is the vehicle traffic route identifier; $Route_{Property_{yi}}$ is the type of vehicle traffic route; $Ident_{SR_i}$ is the identifier of the section of the high-speed mode of movement of motor vehicles; $Coordinate_{First_i}$ is the beginning of the section of the high-speed traffic mode of vehicles; $Coordinate_{End_i}$ is the end of the high-speed vehicle traffic section; $Value_{V_i}$ is the permissible speed of vehicle movement is established; $Date_i$ is the date of setting the vehicle speed; $Base_{Speed_i}$ is the set (base) speed on the vehicle movement section.

Along with the road model, in the information model, there is an opportunity to describe the coordinates of road infrastructure objects. With the help of the map of the monitoring information system, it is possible to record the coordinates and features of the impact on the movement of vehicles under the conditions of operation of bridges, crossings, traffic lights, etc. Also, the features of operating conditions are influenced by existing nearby man-made objects (indicating the type and type of production) or natural territorial systems. All the objects listed above are typified according to their characteristics and contain a parameter for assessing the impact on vehicle movement processes and vehicle operating conditions.

In the processes of researching the vehicle's operating conditions in interaction with the vehicle's driver and the information and software complex, it is possible to solve the problems of road, transport, and atmospheric-climatic conditions. At the same time, the condition of the road, road infrastructure facilities, and the surrounding area is assessed.

The assessment of the type and state of the road surface is carried out as a result of processing the data of the remote survey of the driver using the on-board information complex, and then the assessment of the type and state of the road surface is formed during the operation of the vehicle. The parameters characterizing the road conditions, according to which the classification signs of limiting the permissible speed or closing the movement of vehicles, are: longitudinal profile of the road, height above sea level, width of the carriageway and the condition of the surface, traction of the wheels with the road, etc. [4, 12, 27–30]. The data on the types and sizes of defects are compared with the normative indicators for the vehicle speed, the degree of deviation is determined, and a point assessment of the condition of the road conditions is formed (according to the condition of the road surface) (see Eq. 13):

$$O_p(t)_i = \left(O_{p_1}, O_{p_2}, O_{p_3}\right),\tag{13}$$

where O_{p_1} is the excellent and good condition, O_{p_2} is the satisfactory condition, O_{p_3} is the unsatisfactory condition.

To determine the value of the assessment in the information and software complex of the analyzed section of the road surface, a table of point events is formed, which contains a point assessment of the state of road operating conditions for each detected defect, as follows (see Eq. 14):

$$F_{ts}(Event_{Point})_i = \begin{pmatrix} Ident_{RV_i}, Date_i, Distans_i, \\ Defect_i, Discribe_i \end{pmatrix},\tag{14}$$

where $Ident_{RV_i}$ is the identifier and traffic section of the vehicle; $Date_i$—the date of setting the vehicle speed; $Distans_i$ is the distance from the starting point of the route (linear coordinate); $Defect_i$ is the point assessment of the defect; $Discribe_i$ is the type of defect.

Aggregation of assessments of the condition of the road surface according to the detected defects is carried out for sections of the speed model of vehicle traffic modes according to the expression of the form (see Eq. 15):

$$O_{p_i}(t)_i = \max_{J=1.N} O_{p_{ij}}(t)_i,\tag{15}$$

where $O_{p_{ij}}(t)_i$ is the score of the j-th defect on the i-th section of the vehicle speed mode model, N is the number of defects detected on the section, t is the time factor.

The assessment of the condition of transport infrastructure facilities is determined as a result of inspections and inspections. At the same time, it is determined whether the detected defects affect the safety of the vehicle movement and whether special attention should be paid to the condition of these objects. The assessment is carried out by drivers in the vehicle movement processes, other vehicle monitoring participants and vehicle movement participants. The results of inspections and inspections are entered into the database of measured parameters and objects of the monitoring information system. Each transport infrastructure object is characterized by a vector of parameters (see Eq. 16):

$$F_{ts}\left(RV_{TransInfluence}\right)_i$$

$$= \begin{pmatrix} Ident_{RV_i}, \ Type_i, \ Discribe_i, \ Objectid_i, \\ ObjectId_i, \ DateUpdate_i \end{pmatrix}, \tag{16}$$

where $Ident_{RV_i}$ is the identifier of the i-th traffic section of the vehicle; $Type_i$ is the type of object of transport infrastructure (transport crossings, transport facilities, facilities at highways, etc.); $Discribe_i$ is the description of the transport infrastructure object; $Objectid_i$ is the transport infrastructure object identifier; $ObjectId_i$ is the assessment of the condition (impact indicator) of the transport infrastructure object; $DateUpdate_i$ is the date of assessment.

The assessment of the condition of transport infrastructure objects in accordance with regulatory documents is carried out according to a three-point system (see Eq. 17):

$$O_c(t)_i = (O_{c_1}, O_{c_2}, O_{c_3}), \tag{17}$$

where O_{c_1} is the normal condition, O_{c_2} is the requires attention, O_{c_3} is the needs repair.

A fragment of the map showing the possibility of identifying transport infrastructure objects using (interprocess communication) IPC based on ISM and allowing to highlight objects that have a direct impact is presented in [8].

Unification of assessments of the condition of transport infrastructure objects based on the results of the impact is carried out on sections of the speed model of vehicle traffic modes in accordance with the expression of the form (see Eq. 18):

$$O_{c_i}(t)_i = \max_{J=1.N} O_{c_{ij}}(t)_i, \tag{18}$$

where $O_{c_{ij}}(t)_i$ is the score of the j-th object on the i-th section of the vehicle speed mode model, N is the number of objects detected on the section, t is the time factor.

The assessment of the condition of the adjacent territory of the highway is carried out as a result of the analysis of the level of danger of natural and man-made objects on the condition of the road surface and the movement of vehicles, taking into account their distance from the highway. The level of danger is carried out by drivers in the processes of vehicle movement, other participants in vehicle monitoring and vehicle movement participants as a result of periodic inspections, while the impact assessment is formed on a three-point scale (see Eq. 19):

$$O_t(t)_i = (O_{t_1}, O_{t_2}, O_{t_3}), \tag{19}$$

where O_{t_1} is the low impact, O_{t_2} is the medium impact, O_{t_3} is the high impact.

The third category is assigned to dangerous productions located near the highway, which require repair work or reconstruction, which with a high probability can cause emergency situations on the road in the conditions of operation of the vehicle. These can also be emergency situations of natural origin (landslides, landslides, mudslides).

The presence of such objects requires adjustment of the speed regime on the section of the road.

The results of the analysis of the impact of natural and man-made objects are entered into the database of measured parameters and objects of the monitoring information system. Each object of the transport infrastructure and the surrounding area is characterized by a vector of parameters (see Eq. 20):

$$F_{ts}\left(RV_{TerritiryInfluence}\right)_i$$
$$= \begin{pmatrix} Ident_{RV_i}, Type_i, Discribe_i, Objectid_i, \\ Ident - Sr_i, ObjectId_i, DateUpdate_i \end{pmatrix}, \tag{20}$$

where $Ident_{RV_i}$ is the identifier and traffic section of the vehicle; $Type_i$ object type of the adjacent territory (natural, man-made); $Discribe_i$ is the description of the transport infrastructure object, $Objectid_i$ is the identifier of the transport infrastructure object; $Ident - Sr_i$ is the identifier of belonging to the section of the high-speed model; $ObjectId_i$ is the assessment of the condition (impact indicator) of the transport infrastructure object; $DateUpdate_i$ is the date of assessment.

The final assessment of the impact of natural and man-made objects is built on the basis of such parameters as the number and indicators of the impact of natural and man-made objects, recorded in the "buffer" zone of the highway according to the sections of the speed model of the traffic modes of vehicles according to the expression of the species (see Eq. 21):

$$O_{t_i}(t)_i = \max_{J=1..N} O_{t_{ij}}(t)_i \tag{21}$$

where $O_{t_{ij}}(t)_i$—score of the j-th object on the i-th section of the vehicle speed mode model, N—the number of territorial objects of the section, t—the time factor.

The estimation of vehicle fuel consumption in operating conditions is carried out on the basis of server solutions and a local source of information (LSI) in the process of comparison with the linear norms of vehicle fuel consumption, established by normative indicators [4, 12, 30–32, 39], and determine the degree of deviation (see Eq. 22):

$$O_{Gt_i}(t)_i = \max_{J=1..N} O_{Gt_{ij}}(t)_i \tag{22}$$

where $O_{Gt_{ij}}(t)_i$—is the largest fuel consumption Gt_{ij} of the j-th monitoring object (vehicle) on the i-th section of the vehicle speed mode model, N—is the number of determined fuel consumption at the corresponding sections, t—is the time factor.

Adjustment of vehicle speed depending on operating conditions It is proposed to adjust vehicle speed depending on operating conditions. At the same time, priority is certainly given to ensuring road traffic safety and fuel economy.

The adjustment of the vehicle speed is carried out by sections of the highway speed model according to the matrix of events (see Eq. 23):

$$A_{iK}(t)_i = \left(O_{p_i}(t)_i, O_{c_i}(t)_i, O_{t_i}(t)_i, O_{Gt_i}(t)_i\right), K = 1, 8, \tag{23}$$

where $O_{p_i}(t)_i$ is the assessment of the type and condition of the road surface; $O_{c_i}(t)_i$ is the assessment of the condition of transport infrastructure facilities; $O_{t_i}(t)_i$ is the assessment of the level of danger of natural and man-made objects; $O_{Gt_i}(t)_i$ is the evaluation of the fuel economy of vehicles in operating conditions based on server solutions and a local source of information (vehicle) in the process of comparison with linear rates of fuel consumption of vehicles established by regulatory indicators [4, 12].

The matrix of events establishes the correspondence between the values of the estimates and the degree of emergency of the i-th section of the speed model of the vehicle movement. It is proposed to use 8 degrees of emergency (dangerousness) movement depending on the operating conditions of the vehicle [15, 22].

For each situation $A_{iK}(t)_i$, the adjustment of the vehicle speed should be determined in accordance with the actual route of movement with minimum fuel consumption on the route, depending on the operating conditions.

Then the speed of the vehicle $V_i(t)_i$ on the i-th section can be determined as follows (see Eq. 24):

$$V_i(t)_i = F(V_{ib}(t)_i, V_i(t - \tau)_i, A_{iK}(t - \tau)_i), \tag{24}$$

where $V_{ib}(t)_i$ is the recommended (basic) speed of the motor vehicle, $V_i(t - \tau)_i$ is the set speed of the motor vehicle, $A_{iK}(t - \tau)$ is the established emergency (dangerousness) of movement depending on the operating conditions of the motor vehicle.

The model of the subject area of the vehicle technical condition parameters monitoring system is presented in the form of the following set of components and components of the information system, in terms of technical parameters of the vehicle engine condition, technical parameters of the vehicle condition and operating condition parameters) of the vehicle [16, 17, 34] (see Eq. 25):

$$M_{np.o.} = \langle F, H, P, O, V_{ex.}, V_{eux.}, R \rangle, \tag{25}$$

where $F = \{f_i | i = \overline{1, I}\}$ is the set of user functions (automation functions) performed by the system for monitoring the parameters of the technical condition of the vehicle; $H = \{h_j | j = \overline{1, J}\}$ is the set of data processing tasks of the system for monitoring parameters of the technical condition of vehicles; $P = \{p_k | k = \overline{1, K}\}$ is the set of users (number and composition of personnel), which provides work with the system for monitoring parameters of the technical condition of vehicle; $V_{ex.} = \{v_l | \in L_{ex}\}$ is the set of input information elements; $V_{eux.} = \{v_l | \in L_{eux.}\}$ is the set of initial information elements; $V = V_{ex.} \cup V_{eux.}$ is the complete set of information elements; $O = \{o_m | m = \overline{1, M}\}$—a set of vehicle automation objects, which can be presented as independent parts for blocks of information collection and transmission: from the vehicles engine; from technical specifications about its parameters; about the operating conditions of the vehicle; about the results of identification; about the results of diagnostics; on environmental safety parameters; about fuel consumption; $R = \{r_y | y = \overline{1, Y}\}$ is the set of relations (relationships) between the components of $M_{np.o}$ of the subject area (25) of the system of monitoring the parameters of the technical condition of vehicles.

The process of formation and analysis of graphs of information structures of the model of the system "IdenMonDiaOperCon "HNADU-16"", in the part of the system for monitoring parameters of the technical condition of vehicles, included [12, 16, 18] the following interrelated operations: construction of sets of structural elements based on the model subject area of the system; formation of a matrix of semantic adjacency on a set of structural elements, construction of an oriented graph of its information structure; formation of a matrix of semantic reach on a set of structural elements; definition of informational and group elements of structural sets; arrangement of groups of structural elements by hierarchy levels, selection and formation of a set of keys and attributes in groups of data subsystems; construction of canonical models of database subsystems of the system. A fragment of the matrix of semantic adjacency in the form of a graph of the information structure G, the sets of vertices of which are the structural elements of the corresponding sets, and the arcs correspond to the entry in the matrix [18, 33–36]. The arcs of the organized graph (orgraph) G (see Fig. 3) reflect the presence or absence of semantic connectivity between their structural elements of the model.

4 Experimental Verification and Results of the Information and Software Complex for Monitoring the Vehicles Technical Condition in Operating Conditions

The main object of experimental studies for monitoring the parameters of the technical condition was a C2 segment car (Volkswagen Golf VII GTD 2.0 TDI, Fig. 4) with a CLLA diesel engine, equipped with information monitoring devices between ITS elements.

Figure 5 shows the scheme of information exchange developed by the authors between the elements of the ITS information and software complex of the vehicle for the remote study of the rapidly changing work processes of the vehicle operation in the process of changing operating conditions [19]. Features of the information interaction between the ITS elements of the vehicle and the transport infrastructure and highway infrastructure in the processes of remote management of its performance are described in [19].

The structure and relationship of the functionality of the on-board information complex for obtaining information about the operating conditions of the vehicle within the limits of the virtual enterprise of road transport are shown in Fig. 6.

System interaction is based on the following main functions of the on-board information complex, namely, ensuring the determination of the position of the vehicle

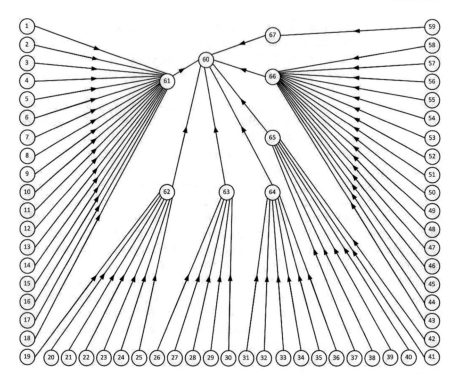

Fig. 3 Orgraph *G* of the information structure of the model systems for monitoring parameters of the technical condition of vehicles

Fig. 4 C2 segment vehicle during parameter monitoring technical condition under operating conditions

(tracking the position of the vehicle), ensuring the monitoring of parameters of the technical condition of the vehicle, solving the task of assisting the driver of the vehicle in the processes of operating the vehicle, ensuring the transport safety of the vehicle tool.

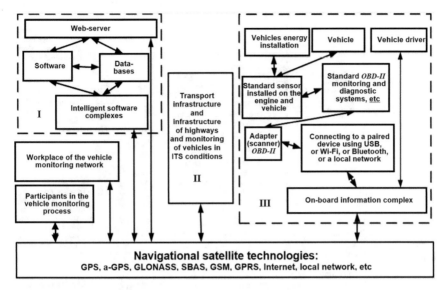

I – the server part of the ITS of the vehicle; II – vehicle ITS monitoring infrastructure in transport infrastructure and highway infrastructure; III – is the on-board part of the vehicle ITS as part of the on-board information complex

Fig. 5 Scheme of information exchange between the ITS elements of the vehicle and transport infrastructure and highway infrastructure in the processes of monitoring parameters of the technical condition in operating conditions

The functioning of the main functions of the on-board information complex is ensured by the performance of the functions assigned to it with the help of the system interaction of the structural features of the vehicle and the components of the ITS, namely (Fig. 6) in the part of: working with information (in the presence of different protocols) received from the sensors of the vehicle, connected by K, L or CAN lines; work with various interfaces of software complexes; vehicle identification in the traffic flow; data transmission and processing with simultaneous interaction between the main functions; engine and vehicle operation with determination of: engine and vehicle parameters in operation, maintenance and repair and their changes, deviations from performance standards, urgent (temporary) engine and vehicle operation conditions, with the formation of geozones regarding vehicle operation parameters; transport safety of vehicles during the performance of surveillance and recording functions (video, photo, audio); navigation when working with maps and services; registration of engine and vehicle condition; logging in and out of the server's software applications; assistance to the driver: with information about errors and malfunctions in work, with elimination of errors and malfunctions in work; with the transfer of information about errors and malfunctions in work to an external information storage, etc.

The results of the monitoring of parameters of the technical condition of the vehicle in the conditions of operation by means of ITS are shown in Fig. 7.

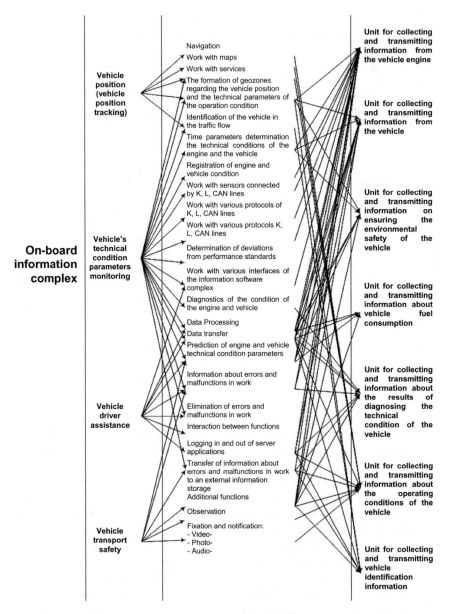

Fig. 6 The structure and relationship of the functional capabilities of the on-board information complex as part of the ITS

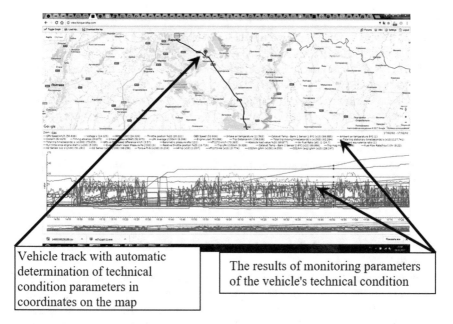

| Vehicle track with automatic determination of technical condition parameters in coordinates on the map | The results of monitoring parameters of the vehicle's technical condition |

Fig. 7 *Torque* (Terascale Open-Source Resource and QUEue Manager) working window with monitoring options parameters of the technical condition of vehicles

The *Torque* working window (Fig. 7) displays the current operating parameters of the vehicle engine, other systems and units, "error codes", "erase errors" from the electronic control unit of the vehicle, automatically send the values of the parameter values controlled by the sensor to the integrated electronic information metaspace, where during the set time it is possible to see not only the values of the controlled values at different times, but also to see on the map the entire route of the vehicle for the set period [20]. In the process of vehicle monitoring, at the same time and in accordance with the received report (Fig. 7), a report was received in MS Excel format on the results of monitoring parameters of the technical condition of the vehicle, when using the *Torque* software module in *csv* and *xlsx* formats. When monitoring a vehicle, the service of the information and software complex provides wide opportunities for generating reports. The report is a sample for a certain vehicle or a group of vehicles, presented in the form of graphical (track) or tabular information. It is the reports that make it possible to find out where the vehicle was located, where and at what time the scanner-communicator of the vehicle could not communicate, etc. The following types of tracks exist in the service: vehicle track tracking on the map; export to *.gpx* format; export to *.xls* format. If the event line is highlighted in red, the possibility of fuel draining is assumed, if it is green—the possibility of refueling.

The process of diagnosing and determining the vehicle malfunction status using DTCs was carried out automatically according to the algorithm developed by the authors using the *Torque* module within the framework of the developed information and software complex in ITS conditions. The final report on the results of diagnosing the technical condition and determining the status of malfunctions in the interaction with the user's computer through the interprocess communication (IPC) can also be available in the form of a consolidated table—vehicle DTCs [21, 34].

The speed of the vehicle has a great influence on the main technical and economic performance indicators. It is known [4] that the determination of the average technical speed of the vehicle is possible if there is a total mileage of the vehicle S according to the speedometer and the time of movement of the vehicle t_{mov}. The relative coefficient of change in the speed of movement of the vehicle, which is accepted as the main criterion when determining the group of operating conditions and subsequent adjustment of maintenance and repair standards, can be determined by the formula (see Eq. 26):

$$K_v = S / (t_{mov} \cdot V_{a1}) \approx 1.43 \cdot S / (t_{mov} \cdot V_{max}), \qquad (26)$$

where V_{a1}—is the speed of movement of this type of vehicle on the road of the 1st group $(0.7 \cdot V_{max})$.

Let's consider the results of the study on the example of one route of the vehicle movement in the sections of the movement distance with the use of a geophone (Fig. 8). In this case, the determination of the speed of the vehicle in operating conditions by means of ITS was carried out in several stages. 8 geozones were formed, with the coordinates of the start and end of the movement. The speed of the vehicle in the geozones was set in accordance with the provisions [4], namely, in geozones 1, 3, 5, 7 a limit of 110 km/h was established (for the conditions of the vehicle movement outside the city), and in geophones 2, 4, 6, 8—50 km/h (for vehicle traffic conditions in the city). To form the final report on the movement of the vehicle and determine the operating conditions of the vehicle by speed, an analysis and determination of technical and economic performance indicators and parameters of the technical condition of the vehicle in the relevant operating conditions were carried out as a result of monitoring by ITS means.

The results of a detailed analysis of vehicle traffic conditions and determination of initial data for further calculation of parameters were presented by means of a comparison of monitoring results (Fig. 8) and as a result of processing the research protocol, a change in vehicle speed was obtained depending on the position of the station, the distance of the journey and the time of travel, which are shown in Fig. 9.

Figure 9 shows the diagram of the change in vehicle speed depending on the section (Fig. 9a) and the graph of the change in the speed of the vehicle depending on the distance (Fig. 9b) and depending on the time of movement (Fig. 9c), which obtained on the basis of the analysis of the report, where for each district they calculated (in the order of calculation) according to the following dependencies (see Eqs. 27–30):

$$V_i = S_i / t_{mov\,i}; \qquad (27)$$

$$V_i = S_i / (t_{mov} + t_{st})_i. \qquad (28)$$

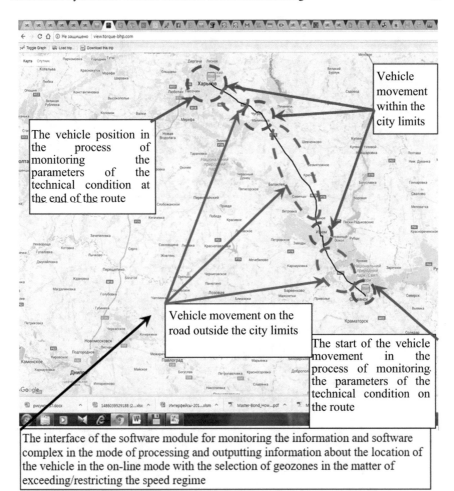

Fig. 8 Formation of the experimental site geozones

$$V_i = \Sigma \, V_{GPSi}; \tag{29}$$

$$V_i = \Sigma \, V_{OBDi}, \tag{30}$$

where V_i is the speed of the vehicle within the boundaries of the precinct; S_i—the distance of the i-th precinct; $V_{GPS\,i}$ is the GPS speed of the vehicle within each section obtained from the report; $V_{OBD\,i}$ is the OBD speed of movement of the vehicle within each section, obtained from the report; $t_{mov\,i}$ is the time of vehicle movement within the i-th precinct; $(t_{mov} + t_{cm})_i$—Σ of the time of vehicle movement and stops, parking within the boundaries of and—precinct.

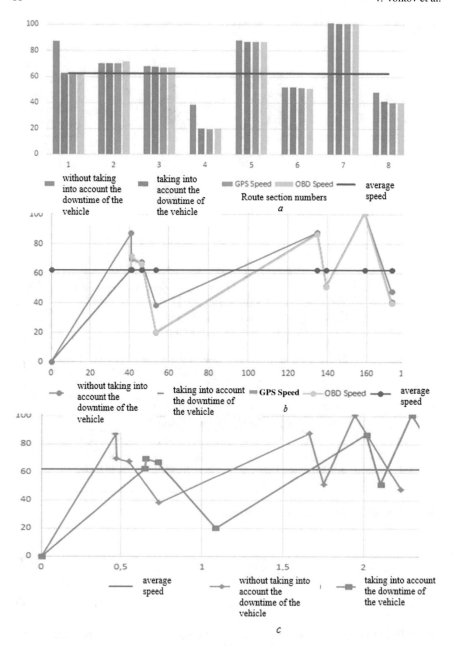

Fig. 9 The results of the study of the change in the average speed of the vehicle at the experimental stations within the distance of the journey: **a** depending on the position of the station; **b** depending on the distance of the journey; **c** depending on the time of movement

All obtained results of changing the parameters in the report, in part $V_{average}$ of vehicle movement, are shown in Fig. 10 and were obtained according to the following dependencies (see Eqs. 31–36):

$$V_{average} = S_{\Sigma i} / t_{\Sigma mov\, i} \tag{31}$$

$$V_{average} = S_{\Sigma i} / (t_{mov} + t_{st})_{\Sigma i} \tag{32}$$

$$V_{average} = \Sigma \, (S_i / t_{mov\, i})_i / n_i \tag{33}$$

$$V_{average} = \Sigma \, (S_i / (t_{mov} + t_{st})_i)_i / n_i \tag{34}$$

$$V_{average} = \Sigma \, V_{GPS\, average\, i} / n_i \tag{35}$$

$$V_{average} = \Sigma \, V_{OBD\, average\, i} / n_i \tag{36}$$

where $V_{average}$ is the average speed of the vehicle within the driving distance; $S_{\Sigma i}$ is the sum of the distances i-th stations; $t_{\Sigma movx\, i}$ is the sum of the vehicle's movement time at i-th stations within the distance of movement; $(t_{mov} + t_{st})_{\Sigma i}$ is the sum of the time of vehicle movement and stops, parking at i-th stations within the distance of movement; n_i is the number of precincts; $V_{GPS\, average\, i}$ is the average GPS speed of the vehicle within each i-th stations obtained from the report; $V_{OBD\, i}$ is the average OBD speed of the vehicle within each i-th stations, which were obtained from the report.

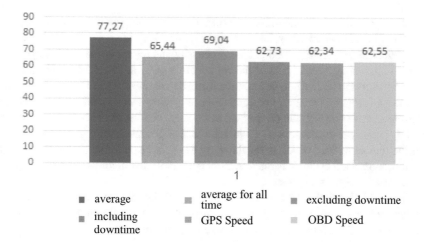

Fig. 10 The results of determining the change in the average speed of the vehicle based on the results of processing the report

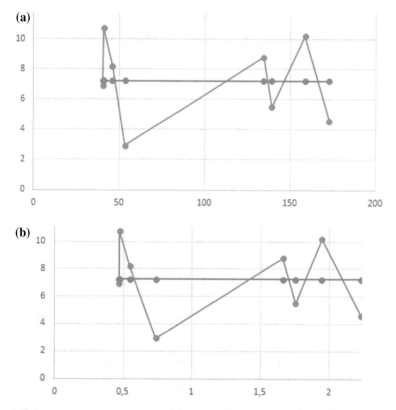

Fig. 11 The results of the study of vehicle fuel consumption at experimental sites within the distance of the journey: **a** depending on the distance of the journey; **b** depending on the time of movement

The results of the change in fuel consumption depending on the distance traveled and the time of movement of the truck are shown in Fig. 11.

It can be seen from Fig. 12 that after processing the results of the obtained parameters of the technical condition of the vehicle, it is possible to calculate the value of fuel consumption taking into account the formed geozones (Fig. 11). As a result, the average vehicle fuel consumption for the entire driving distance, taking into account geozones, was obtained, which is equal to $G_{average.} = 7.23$ l/h. This value is chosen as the actual fuel consumption of the vehicle. For comparison, Fig. 12 also shows the fuel consumption, which is regulated by the vehicle manufacturer in the urban/extra-urban/mixed cycle (information from the factory instructions) and the value of the average fuel consumption.

Figure 13 shows the determination and study of the relative coefficient of speed change of vehicle movement. The peculiarity of the determination of the relative coefficient of change of the movement speed was as follows:

– 2 V_{max} speeds were chosen as restrictions on the maximum speed on the relevant section, taking into account the geozones formed, respectively: *a* is the maximum

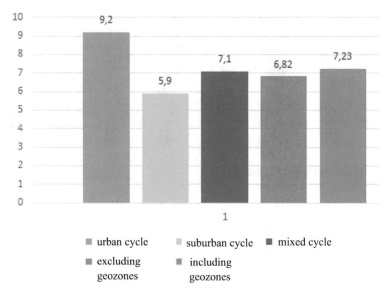

Fig. 12 The results of determining the change in the fuel consumption of vehicles during the movement based on the results of processing the report

possible V_{max} speed of the experimental vehicle on the road of the 1st group (outside the city)—110 km/h; b is the maximum possible speed V_{max} of movement (in accordance with the traffic rules) of the experimental vehicle on the road of the 1st group (in the city)—50 km/h;

- the relative coefficient of speed change of vehicle movement for each site, taking into account the geozones: was determined by dependence (see Eq. 37):

$$K_{vp} = S_{section\ i} / (t_{mov} \cdot V_{a\ aver.\ section\ i}), \tag{37}$$

- the average value of the relative coefficient of speed change of vehicle movement was determined as the average for the entire distance of movement of the vehicle. As a result of the study, the value of $K_{vp} = 0.94$ was obtained, which refers to the first group of operating conditions [4], K_{vp} changed at the sections of the vehicle movement path within the limits $K_{vp} = 0.69$–1.25.

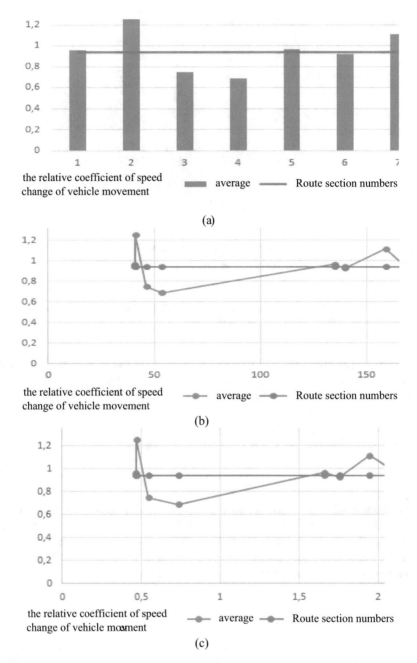

Fig. 13 The results of the study of the relative coefficient of speed change of vehicle movement at experimental stations within the driving distance: depending on the position of the station (**a**); depending on the driving distance (**b**); depending on the time of movement (**c**)

5 Conclusion

In connection with the conducted analysis of the state of technical operation of cars, it can be concluded that the traditional system of maintenance and repair, which has been formed on public car transport for many years, no longer meets modern requirements in general.

A new technique for public road transport in the field of technical control of rolling stock is the creation of information systems for organizational and functional support of rolling stock operation processes. Implementation of the new basic principles of the "adaptive" system for managing the technical condition of the vehicle in the technical operation of cars, the key point of which is the development of an information and communication system and a database of predictive models that ensure, through monitoring, the remote acquisition of the necessary current information from the rolling stock and its processing, as well as production of corrective influences.

An information system for monitoring the condition and operating conditions of vehicles was formed, the general information support of the system was proposed, and the following were described: research processes and evaluation of the operating conditions of vehicles; information model of the position of vehicles; speed model of vehicle traffic modes. In addition, evaluation tools are presented: the type and condition of the road surface; state of transport infrastructure facilities; condition of the adjacent area of the road; fuel economy of vehicles in operating conditions, as well as an algorithm for adjusting the speed of vehicles.

A mathematical model of the information system for evaluating parameters of the vehicle's technical condition in operating conditions has been developed. For its implementation, the features of the information system for evaluating the parameters of the technical condition of the vehicle in operating conditions are described, and the structured information model of the information software complex "IdenMonDiaOp-erCon "HNADU-16"" has been verified, which actually ensures the operation of the information system for evaluating the parameters of the vehicle technical condition.

An assessment of the technical and economic indicators of work and parameters of the technical condition of vehicles in operating conditions was carried out, which was carried out based on the average speed of the vehicle and fuel consumption. The operating conditions were evaluated based on the relative coefficient of speed change of vehicle movement in accordance with the provisions of the theory of car operation.

References

1. Transport and Communication of Ukraine for 2014, 222 pp. Consultant, Kyiv (2015) (in Ukrainian)
2. Automobile Transport of Ukraine: State, Problems, Development Prospects: Monograph/ State Motor Vehicle Research and Design Institute, 400 pp. In: Redjuka, A.M. (ed.). SE "Derzhavtotrans NDIproekt", Kyiv (2005) (in Ukrainian)

3. National Center for Statistics and Analysis. Distracted Driving: 2016. Traffic Safety Facts Research Note. Report No. DOT HS 812 517, 6 pp. National Highway Traffic Safety Administration, Washington, DC (2018)
4. Hovorushchenko, N.: Technical Operation of Cars, p. 312. Vysh. shkola Publ, Kharkiv (1984)
5. Hovorushchenko, N., Varfolomeev, V.: Technical Cybernetics of Transport: Textbook, 271 pp. KhNADU, Kharkov (2001) (in Ukrainian)
6. Kuznetsov, E.: Technical Operation of Cars in the USA, 168 pp. Transport, Moscow (1978) (in Russian)
7. Volkov, V., Mateichik, V., Nikonov, O., et al.: Integration of the Technical Operation of Vehicles into the Structures and Processes of Intelligent Transport Systems, 398 pp. Under the editorship of Volkov V. Donetsk, Publishing House "Knowledge" (2013) (in Ukrainian)
8. Volkov, V., Mateychyk, V., Hrytsuk, I.: Monitoring of the Technical Condition of the Car in the Life Cycle, 301 pp. In: Volkov, V. (ed.). KhNADU, Kharkiv (2017) (in Ukrainian)
9. Troitsky-Markov, T., Sennovsky, D.V.: Principles of construction of energy efficiency monitoring system. Monitoring. Sci. Secur. **4**, 34–39 (2011)
10. Dmytrychenko, M., Mateichyk, V., Hryschuk, O., et al.: Methods of System Analysis of the Properties of Automotive Equipment: Training, 168 pp. Manual. K. NTU (2014) (in Ukrainian)
11. Hrytsuk, I.: The Concept of Ensuring the Optimal Temperature State of Engines and Vehicles in Operating Conditions, 40 pp. Doctoral thesis in the technical sciences 05.22.20; Kharkiv National Automobile and Road University, Kharkiv (2016) (in Ukrainian)
12. Hovorushchenko, N.: System Technic of Vehicles (Calculation Methods of Research), 292 pp. KhNADU Publ., Kharkiv (2011) (in Ukrainian)
13. Mateichyk, V., Volkov, V., Hrytsuk, I., et al.: Peculiarities of monitoring and determining the status of vehicle malfunctions as part of the on-board information and diagnostic complex. Bull. Natl. Transp. Univ. **30**, 51–62 (2014)
14. Volkov, V., Mateychyk, V., Komov, P., et al.: Organization of the technical operation of cars in the conditions of the formation of intelligent transport systems. Bull. Natl. Tech. Univ. "KhPI" Collect. Sci. Papers Ser.: Automob. Tractor Constr. **29**(1002), 138–144 (2013)
15. Alekseev, V., Korolev, P., Kurakina, N., et al.: Information-measuring and control systems for monitoring the state of distributed technical and natural objects. Pribory **10**, 28–42 (2009)
16. Atroshchenko, V., Shevtsov, Yu., Yatsynin, P., et al.: Technical Possibilities of Increasing the Resource of Autonomous Power Plants of Energy Systems. Monograph, 192 pp. South Publishing House, Krasnodar (2010) (in Russian)
17. Electronic Resource. Access mode: https://www.scania.com/group/en/boosting-uptime-scania-remote-diagnostics/
18. Hrytsuk, I., Hrytsuk, Yu., Volkov, Y., Kalashnikov, E.: Peculiarities of the development of mathematical models for evaluating the current and forecasting parameters of the vehicle's technical condition. Interuniv. Collect. Sci. Notes. **45**, 44–51 (2017). Lutsk National Technical University, Lutsk
19. Hrytsuk, I., Hrytsuk, Yu., Volkov, Yu.: Peculiarities of the information system for monitoring and forecasting parameters of the technical condition of the engine and the vehicle in its conditions. Mod. Technol. Mech. Eng. Transp.: Sci. J. **2**(6), 43–49 (2016). Lutsk National Technical University, Lutsk
20. Sytnyk, M., Krasnyuk, T.: Intelligent Analysis of Data (Data Mining): Training. Manual, 376 pp. KNEU, V.F. Kharkiv (2007)
21. Volkov, V., Mateychyk, V., Komov, P., et al.: Technical regulation of the software product "Service Fuel Eco "NTU-HADI-12"" during normal operation (Essay of a scientific and practical nature). Applicant and patent owner V. Volkov and KhNADU. Certificate of copyright registration for the work No. 53292 dated January 24, 2014. Application dated November 22, 2013 No. 53604, 3 pp. (2014)
22. Hrytsuk, I., Volkov, V., Khudyakov, I., Volkova, T., Kuzhel, V.: Operational Control of the Technical Condition of Vehicles, 197 pp. Edelweiss and K, Kharkiv–Kherson–Vinnytsia (2022). ISBN 978-617-7417-00-1 (in Ukrainian)

23. Hrynkiv, A.: Using of forecasting techniques to manage the technical condition of aggregates and systems of vehicles. Tech. Agric. Ind. Mach. Build. Autom. **29**, 25–32 (2016)

24. Volkov, V., Gritsuk, I., Gritsuk, Yu., Volkov, Yu.: Specifically, formulating the information systems of the classification of the minds of transport interventions. In: Proceedings of the State Economic and Technological University of Transport Ministry of Education and Science of Ukraine Series "Transport Systems and Technologies", vol. 30, pp. 84–94 (2017)

25. Template: Real-Time Operating Systems/Wikipedia, The Free Encyclopedia. wikipedia.org. https://en.wikipedia.org/wiki/Template:Realtime_operating_systems

26. Gritsuk, I., Bilousova, T., Gritsuk, Yu., Volkov, Yu.: Features of the formation of the subject area and the information system for evaluating the parameters of the technical condition of the vehicle in the conditions of exploitation. Vestnik Kherson Natl. Tech. Univ. **3**(32), 302–306 (2017)

27. Volkov, V., Gritsuk, I., Gritsuk, Yu., Shurko, G., Volkov, Yu.: Features of the formation of a methodology for the classification of conditions of vehicle operation in the information conditions ITS. Bull. Natl. Tech. Univ. «KhPI» Collect. Sci. Works Ser. Transp. Eng. **14**(1236), 10–20 (2017)

28. Hrytsuk, I., Volkov, V., Ukrainskyi, E., Volodarets, M., Kuzhel, V., Volkova, T., Ryzhova, V.: Information system for operational provision of normalization of vehicle operation indicators. Bull. Mech. Eng. Transp. **2**(16), 16–22 (2022). https://doi.org/10.31649/2413-4503-2022-16-2-16-22

29. Han, J., Kamber, M.: Data Mining: Concepts and Techniques, 800 pp. Morgan Kaufmann (2006)

30. Amberg, B., Vetter, T.: Optimal landmark detection using shape models and branch and bound. In: Proceedings of the International Conference on Computer Vision, pp. 455–462 (2011)

31. Volodarets, M., et al.: Optimization of vehicle operating conditions by using simulation modeling software. SAE Technical Paper 2019-01-0099 (2019). https://doi.org/10.4271/2019-01-0099

32. Kuric, I., Mateichyk, V., et al.: The peculiarities of monitoring road vehicle performance and environmental impact. In: MATEC Web of Conferences Volume 244, 5 December 2018, № 03003 3rd Innovative Technologies in Engineering Production (ISSN: 2261236X), ITEP 2018; Bojnice; Slovakia; 11 September 2018–13 September 2018; № 143364 (2018). https://doi.org/10.1051/matecconf/201824403003

33. Savostin-Kosiak, D., Sakhno, V.: Study of fuel consumption of city buses with diesel by means of mathematical modeling. Wybrane Zagadnienia: monografia pod redakcja naukowa Kazimierza Lejdy. Series Transport, No. 9, S. 95–103. Politehnika Rzeszowska, Rzeszow (2017)

34. Volkov, V., Hrytsuk, I., Hrytsuk, Yu., Volkova, T., Kuzhel, V., Volkov, Yu.: A general approach to the formation of models for evaluating the technical condition of a car in operating conditions. Bull. Mach. Build. Transp. **1**(9), 27–37 (2019). https://doi.org/10.31649/2413-4503-2019-9-1-27-37

35. Volkov, V., Pavlenko, V., Kuzhel, V.: Investigation of the agent approach to control the technical condition of vehicles. Bull. Mach. Build. Transp. **2**(10), 16–23 (2019). https://doi.org/10.31649/2413-4503-2019-10-2

36. Volkov, V., Nikonov, O., Volkov, Yu.: Prospects for the introduction of an adaptive system for car maintenance. In: Collection of Reports of the XX Scientific and Technical Conference with International Participation "Transport, Ecology-Sustainable Development". Varna, Bulgaria, pp. 404–409 (2014)

37. Boreiko, O., Teslyuk, V., Kryvinska, N., Logoyda, M.: Structure model and means of a smart public transport system. Procedia Comput. Sci. **155**, 75–82 (2019). https://doi.org/10.1016/j.procs.2019.08.014

38. Sajid, F., Javed, A.R., Basharat, A., Kryvinska, N., Afzal, A., Rizwan, M.: An efficient deep learning framework for distracted driver detection. IEEE Access **9**, 169270–169280 (2021). https://doi.org/10.1109/ACCESS.2021.3138137

39. Gritsuk, I., Zenkin, E., et al.: The complex application of monitoring and express diagnosing for searching failures on common rail system units. SAE Technical Paper, 2018-01-1773 (2018). https://doi.org/10.4271/2018-01-1773

Model for New Innovation Knowledge Spreading in Society

Anatolii Shyian and **Liliia Nikiforova**

Abstract Innovation knowledge begins with the help of objective characteristics of the manifestations of human activity (both in its activity or behavior in communication with a social group). This made it possible to introduce a tuple of quantitative and qualitative parameters as a characteristic of knowledge. It shows how a metric can be entered into a tuple of knowledge (individual, social group, or society). The presence of the metric allows you to set the task of knowledge spreading as a change in the number of people who have a certain tuple of knowledge. The general model for the diffusion mechanism of knowledge spreading is considered. A method of identifying the tuple of knowledge possessed by an individual, social group, or society has been developed. The discussion is an example of the application of this method to important processes that accompany the development of society. It is shown that the obtained results can increase the efficiency of the management of migration processes. The method of determining a set of tuples, the presence of which is necessary for the functioning of a developed country, is described. The possibility of applying the obtained results to increase the efficiency of regional development management is shown.

Keywords Innovation knowledge · Diffusion spreading · Society · Development

1 Introduction

Innovation knowledge is a reason for the country's development in the modern world. Universities are one of the main drivers for the development of innovation knowledge and its transformation into innovation. Universities in modern society simultaneously perform two functions [1]. Firstly, new innovation knowledge is formed in their

A. Shyian (✉) · L. Nikiforova
Vinnytsia National Technical University, Khmelnytske Shose 95, Vinnytsia 21021, Ukraine
e-mail: anatoliy.a.shiyan@gmail.com

L. Nikiforova
e-mail: nikiforova@vntu.edu.ua

© The Author(s), under exclusive license to Springer Nature Switzerland AG 2024   97
A. Semenov et al. (eds.), *Data-Centric Business and Applications*, Lecture
Notes on Data Engineering and Communications Technologies 195,
https://doi.org/10.1007/978-3-031-54012-7_5

environment. Secondly, they prepare a new generation for innovative knowledge implementation in the social and economic life of society.

A large number of scientific studies are dedicated to the study of the role and influence of innovation in social, economic, and other areas of social development. As a rule, these works were carried out within the framework of separate scientific branches. However, a unified approach to describing such problems has not yet been built.

In [2], it is emphasized that "Science is considered essential to innovation and economic prosperity. Understanding how nations build scientific capacity is, therefore, crucial to promoting economic growth and national development." As a result of studying this issue, the authors concluded that "revealed disciplinary clusters inform economic development, where the number of publications in applied research centered cluster significantly predicts economic growth." For example, the influence of universities through knowledge and science is a powerful driver of the development of society and the state.

A recent study [3] examined a wide class of concepts and methods of society development, including the impact of innovation. However, the impact on the characteristics of its development from the spreading of knowledge in society is not considered in detail. Similarly, the role of universities as centers for spreading knowledge and forming innovation is also not considered.

Today, there is a rapid development of communication tools, primarily in the field of new innovative knowledge spreading. Therefore, more people throughout their lives are often forced to master a fairly large array of new knowledge for themselves. However, the knowledge spreading throughout society has not yet been sufficiently studied despite recently growing interest.

In several recent studies [4, 5] knowledge spreading using social networks is modeled and simulated. This is often done using different random walk models [6, 7]. However, such network processes with a large number of agents can be regarded as diffusion processes. In this case, the diffusion coefficient can be calculated using the characteristics of the random walk [8]. This opens up the possibility of finding the diffusion coefficient of knowledge spreading directly from experiments with the time dynamics of social networks.

In [9], the study of the process of knowledge spreading from teacher to student was carried out, and quantitative data was obtained by a student survey. In the paper, the authors' attention was focused primarily on the individual trajectories of student learning. However, the process of transition to describing the activities of students after graduation from the university, as well as to changing their knowledge in the process of adaptation to society, still requires detailed study.

In recent years, there has been an increasing interest in the connection between clusters in social networks and the regional distribution of knowledge (information) across geographic regions. The importance of such studies is emphasized by their direct influence on the political and economic processes in the country. For example, in [10], a set of characteristics of a geographic region that could influence an electoral process was obtained. However, this article did not compare the knowledge diversity between geographic regions and the network. At the same time, the following example can demonstrate the importance of the problem in the development of correct methods for the quantitative transition from the network diversity of knowledge (opinions) to geographic diversity. For example, let's suppose there is a localization of a relatively small number of people on a national scale, which are configured for autonomy. On a national scale, such a number of people can be quite small. However, if they are localized in a region with a small quantity of population, then their relative number may be large. As a result, separatist processes can be carried out using democratic mechanisms and procedures. The situation with Great Britain's Brexit, "democratic voting" in Crimea, events in the East of Ukraine, and many other events in recent years in different countries convincingly demonstrate the importance of modeling and forecasting such incidents.

A recent book [11] examines errors, mistakes, and failures that arise in the communication of people and societies that belong to different cultures (e.g., different knowledge). Such errors, mistakes, and failures occur at all levels, from the activities of individuals in society to large-scale economic and political processes. However, culture can be viewed as a set of rules and norms of behavior. In developed countries, the main pool of rules and norms of behavior and activity is formed in universities [1]. In [12], the authors note: "Distrust in scientific research has important societal consequences, from the spread of diseases to hunger in poorer regions. However, these scientific beliefs are hard to change because they are entrenched within many related moral beliefs and perceived beliefs of one's social network." Thus, in a society, there may be active or passive resistance to knowledge spreading, which is necessary for the development of society. Such processes lead to a decrease in the intensity of the knowledge spreading from a university throughout society.

In [13], attention is drawn to the fact that consideration of the influence of science requires an obligatory analysis of changes in social behavior, which in practice is rather difficult to implement. In [14], it is proposed to take into account the influence of journalists as a quantitative characteristic of scientific influence. However, the question of the impact on society remains open. In [15], the authors studied the impact on society of two discoveries in the field of physics: "(i) the discovery of the exoplanet Proxima b in 2016; (ii) the Cassini-Huygens mission, which culminated in a spectacular Grand Finale in 2017". They stressed that "We found convincing evidence for increased public awareness of astronomy topics; discussion and debate of advanced scientific concepts; and a sustained legacy of engagement with our chosen research results." Thus, the information spreading about "high-field research projects" increases interest in the content of scientific knowledge produced by universities. As a result, interest in research on the impact of knowledge on society is increasing, and one might expect a concomitant increase in the number of methods addressing indicators of this process. However, this poses a new challenge: the necessity to identify characteristics and processes in need of measurement.

In [16], there exists "the need to capitalize on diversity to develop students' entrepreneurial intentions and drive entrepreneurship." This means that there are non-linear effects for innovation increasing that are related to the total amount of knowledge from the university.

Ruoslahti in [17] asks the question: "How does complexity affect the co-creation of knowledge in innovation projects, according to project participants?" He concludes that "although this study demonstrates that the elements of complexity can be used to gain insight into innovation projects, the results show that not all elements of complexity are equally important in this context and that they appear in a certain order." That is, innovative projects are influenced by a lot of knowledge factors of the stakeholders, which have different strengths of influence. Therefore, the identification of those elements of knowledge tuples, which are a place for the creation and implementation of innovative projects, is an element of the development of society. For this, there must be powerful analytical structures in society, which today traditionally function within the framework of universities.

Bhattacharya and Packalen in [18] note that: "New ideas no longer fuel economic growth the way they once did." This, in their opinion, was since "emphasis on citations in the measurement of scientific productivity shifted scientist rewards and behavior on the margin toward incremental science and away from exploratory projects that are more likely to fail, but which are the fuel for future breakthroughs." Park et al. in [19] also pay attention to the following: "Although the number of new scientific discoveries and technological inventions has increased dramatically over the past century, there have also been concerns of a slowdown in the progress of science and technology. We analyze 25 million papers and 4 million patents across 6 decades and find that science and technology are becoming less disruptive of existing knowledge, a pattern that holds nearly universally across fields. We link this decline in disruptiveness to a narrowing in the utilization of existing knowledge." Thus, today, there is a growing understanding in society that the modern system of organizing the connection between science and the economy needs significant reformation. Such reforming should be carried out primarily in order to motivate employees to increase the knowledge tuples that they use in their activities.

The main purpose of this paper is to develop a general approach to modeling the influence of a university's impact on the spreading of knowledge for society's development.

2   Baseline Model

2.1   How Is the Knowledge Measured?

Let's define, following [1, 20], knowledge (information) as norms of behavior, methods, results of activity, algorithms, and programs of activity of individuals, as well as ways of communicating with people in their joint behavior and joint activities.

As a rule, today the population receives the main body of knowledge related to the implementation of activities while studying at the university. Therefore, the presented results will be applied primarily to tertiary education.

The knowledge of a person (or a group of people) will be determined by the tuple K, which includes both quantitative and qualitative characteristics for individual components of knowledge. Individual components of the tuple K can be considered as components in some local coordinate system localized on a person or a separate group of people. The tuple includes methods and technologies that are necessary to maintain human life in society. It may also include, for example, social, economic, moral, ethical, religious, and other characteristics that determine the behavior of an individual or group. The tuple also necessarily includes the characteristics that describe the social and professional activities of an individual (or a group of people if it is considered as a single and indivisible whole).

There typically are many tuples for people in a society that are all tied together into a single entity. They are necessary for the existence and maintenance of the stability of society.

Each tuple can be represented as $K_s = (k^s{}_1, k^s{}_2, ..., k^s{}_i, ...)$. Here $k^s{}_i$ is a separate method, technology, or way of doing an activity (either for a person or in collaboration with other people). Such a tuple can be viewed as a knowledge piece that is localized on an individual. Note that many people can have the same tuple. In this sense, learning (education) is a way of localizing a tuple on a person. The totality of society's knowledge is a collection of individual tuples (see Eq. 1):

$$K = U_{s=1}^{s=n} K_s. \tag{1}$$

Here n is a total quantity of separate tuples.

Each tuple is a separate point K_s in the set K. The metric $d(K_i, K_j)$ in space K of the society knowledge can be introduced as follows (see Eq. 2):

$$d(K_i, K_j) = \| K_i - K_j \| = \| \Delta K_{ij} \|. \tag{2}$$

Here the indices i and j correspond to the tuple of knowledge.

The specific type of metric can be chosen depending on the task at hand. In the general case, it is necessary to compare the components of tuples with each other. For example, one can choose the following metric (see Eq. 3):

$$d(K_i, K_j) = w_1 \left| k_1^i - k_1^j \right| + w_2 \left| k_2^i - k_2^j \right| + \cdots + w_n \left| k_n^i - k_n^j \right|, \sum_{s=1}^{s=n} w_s = 1. \tag{3}$$

Here w_s are the corresponding weights for the characteristics. They are selected from the conditions of the problem being considered (for example, they can be selected when processing the opinions of experts). One can use other metrics, which are similar to (3) (for example, instead of (3) go to the Cartesian difference for a set of characteristics).

Thus, the presence of a metric makes it possible to order the set K. This is especially easy to implement in the case when the components of the tuple are presented in the form of numerical characteristics.

Note that due to the definition of knowledge, knowledge tuples K can be formed, for example, for certain professions, certain areas of education (for example, secondary, tertial, economic, engineering, etc.), certain geographical regions, different types of activities (which include several professions), for religious or ethnic groups, individual countries, collections of countries, and so on. Then, for these tasks, metric (2) or similar can be used to describe differences in knowledge (more precisely, in tuples (1), which are used, as a rule, to characterize groups of people).

The order of the tuple of knowledge should also take into account the time (averaged for the society) that is required to acquire certain knowledge. This time can be calculated, for example, as the time needed to master certain professional skills. Part of the tuple of knowledge of society is established in secondary (general) school when the new generation is taught "general conditions of life" in society.

Thus, metrics (2) and (3) refer to innovative knowledge that exceeds the tuple of knowledge that is necessary for the existence of a person in society. Part of such knowledge is often associated with the culture that is localized in the society.

As for migrants and expats, there is a need to teach them the necessary elements of a tuple. Here, additional tuple elements and additional time intervals already appear for them. Therefore, the results obtained within the framework of the proposed model can be applied to the task of integrating migrants and expats into society.

The set of people n who know the tuple K_s will be denoted as $n(K_s)$. In what follows, the notation $n(K)$ will be used. Here it is emphasized that the knowledge of K is a variable, and n is a function.

To consider knowledge change problems, one can explicitly use the time variable t. Then the desired function that characterizes the knowledge of an individual, social group, or society as a whole will be the function (distribution) $n(K, t)$.

As an example, consider the university as a structure that knowledge changes. It can be considered as a point source for the components change (some or all at once) in the tuple K. Since there are no conservation laws for knowledge (information), the number of carriers of "university knowledge" (e.g., innovation knowledge) will increase with time t. Thus, the university will introduce knowledge with the K_U tuple into society. The number of people with "university knowledge" can be written as follows (see Eq. 4):

$$n_U(K, t) = f_U(K, t) \tag{4}$$

Universities are considered as precisely those structures that disseminate (socialize) new knowledge in society.

For example, if knowledge is accumulated within the framework of one paradigm, then $f(K_i, t)$ for a specific component of the knowledge tuple Ki can, in some approximation, be represented by a logistic curve. Then the simplest time equation for variation of the function $f(K_i, t)$ takes the following form (see [18, 21]).

$$f_U(K_i, t) = K_{i\infty} \cdot \frac{K_i(t = 0)}{K_i(t = 0) + \{K_{i\infty} - K_i(t = 0)\} \cdot \exp(-at)} \tag{5}$$

Here $K_{i\infty}$ is the maximum possible knowledge within the given paradigm. a is some constant that characterizes the rate of change of a given component of knowledge over time.

Note that formula (5) can only be used within one paradigm, that is, at the stage of knowledge accumulation. The change of knowledge during the transition to a new paradigm is carried out quite quickly, often abruptly.

One of the ways to measure knowledge is also financial. Knowledge in the form of innovation can be capitalized through use in the economy. The use of knowledge stimulates its spreading when an increasing number of the population is involved in the use of new methods, ways, and technologies of activity. The possibility of capitalization is determined by the institutional structure of society. Developed societies have created the most powerful ways to spread knowledge through economic benefits.

In developing societies, the capitalization of knowledge is quite difficult due to the lack of appropriate special institutions, mainly of a political and economic nature [22]. In such societies, new knowledge can even be harmful to its bearer (that is, brings negative benefits). Examples are authoritarian and totalitarian societies.

Thus, some part of the measurable characteristics of knowledge will depend on society as a whole. In this sense, the problem of knowledge spreading in society is self-coordinated and self-organized.

2.2 Knowledge Spreading by Diffusion

The quantity of people is a constant value (this quantity can only change due to the balance of birth and death). Knowledge is spread in such a way that the quantity of people with one knowledge tuple decreases, and the quantity of people with another knowledge tuple (other tuples) increases by the same amount. As a result, the knowledge spreading (transformation, distribution, diffusion) can be described as a change in the number of people in the knowledge space K.

In the general case, the dynamics of changes in the distribution of the people quantity by knowledge can be described by such a differential equation (see Eq. 6):

$$\frac{\partial n}{\partial t} = O(n, K, t) + \sum_{i=1}^{N} U(K_i, t). \tag{6}$$

The initial and boundary conditions for this differential equation are set by the conditions of the problem.

In (6) $O(n, K, t)$ is an operator describing the processes of spreading and assimilation by a society of new tuples of knowledge K (that is, changes in the people quantity $n(K, t)$ depending on the conditions in society). The operator $U(Ki, t)$ describes the influence of universities on changes in the distribution $n(K, t)$. N is the total number of universities.

Equation (6) with appropriate initial and boundary conditions describes the influence of universities that have an impact on the considered society.

In general, since K is a tuple, Eq. (6) will be a system of non-linear differential equations that are related to himself.

The simplest case is when knowledge spreads in society according to the diffusion law (for example, the term "diffusion of innovations" is already quite widespread) [23].

If earlier the knowledge spreading was carried out in a geographical space, when the physical carriers of knowledge (for example, people or books) migrated from point to point, today the situation has changed dramatically. The Internet contains all the scientific knowledge that universities have [24]. Sometimes access to this knowledge (scientific articles, textbooks, lecture courses, etc.) is difficult due to financial difficulties. However, the presence of e-mail, as well as the availability of web archives of scientific papers with free access to scientific information, has significantly changed the situation with knowledge spreading. As a result, today the limiting factor for knowledge spreading is not the physical spreading of its carriers [25]. Today, the limiting factor is the readiness (ability) of a person to perceive the new knowledge that he needs. Therefore, the processes of knowledge diffusion do not occur in the "physical" space, but in the space of knowledge K.

Let the operator $O(n, K, t)$ describe the diffusion process with the diffusion coefficient $D(n, K, t)$. This will simulate the propagation of characteristics of a tuple K from a university through society. The heterogeneity of society will occur through the diffusion coefficient D dependence from n, K, and t.

Then the influence of universities on a given society can be described as a nonlinear problem in this way (∇_m is the gradient operator for the considered tuple K).

$$\frac{\partial n}{\partial t} = (\overrightarrow{\nabla}_K D(n, K, t), \overrightarrow{\nabla}_K n) + \sum_{i=1}^{i=N} U(K, t) \tag{7}$$

Here, (…, …) denotes the scalar product of vectors on tuple space K. The initial and the boundary conditions can be separately set.

In general, Eq. (7) is a system of nonlinear differential equations. These equations have been studied in sufficient detail in recent years. For example, [5] studied a coupled set of nonlinear integral-deferential equations. It was found that the system has one or more fixed points. That is, such equations can admit stationary solutions.

In the general case, D depends on n as well as on K and t. The greater the proportion of people n who possess a certain level of knowledge K, the greater will be the diffusion coefficient. That is, $\partial D/dn > 0$ at a constant K. On the contrary, at a constant n, the greater the level of knowledge, the smaller the diffusion coefficient D will be. That is, $\partial D/\partial K < 0$.

However, another situation is possible. For example, certain knowledge becomes important for a certain group of society. Then this group of people will have a social or economic motivation to study them. Such cases can occur in the diffusion of innovations [26], when the possession of the knowledge necessary for the release of the innovation gives significant advantages to its bearers. It is also manifested in the phenomena of the so-called mainstream in scientific research and development, when some areas of scientific research become "fashionable". For example, when the scientific paradigm changes [21]. It also manifests itself in cases of "infection" of people with certain information [27], which stimulates them to change the methods and technologies of their behavior. The latter is characteristic of the development of culture. In this case, $\partial D/\partial n < 0$ and $\partial D/\partial K > 0$ will be observed. This happens, for example, when the scientific paradigm changes [21] or about the differences of different social groups in society [28]. Digitization of knowledge and the availability of a large number of educational programs on the Internet (see, e.g., [29]) are also a typical example of such a situation.

As for the dependence on time t, the diffusion coefficient decreases over time due to the fact that the distribution of knowledge comes to equilibrium. That is, $\partial D/\partial t < 0$.

Thus, the coefficient of diffusion of knowledge in society will depend on the institutional features of the structure of this society. The presence of special institutions, that is, social structures and strata of society, which have a high level of K, and to which the possibility of wide access of other groups of society is organized in society, leads to an increase in the coefficient of knowledge diffusion.

Therefore, open knowledge leads to the acceleration of establishing the stability of society as a whole. For example, the movement for Open Innovation in Science leads to an increase in the diffusion coefficient of knowledge.

3   Example: One University Impact

Let the knowledge (information) be characterized by only one variable, that is, K is a scalar stationary function.

Let the university have knowledge advantages over the surrounding society. This excess is denoted as $\Delta K = K_U - K_s$. K_s is the upper boundary of society's knowledge, and K_U is the upper boundary of society's university. Using the metric (2), we can assume that this knowledge is a "higher" level compared to society knowledge. Knowledge ΔK is available only to employees of the quantity n_U, and $n_U \ll n_s$.

In this case, the diffusion spreading of knowledge from university to society can be described by such an equation with such initial and boundary situations (see Eq. 8):

$$\begin{cases} \frac{\partial n}{\partial t} = D\frac{\partial^2 n}{\partial K^2} \\ n(K, t = 0) = n_U, \, K \in [K_s, K_U] \\ n(K_s, t) = n_s \\ n(K_U, t) = n_U \end{cases} \tag{8}$$

When writing Eq. (8), the following circumstances were taken into account:

(1) The knowledge diffusion coefficient is assumed to be independent of the task variables.

(2) Due to the inequality $n_U \ll n_s$, the impact of the university on the distribution of knowledge will cover a fairly small part of society. First of all, those who have knowledge close to K_s.

(3) Considering that new knowledge learning requires a certain time, a relatively small number of people will acquire new knowledge. These will be those people whose knowledge is close enough to the K_s boundary.

The qualitative behavior of solutions to such a problem in time is shown in Fig. 1. Straight line correspond to steady-state solutions at $t \to \infty$.

The steady-state solutions for problem (8) are described by such a formula (see Eq. 9).

$$n(K, t \to \infty) = n_s - \frac{n_s - n_U}{K_U - K_s}(K - K_s) \tag{9}$$

Formula (9) does not depend on the diffusion coefficient and therefore has a universal character. Such a solution is established in a time that satisfies this condition:

$$t \gg \frac{(\Delta K)^2}{D}. \tag{10}$$

The higher the rate of diffusion D for knowledge, the faster the established regime sets in (9).

Figure 1 shows that at first society acquires lower knowledge, and only then moves on to knowledge of a higher level.

This applies to people studying at university.

There is forgetting, but little depends on the number of people. "Old" knowledge is used all the time, so repetition does not allow you to forget much.

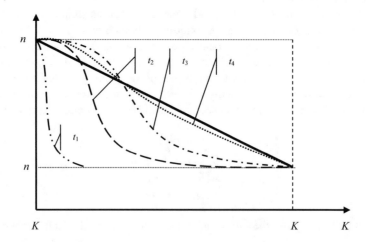

Fig. 1 Qualitative representation of the knowledge dynamics. $t_1 < t_2 < t_3 < t_4$

Diffusion from teachers is constant, teachers have the same level of knowledge. The value of the diffusion coefficient is directly proportional to the "intensity of learning". Students are also working on increasing the diffusion coefficient.

The boundary conditions apply to senior students who, together with teachers, work at grants.

The second boundary condition is junior students/applicants who come to the university after school (with school knowledge).

How long do you need to study at university to get to the "advanced level"? This is determined by relation (10). This ratio determines the amount of knowledge that a community of students can master. Generally speaking, the real time will be somewhere between $t_2 - t_3$. This is due to the fact that students who "don't have time" are excluded from university. That is why the following system is effective: 2 years of college studies + 2 years to obtain a "bachelor's" + 2 years to obtain a "master's". Because the university/college still "raises the level of knowledge" (that is, the amount of knowledge), and it may be enough for more effective participation in the socio-economic development of the country.

The results obtained below can serve as the first stage for the optimization of educational programs for university students. They can also serve as a direction for further improvement of the model by taking into account the non-linear dependence of the diffusion coefficient on the peculiarities of students' perception of new information. It is also possible to experimentally determine the distribution of students according to the diffusion coefficient. Also, the diffusion coefficient depends on the amount of new knowledge for the student, which allows to manage educational programs. The time factor necessary for obtaining the optimal values of the learning terms can also be obtained as a result of both the development of the model and the obtaining of experimental values of the dependence of the diffusion coefficient on forgetting, the speed of mastering the material, the number of people who possess certain knowledge (because here the "social training"), etc.

Note that the same equation will describe the diffusion of innovations. This is because innovations are the same knowledge on the use of new methods and technologies to achieve the goal in people's activities.

The extension of the model to the case of many universities with different knowledge usually requires computer simulation. However, the general trends of this model results will not change.

It is interesting that if the knowledge of universities does not differ much, then in some approximation it is possible to build a "cells" model. That is, one university is considered, around which there is a certain number of people. The whole country is a collection of "identical cells". (Note, that tertiary education in totalitarian or authoritarian states has much in common with this model.)

The knowledge of the university will correspond to the knowledge averaged over all universities (one gets a kind of "average university"). The number of people is obtained by dividing the population quantity by the university's quantity (one gets a kind of "average number of people who fall on one university"). Of course, the results obtained in this case will only be indicative. However, for many tasks (for example, determining the resources that are needed for the development of tertiary education in the country), this may be quite enough.

Given and Fig. 1 qualitative behavior of the distribution of university graduates by knowledge corresponds to the realities of social life. Indeed, over time, the knowledge acquired by graduates at the university, which is redundant for use in work and current life, will be forgotten. As a result, the number of people with such knowledge will decrease. That is, the trajectory of such a graduate will go down to the straight line in Fig. 1. We note that the quantitative results that will describe the real situation can be important for the company's planning of the career development of young people. This will be important so that the company can use the knowledge of graduates for its development with maximum benefit.

Figure 1 shows the importance of implementing two strategies for managing the innovative development of the country's economy and social life.

The first strategy consists of creating conditions for identifying those graduates who have a tuple of knowledge that exceeds the straight line in Fig. 1. These graduates should be given the opportunity to use their knowledge almost immediately after graduation. That is, they need to create conditions for their career to develop more intensively.

The second strategy is to create economic and social motivation in society for graduates who have a tuple of knowledge smaller than the straight line in Fig. 1. As a result, they will have the motivation to "rise" because of the straight line. This will significantly increase the speed of economic and social development of the country as a whole. For example, if a university graduate has insufficient knowledge, then the need to perform work will force him to increase his knowledge level on his own. That is, the trajectory of such a graduate will go upwards to the straight line in Fig. 1.

Quantitative calculations will allow to determine the time, which, on average, is necessary for the studied community to reach an equilibrium state.

The proposed model also describes a situation when an employee of the firm will be forced to improve his qualifications. Such information is important for many participants in the economic process. For companies, it will be important to optimize human resource management. This will be important for training companies, as it will allow them to predict the direction of development of their activities. It will be important for employees to manage their career development. This will be important for the state, as it will allow forecasting the development of the economy.

4   Methods of Model Using

4.1   Method for Knowledge Tuple Formation

To obtain numerical characteristics using the obtained model for specific situations, it is necessary to have numerical knowledge for the diffusion coefficient D and the initial values of knowledge $n(K)$. In this case, it should be taken into account that, in the general case, knowledge in society forms a tuple $K = (k_1, k_2, ..., k_i, ...)$. Here k_i is a particular method, technology, or way of doing an activity (either for a person or in collaboration with other people). Therefore, for many tasks, the initial distribution of knowledge can be given by the following vector (see Eq. 11):

$$\vec{n} = (n_1(k_1), n_2(k_2), \ldots, n_i(k_i), \ldots) \tag{11}$$

Similarly, the diffusion coefficient D may depend on the variable k_i. If the diffusion coefficient is also a vector of the form (11), then the diffusion equation of the form (7) can be reduced to the form (8). In essence, this means that all components of knowledge, i.e., all methods or technologies for carrying out an activity are independent. The validity of such an assumption has a limited scope. However, it can be used as a first approximation to obtain preliminary estimates.

This allows using (8)–(11) for each of the knowledge components independently.

The numerical value of the coefficient D of knowledge diffusion in society depends on many factors. For example, it depends on the presence of appropriate social, legal, and economic institutions, the level of culture, diversity, and tolerance in society, and many others. As a quantitative estimate of the coefficient D, we can take the values of the diffusion coefficient for innovations in a given society. These can be political processes, the spreading of new opinions (including in social networks), etc. One can also use the information on the spreading of scientific theories and new paradigms, quantitative patterns on the influence from Elon Musk, Steve Jobs, Bill Gates, CNN, Google, Amazon, Chinese companies, etc.

The study of knowledge diffusion in society has been studied quite actively in recent years. For example, in [30] the comparative characteristics of the processes of spreading scientific and fake knowledge are studied. Models for describing the diffusion process in social networks are being actively studied both theoretically and experimentally [31, 32].

The formation of a tuple of knowledge in society is studied much less actively. First of all, not all researchers understand the same thing by the term "knowledge". For example, the term "mathematical knowledge" can include different meanings: for "pure" mathematicians it will have one meaning, and for "applied" mathematicians (for example, for physicists) it will have a different meaning. The definition of knowledge used in this article is based on the description of the activities of an individual and a group of people. Therefore, it can be measured using the results of their activity. Although this requires special studies, the formation of a knowledge tuple needs to be done only once. Then it will be possible to use it in solving various problems of the development of society.

For example, the formation of a knowledge tuple for an arbitrary social group or society as a whole can be carried out according to such a method.

1. Database of all methods and technologies of human activity and its joint activity in a social group is formed. To do this, one can, for example, uses the classifiers of professions available in a society (or social group), requirements for professions

and functional duties, formalized descriptions of knowledge bases of elementary, secondary, and higher schools, legislation, and relevant databases, informal requirements for the behavior of a person or group people in society and family (including also accepted religious rules in society), and the like.

2. Formation of a single ordered database from individual fragments of knowledge. To do this, duplicates and analogs will be deleted from the databases formed at stage 1. For example, the ability to use a computer and the Internet, or the ability to use mathematical statistics will be repeated many times in different databases.

3. Each method in the Stage 2 database should/desirably be associated with having a tool to quantify how successful an individual/social group is in mastering that method. It is also necessary to indicate the interval of change for every indicator, which is necessary for each method from the entire body of knowledge.

4. A set of relatively independent fragments of knowledge (for example, professions, norms of behavior, etc.) that society possesses is formed.

5. For each individual (relatively independent) piece of knowledge, a quantitative indicator is formed. This indicator includes the required degree of mastery of this piece of knowledge, as well as the time and effort required for the individual. At the same time, many methods, technologies, algorithms, and so on, which are necessary for a person to survive in this society, will also be revealed.

6. Using the results of measurements from stages 4 and 5, it is possible to form for a given society both K_{min} and maximum K_{max} tuples, which determine the need for an individual to successfully survive in modern society. Such maximum and minimum tuples can be created for different countries of the world both developed and developing.

7. On the interval $[K_{min}, K_{max}]$, the numerical values of the tuples must be ordered from the minimum to the maximum. Depending on the considered problem, ordering can be done here within the framework of various knowledge. For example, for a given person, for a particular profession, for a group of people within a given profession, for a specific context of activity (for example, for a user of a medical system, for a traveller, etc.), for a country as a whole. For many tasks, ordering can be done, for example, using the dichotomy "available (1)—not available (0)". Perhaps, in this case, it will be necessary to take into account the "importance" of a different piece of knowledge. This can be done, for example, by using weights that are either determined experimentally or specified by experts.

8. The use of points 6 and 7 makes it possible to compare the sets of knowledge that have been accumulated as a result of the historical process in each of the countries of the world. On this path, those fragments of knowledge that are necessary for the further development of a given country can be identified. Fragments of knowledge that hinder the development of a given country can also be found.

Note that many stages from the method written above already exist for different fragments of knowledge. For example, there is PISA (Program for International Student Assessment [33]) that measures 15-year-old students' reading, mathematics, and science literacy every 3 years. There are also many different tools related to training or required to gain access to certain professional duties.

For the quantitative characteristics of a certain piece of knowledge (that is, certain methods of carrying out activity), one can also use the time that is necessary to study it. In this case, the metric (distance) between two fragments can be calculated as the time it takes to retrain (or train) a human on the missing difference between the fragments. Note that this metric can only be used effectively for societies that have the same (or similar) structures for learning.

Finally, it must be taken into account that each individual has a finite capacity for knowledge [1, 20]. Therefore, the capacity of knowledge for an individual is limited. The question of talents and geniuses and their role in society will be considered in a separate work.

Probably, when building a common tuple of knowledge for a given country, it will be possible to use ontology technologies [34].

4.2 Using the Knowledge Tuple to Analyze the Process of Migration to Developed Countries

Let's give one example of using the obtained method.

Let's consider the problem of the influence of migrants from a developing country on the level of the knowledge system of a developed country. Since people come from a developing country, they do not have all the necessary fragments of knowledge to carry out activities and life in a developed country. Conversely, some fragments of knowledge cannot be used by them in a developed country. And even more than that, certain fragments of migrants' knowledge may even be unacceptable for a developed country.

In this case, two options are possible.

The first variant is presented in Fig. 2. In this case, migrants from a developing ("less developed", will be marked as "L") country have separate fragments of knowledge that can be used in the developed country (will be marked as "D"). This means that migrants with knowledge in the interval $[K^D\text{min}, K^L\text{max}]$ can adapt to the conditions of life and work in a developed country fairly quickly and with relatively little education costs. It emphasizes that coverage training from developed country knowledge should be mandatory since most likely, migrants will not have some fragments of developed country knowledge. These are, first of all, those fragments of knowledge that are useful for life and joint activities in a developed country. Note that, as a rule, in developing countries, the total quantity of people who have knowledge in the interval $[K_{\min}, K_{\max}]$ is a small fraction of the total population.

An important circumstance is also the fact that motivational conditions should be created in the country so that migrants stop using the knowledge from the interval $[K^L\text{min}, K^D\text{min}]$. In fact, this is the knowledge that makes the country from which they emigrated developing. This is an important circumstance because such knowledge will be harmful to a developed country. We will remind that in the framework of this article, by knowledge, we understand norms of behavior, methods, results

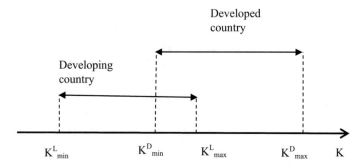

Fig. 2 Knowledge tuples of migrants and natives have common fragments

of activity, algorithms, and programs of activity of individuals, as well as ways of communicating with people in their joint behavior and joint activities. That is, in a developed country, the relevant behavior of migrants will be destructive.

Note that for ex-pats from a developed country to a developing country, the situation in Fig. 2 may also occur. However, in this case, they may well learn the missing fragments of knowledge on their own.

Let's consider an example from China. The intensive development of the Chinese economy was accompanied by an intensified return to Chinese universities of established scientists and those who received education and Ph.D. in developed countries. This allowed the creation of several world-class universities in China, which significantly increased the number of people who had modern knowledge. Shen [35] provides an "own inside view" of the universities in China. The author emphasizes that "compared with older generations of scientists, Chinese scholars born after the 1970s, especially Chinese scholars who have obtained doctorates abroad ... seem to be less affected by nationalism and patriotism, showing a stronger cosmopolitan orientation." The author also draws attention to the fact that "the current academic evaluation and talent policies may restrict the scientific research potential of young scholars in China." In particular, "the Ph.D. returnees are disadvantaged in terms of accruing social capital in the domestic academic community—so they may fail in local competitions for resources". This happens for the reason that "even well-trained young scholars face many difficulties in their professional development when they are required to produce world-leading research in an academic system which prioritizes support to a very small number of senior scholars who have made outstanding achievements." These phenomena testify to the growing conflict between the knowledge tuples in China and the knowledge tuples in developed countries. These phenomena will likely determine the development of China in the future.

The second variant corresponds to the case in Fig. 3, when the knowledge tuples of migrants and natives do not have common fragments (or their influence is negligible). In this case, it is necessary to carry out significant and varied general training for migrants. At the same time, it must be taken into account that only a very small number of people can acquire an additional very significant amount of new knowledge. Then,

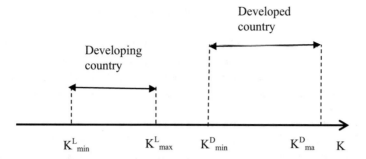

Fig. 3 The case when knowledge tuples of migrants and natives do not have common fragments

the selection of migrants should be carried out solely on their ability and desire to learn, as well as their ability to acquire the minimum necessary knowledge.

The problems of organizing the training for migrants to the developed countries are in detail considered [36].

Note that in a situation where people from a developed country get into a developing country and are forced to live "according to its laws", the situation is the opposite. Already, people from developed countries have to learn new pieces of knowledge. At the same time, it should be taken into account that in developing countries, as a rule, the education system does not include teaching fragments of knowledge that are "familiar from early childhood" or "obvious" to the indigenous people. There are many examples from Vietnam, Afghanistan, Iraq, and many other countries, which convincingly testify to the existing misunderstanding between people from developed countries (usually the military contingent) and indigenous people.

4.3   Using the Knowledge Tuple to Analyze the Regional Heterogeneity of Society

The proposed method of forming tuples and the model for describing the temporal dynamics of the distribution of the population by knowledge can be applied not only at the macro level but also at the micro level. For example, considering the regional heterogeneity of society. At the same time, when comparing different regions, the situations described in Figs. 2 and 3 may take place. Using the results of this work, it is possible to identify those fragments of knowledge tuples that need to be learned by people in a given region. The use of the obtained mathematical model will allow us to develop an optimal model for teaching the population. For example, from the mathematical model can, be found that it required time to learn people.

The obtained results can be important for a wide range of problems in territorial development management. Territories here mean not only the regions of one country but also individual countries, which sometimes have quite distinct territorial differences. The lives of people in these territories will require different knowledge. The difference in knowledge, as a rule, can be localized in individual elements of the tuple.

The importance of successfully solving this problem lies in the fact that the free migration of people is one of the crucial elements of a market economy. People should be free to choose their place of residence. They should also be free to choose to change their place of work and profession. The development of the country is based on this.

However, there are many obstacles on this path. One of the important ones is the regional specificity of the language. As a rule, within the framework of one country, it manifests itself in the presence of regional dialects (although, for a number of countries, it has several quite different languages). For different countries, the issue of language can often be quite a serious challenge, especially in the case when the languages belong to different language groups.

People's way of life also has regional features. For example, life on the equator, in the tropics, and in the north requires completely different knowledge. Note that some of such knowledge is not sufficiently formalized, and some knowledge may even be "obvious" to local residents. But they will be incomprehensible to migrants.

Thus, the presence of special infrastructure in countries, which can consist of organizations of various forms of ownership, is an important guarantee of the development of society. These infrastructures should be aimed at teaching internal migrants and ex-pats precisely those elements of the knowledge tuple that they need for a comfortable life and successful work. Such infrastructures should actively cooperate with universities that have extensive experience both in researching the problem and in developing and applying effective learning technologies.

The situation somewhat resembles the one described in the previous chapter, but it also has important differences. They consist in the fact that the situations Figs. 2 and 3 will now not refer to the general knowledge tuple of society. Now, they will describe the situation only for some fragments of knowledge tuples of local residents and migrants. This is a fundamental difference because these fragments will not, as a rule, relate to the basic elements of a person's picture of the world. Moreover, in this case, migrants will have sufficient knowledge about how to study and, often, how to organize the educational process.

5   Conclusion

The expansion of innovative knowledge is actively studied both theoretically and experimentally. The rapid development of the Internet and social networks, educational platforms, distance education, etc., requires a transition to several characteristics of the concept of "knowledge". This allows the development of new experimental methods.

Knowledge, including innovative knowledge, begins with the help of objective characteristics of the manifestations of human activity (both in its activity and in communication with a social group). This made it possible to introduce a tuple of quantitative and qualitative parameters as a characteristic of knowledge. It shows how a metric can be entered into a tuple of knowledge (individual, social group, or society). The presence of the metric allows you to set the task of knowledge spreading as a change in the number of people who have a certain tuple of knowledge.

The diffusion mechanism of knowledge spreading is considered. The example of a university as an important source of innovative knowledge is considered. It is the presence of new knowledge that allows us to talk about the development of society. The model of the spread of knowledge from one source in society is shown in detail. The obtained results made it possible to propose a method for more effective use of universities as a driver for the development of the economy and society.

A method of identifying the tuple of knowledge possessed by an individual, social group, or society has been developed. The discussion is an example of the application of this method to important processes that accompany the development of society. It is shown that the obtained results can increase the efficiency of the management of migration processes. The method of determining a set of tuples, the presence of which is necessary for the functioning of a developed country, is described. The possibility of applying the obtained results to increase the efficiency of regional development management is shown.

References

1. Shiyan, A.A., Nikiforova, L.O.: Typology of institutions—theory: classification of institutions via the methods for transmission and modification of knowledge. Soc. Sci. Res. Netw. (2013). https://doi.org/10.2139/ssrn.2196300. Last accessed 17 Jan 2023
2. Miao, L., Murray, D., Jung, W.-S., et al.: The latent structure of global scientific development. Nat. Hum. Behav. **6**, 1206–1217 (2022)
3. Servaes, J. (ed.): Handbook of Communication for Development and Social Change. Springer Nature Singapore Pte Ltd. (2020)
4. de Arruda, H.F., Silva, F.N., da Costa, L.F., Amancio, D.R.: Knowledge acquisition: a complex networks approach. Inf. Sci. **421**, 154–166 (2017)
5. de Arruda, H.F., Silva, F.N., Comin, C.H., Amancio, D.R., da Costa, L.F.: Connecting network science and information theory. Phys. A **515**, 641–648 (2019)
6. Batista, J.B., da Costa, L.F.: Knowledge acquisition by networks of interacting agents in the presence of observation errors. Phys. Rev. E **82**(1), 016103 (2010)
7. Lima T.S., de Arruda, H.F., Silva, F.N., et al.: The dynamics of knowledge acquisition via self-learning in complex networks. Chaos Interdiscipl. J. Nonlinear Sci. **28**(8), 083106 (2018)
8. Pitaeskii, L.P., Lifshitz, E.M.: Physical Kinetics. Butterworth-Heinemann, Oxford (2012)
9. Velasquez-Rojas, F., Laguna, M.F.: The knowledge acquisition process from a complex system perspective: observations and models. Nonlinear Dyn. Psychol. Life Sci. J. **25**(1), 41–67 (2021)
10. Campisi, M., Ratliff, T., Somersille, S., Veomett, E.: Geography and election outcome metric: an introduction. Elect. Law J. Rules Polit. Policy **21**(3), 200–219 (2022)
11. Vanderheiden, E., Mayer, C. (eds.): Mistakes, Errors and Failures Across Cultures. Springer Nature, Switzerland AG (2020)
12. Dalege, J., van der Does, T.: Using a cognitive network model of moral and social beliefs to explain belief change. Sci. Adv. (8), eabm0137 (2022)
13. Grant, M., Vernall, L., Hill, K.: Can the research impact of broadcast programming be determined? Res. All **2**(1), 122–130 (2018)

14. Williams, S.M.: The research impact of broadcast programming reconsidered: academic involvement in programme-making. Res. All **3**(2), 218–223 (2019)
15. Chris, D., White, P.A., Sajonia-Coburgo-Gotha, B.: What Happened, and Who Cared? Evidencing Research Impact Retrospectively. arXiv.org (2021). https://arxiv.org/abs/2103.06778. Last accessed 17 Jan 2023
16. Haddad, G., Nagpal, G.: Can students' perception of the diverse learning environment affect their intentions toward entrepreneurship? J. Innov. Knowl. **6**, 167–176 (2021)
17. Ruoslahti, H.: Complexity in project co creation of knowledge for innovation. J. Innov. Knowl. **5**, 228–235 (2020)
18. Bhattacharya, J., Packalen, M.: Stagnation and Scientific Incentives. National Bureau of Economic Research. Working Paper 26752 (2020). https://doi.org/10.3386/w26752. Last accessed 17 Jan 2023
19. Park, M., Leahey, E., Funk, R.: Dynamics of Disruption in Science and Technology. arXiv.org (2021). https://arxiv.org/abs/2106.11184. Last accessed 17 Jan 2023
20. Petrov, M.K.: Language, Symbol, Culture. Nauka, Moscow (1991) (in Russian)
21. Kuhn, T.S.: The Structure of Scientific Revolutions. University of Chicago Press, Chicago (1962)
22. Acemoglu, D., Robinson, J.A.: Why Nations Fail? The Origin of Power, Prosperity, and Poverty. Random House, Inc., New York (2012)
23. Stoshikj, M., Kryvinska, N., Strauss, C.: Service systems and service innovation: two pillars of service science. Procedia Comput. Sci. **83**, 212–220 (2016). https://doi.org/10.1016/j.procs.2016.04.118
24. Rauer, J.N., Kroiss, M., Kryvinska, N., Engelhardt-Nowitzki, C., Aburaia, M.: Cross-university virtual teamwork as a means of internationalization at home. Int. J. Manag. Educ. **19**, 100512 (2021). https://doi.org/10.1016/j.ijme.2021.100512
25. Abbasi, A., Javed, A.R., Iqbal, F., Jalil, Z., Gadekallu, T.R., Kryvinska, N.: Authorship identification using ensemble learning. Sci. Rep. **12**. https://doi.org/10.1038/s41598-022-13690-4
26. Rogers, E.M.: Diffusion of Innovations. The Free Press, New York (2003)
27. Kreindler, G.E., Young, H.P.: Rapid innovation diffusion in social networks. Proc. Natl. Acad. Sci. **111**, 10881–10888 (2014)
28. Berliant, M., Fujita, M.: Culture and diversity in knowledge creation. Reg. Sci. Urban Econ. **42**(4), 648–662 (2012)
29. Coursera. https://www.coursera.org/. Last accessed 17 Jan 2023
30. Del Vicario, M., Bessi, A., Zollo, F., et al.: The spreading of misinformation online. Proc. Natl. Acad. Sci. **113**(3), 554–559 (2016)
31. Bago, B., Rand, D.G., Pennycook, G.: Fake news, fast and slow: deliberation reduces belief in false (but not true) news headlines. J. Exp. Psychol. Gen. **149**(8), 1608–1613 (2020)
32. Furioli, G., Pulvirenti, A., Terraneo, E., Toscani, G.: Fokker-Planck equations in the modeling of socio-economic phenomena. Math. Models Methods Appl. Sci. **27**(01), 115–158 (2017)
33. Program for International Student Assessment (PISA). https://nces.ed.gov/surveys/pisa/. Last accessed 17 Jan 2023
34. Staab, S., Studer, R. (eds.): Handbook on Ontologies, International Handbooks on Information Systems. Springer, Berlin (2009)
35. Shen, W.: The achievement, limitation and potential of Chinese Universities in STEM fields: a generational perspective. Univ. Intellect. **1**(1), 49–56 (2021). https://cerc.edu.hku.hk/universities-and-intellectuals/1-1/the-achievement-limitation-and-potential-of-chinese-universities-in-stem-fields-a-generational-perspective/?fbclid=IwAR3KvG2OsOWVyXKQCJd9bqPYrT0-b4gdKthj_lK2MHHMLbaljqvMfuvgj1g. Last accessed 17 Jan 2023
36. Shyian, A., Nikiforova, L.: The role of the state in optimizing communication between generation Z and migrants in the human resources management. In: Leon, R.-D. (ed.) Strategies for Business Sustainability in a Collaborative Economy, pp. 17–36. IGI Global, Hershey, PA (2020)

Informational and Digital Business Security in Tourism as a Component of the Coastal Region Competitiveness

Maryna Byelikova⬤, **Anastasiia Bezkhlibna**⬤, **Yuriy Polyezhayev**⬤, **Valentyna Zaytseva**⬤, and **Hulnara Pukhalska**⬤

Abstract Three types of information and digital tools for improving the competitiveness of coastal regions are analyzed: the concept of SC, digital platforms for communication and information dissemination, applications, and the Big Date system. The world experience of implementing the Smart City program in coastal regions (Singapore, Switzerland, Denmark, Norway, etc.) was studied. It was established that the most popular areas of use of digital technologies are technologies in the fields of transport, logistics, and lighting for the purpose of saving energy resources. In the development of the Smart City concept, coastal regions should rely not only on foreign experience but also on the peculiarities of their own development of elements of competitiveness. The functioning of the big data infrastructure creates the image of a modern digital city, which is based on the use of modern data processing technologies, which are constantly updated and replenished. Using the results of digital analytics data is the basis of timely response and flexibility of the management system of a competitive region. The article is devoted to the issues related to the organization of the activities of travel agencies in the conditions of a global pandemic, which is manifested in changes in the work of travel agencies (transition to remote service, greater flexibility in changing the conditions of booking tours, constant control of the conditions of entry and exit from countries, the use of new documentation forms for tourists). The study was carried out using the analysis of Digital marketing of Zaporizhzhia travel agencies. The methods of induction and deduction made it possible to

M. Byelikova · A. Bezkhlibna (✉) · Y. Polyezhayev · V. Zaytseva · H. Pukhalska
National University "Zaporizhzhia Polytechnic", Zaporizhzhia, Ukraine
e-mail: bezkhlibna@zp.edu.ua

M. Byelikova
e-mail: belikova@zp.edu.ua

Y. Polyezhayev
e-mail: polyezhayev@zp.edu.ua

V. Zaytseva
e-mail: vnz@zp.edu.ua

H. Pukhalska
e-mail: puhalska@zp.edu.ua

© The Author(s), under exclusive license to Springer Nature Switzerland AG 2024   117
A. Semenov et al. (eds.), *Data-Centric Business and Applications*, Lecture
Notes on Data Engineering and Communications Technologies 195,
https://doi.org/10.1007/978-3-031-54012-7_6

study the competitors of the Zaporizhzhia travel agency "Drive Tour" on the Internet, namely their websites, social networks, advertising messages, visual images and intonations, functional and emotional advantages for online tour sales, target audience and needs that are covered. Summarizing existing information made it possible to come to the conclusion that the competitive advantage of the researched travel agency in online sales is the use of its own Viber community with 7.5 thousand participants—potential customers, providing daily up-to-date advertising messages. The target audience consists of active Internet users interested in fast remote online services. On Instagram, the travel agency positions its readiness to work quickly and remotely with clients from all over Ukraine, from application processing to issuing electronic documents, not limited to Zaporizhzhia. It should be noted that in recent years, as a result of the pandemic and quarantine restrictions, the number of electronic documents for entering other countries has increased, the details of which are described by the authors for the first time.

1   Introduction

The development of the regions' study is an urgent issue of the modern science of regionalism. It is necessary to highlight the research of Obolentseva [1], Tulchynska et al. [2], Pushkarchuk [3], Gudz et al. [4], Oliinyk Y., Shkurupska I., Ivanchenkov V., Petrenko O. among the works of scientists [5]. The works of Costanza [6], Costello and Ballantine [7], Morrisey [8], Pontecorvo [9], Witte et al. [10] are focused on the influence of natural and geographical conditions on the peculiarities of the economic development of coastal regions. However, considering the peculiarities of development problems, the digital component of the coastal region competitiveness still remains insufficiently researched.

Geographically, the seaside region has access to the sea coast with its own water area. It specializes in seaside types of economic and ecosystem activities, creating its own maritime complex of industries with appropriate infrastructure, has specific social problems and environmental requirements, and cultural and historical heritage, which reflects the maritime epoch. Hence, the coastal region is an administrative and territorial unit that has natural access to the sea coast and water area is distinguished from other regions by the coastal profile of types of economic activity in the production of ecosystem services, recreation and tourism, industries and productions of the maritime complex, and the network of industrial relations for their maintenance in terms of seasonality, ecological restrictions and requirements for the protection and preservation of the natural and conservation fund, cultural and historical heritage. That is why it requires systematic management and administration.

The competitiveness of the coastal region is the ability of regional authorities, dwellers, and the housing management to use the internal potential of the maritime complex through the formation and development of the adaptive structure of the regional economy, the preservation of the cultural and mental identity of the regional society, and the ability to use opportunities and threats to solve the social and environmental problems of the region, improving the life standards of residents. The ability to produce specific eco-services on the territory of the coastal zone, creating a kind of "growth poles" in spatial development, distinguishes the seaside region from others. The focus on economic profit from ecosystem services allows for the use of the strengths and unique advantages of the region in economic activity. At the same time, the ecosystem approach provides measures not only for commerce but also for the protection, preservation, and restoration of biogeocenosis.

Increasing the digital and information service volume in the coastal region is a strategic direction enshrined in the State Regional Development Strategy. It foresees measures which are necessary for the prosperity of coastal regions in particular:

– increasing of community information resources available to a wide range of users;
– community participation in local self-government via the use of digital platforms;
– creation of a wide internet application platform for smartphones, which relates to the services and enterprise activities, institutions, and organizations of the seaside region. Moreover, special attention is paid to the tourism business;
– the use of the Big Date platform in the implementation of the complex tourist product;
– the use of Smart Cities infrastructure for the purpose of ecosystem development and restoration.

Digital technologies are becoming everyday components of municipal, social, information, business, and other services [11]. Regions, where cities are implementing the Smart City concept, have rival advantages over others [12]. The transition to fully automated production, the management of which is carried out by intelligent constantly improving systems in the mode of constant interaction with the external environment is a characteristic of the Fourth Industrial Revolution [13].

The management of travel agencies is related to the organization of online sales activities—digital marketing and the organization of office management in tourism, in which there is an increase in the number of electronic documents. The global pandemic of 2020–2021 has changed the business processes of travel agencies and affected the organization of business administration in tourism. Effective management of travel agencies as a whole depends on the correct organization of online sales and administration. The tourism business received a challenge in the form of a global pandemic and "digitalized", technologies that were previously in their infancy began to gain momentum and reveal their potential in an accelerated mode. Inadequate development of the issue of the practical application of digital marketing tools by travel agencies and the current relevance of the issue led to the choice of the purpose of the article: the analysis of digital marketing of travel agencies and electronic documents for departure and return to Ukraine due to quarantine restrictions. Digital marketing is considered a comprehensive approach to the promotion of a travel agency and its services in a digital environment, in particular with the use of mobile applications in phones and other digital means of communication.

2 Information and Digital Support for the Implementation of the Strategy of Competitiveness of the Seaside Region

The largest cities in the world spend more than $300 million a year on Smart City projects, while in 2019 the number of such cities has reached 600 units, increasing annually. The International Management Development Institute annually analyzes the largest cities and creates an international ranking of Smart Cities [14]. According to the survey data, in 2021, Kyiv will on the 82nd city in the global ranking of Smart Cities, rising from 98th in 2020. The first three leading cities are Singapore, Zurich, and Oslo.

Table 1 includes cities with coastal zones that have progressive approaches to the implementation of the Smart City program.

According to Table 1, it is necessary to conclude that the most popular areas of digital technology use are technologies in the fields of transport, logistics, and lighting for the purpose of saving energy resources.

It's worth defining that the scope of technology use depends on the state of the city's digital infrastructure. If we examine the experience of countries, we will note that in addition to the use of IoT sensors, almost every country has a national system of digital infrastructure support. For example, Virtual Singapore is a three-dimensional city model developed by the National Research Foundation of Singapore. Oslo has a special innovative park, the goal of which is to ensure carbon neutrality. This park includes, for example, the implementation of a special traffic lane for electric cars, and free parking for them, with the help of digital initiatives. Copenhagen, which has declared carbon neutrality by 25 years, has its own Copenhagen Solution Lab, namely, a laboratory for the development of SC, and special 3W applications for cyclists in addition to calories, they display the pedaling speed for light traffic. New York has its own smart grid sensor system. Amsterdam also has its own IoT Living Lab laboratory, devices, and applications such as LoRaWan, Bluetooth Low Energy, Green Generation, and Carbon Neutrality applications. London is a city that accumulates data; private investors develop products, and business incubator Civic Innovation City program Connected London (5G). It is becoming clear that digital technologies, digital infrastructure, and citizens' digital skills are closely related and mutually dependent. The sustainable development of the administrative and territorial entities of the coastal territories is currently laid down in international standards for the construction of intelligent networks, the so-called Smart City and Smart Communities, which create additional favorable opportunities for the development of territorial communities. In the development of the Smart City concept, coastal regions should rely not only on foreign experience but also on the peculiarities of their own development of elements of competitiveness, which could be reflected in the Smart Coastal City concept. The future architecture of the smart city of the maritime region should include tracking and intelligent control.

In continuation of the development of digital services of coastal regions, SC services are proposed in Table 2 according to the elements of competitiveness of coastal regions.

Table 1 Cities' experience of the Smart City program implementation (based on [14, 15])

Rating place among Smart cities	City implementing the Smart City program (country)	Availability of coastal areas of the city exits to rivers, lakes, seas, oceans	Scope of technology use
1	Singapore (Singapore)	The South China Sea	"Smart houses" (electricity, water resources management system), intelligent energy-saving systems, solar panels, car sharing, toll roads (the cost depends on the load, the area, the day of the week, the time)
2	Zurich (Switzerland)	Lake Zurich	Transport system and management of buildings, traffic, urban infrastructure, and utilities
3	Oslo (Norway)	There are 343 lakes within the city limits, two small rivers flow through the territory: Akeshelva and Alna	Green technologies of vehicles and energy efficiency of buildings, control of lighting, heating and cooling, network of electric car charging stations, smart lighting, city safety, waste recycling
7	Copenhagen (Denmark)	The Baltic Sea	Smart lighting, parking, a network of charging stations, garbage collection
8	Geneva (Switzerland)	Lake Geneva	System of intelligent urban lighting, parking
12	New York (the USA)	The Atlantic Ocean	Garbage removal, waste removal, security (crimes), street patrolling
17	Amsterdam (the Netherlands)	The Amstel and Ey rivers, connected by the Nordze Canal to the North Sea	Sensors on streets and houses, smart transport, clean air, car sharing, electric transport, bicycles, smart LED
22	London (the United Kingdom)	River Thames near the North Sea	Creation of high-speed infrastructure
29	Dubai (UAE)	Persian Gulf	Transport, communication, communal economy, paperless document flow, road management, traffic accidents, autonomous police stations
41	Hong Kong (China)	South China Sea, Pearl River	Road congestion, car traffic, parking spaces, public transport and public digital services
58	Barcelona (Spain)	Mediterranean Sea	Bicycle transport, traffic control sensors, air quality, noise level, smart trash bins

According to the data in the Table 2, we will analyze the state of development of the concept of Smart Cities and digital services in the coastal regions (see Table 3).

Both, the state and local authorities create conditions for the development of digital services and the concept of digitization and digitization in the coastal regions. This happens most efficiently thanks to Smart City (SC) technology.

Table 2 Implementation of the Smart City concept in coastal regions in relation to elements of competitiveness (authors' research)

Element of competitiveness	Smart City service
Regulation	Introduction of the resident's electronic cabinet; Access to information; Involving citizens to participate in city management through electronic systems Provision of electronic administrative and social services
Business-processes	Integrated intelligent transport, including marine, and utility management systems, i.e. intelligent lighting, provision of meter readings; Tourism; Energy; Construction management and supervision system; Financial services
People	The health care system, including the development of electronic medical resources Education system, including automation of the educational process Creating conditions for residents to acquire competencies related to digital technologies
Territory	Video surveillance and geolocation systems, including data collection and analysis centers Collection and processing of data on forecasting of weather conditions and the state of the water area Ecology

We are to consider functional components of the SC mechanism in coastal regions mentioned in Fig. 1.

It can be seen from Fig. 1 that as an intelligent digital mechanism, Smart City functions not only thanks to the infrastructure. Devices and sensors, smartphones, information systems, platforms, applications, Internet coverage, IOT devices, analytical centers and research laboratories, and innovation funds can be listed as additional resources. It can also work thanks to the interest of the state, embodied in the legal component such as normative acts or executive documents on the basis of which Smart City is developed and launched, the organizational component, effective management aimed at the result, not the process, namely, and the financial component as availability of budget, sponsorship or donor funds.

Standards used in the development of an intelligent digital network of the SC:

1. ISO 37120:2014 Sustainable development of communities—Indicators for city services and quality of life.
2. ISO 37122:2019 Sustainable cities and communities—Indicators for smart cities.
3. ISO 37101, Sustainable development in communities—Management system for sustainable development—Requirements with guidance for use.
4. ISO 37151:2015 Smart community infrastructures—Principles and requirements for performance metrics.

Table 3 The state of development of the concept of Smart Cities and digital services in the seaside regions of Ukraine as of 2021 (*source* [16])

Region	State digital services and events in the regions initiated by the Ministry of Digital Transformation	The state of Smart City development in the seaside region
Odesa	In June, the Diya Business Center was opened in Odesa for consulting, holding conferences, renting space for business events, sharing experiences, and a platform for finding investors	The integrated system of analytics and video surveillance
	The Mayor of Chornomorsk initiated the passing of the national test "Digit numbers" and the viewing of educational series on digital literacy and digital competencies among people of all age categories	Odesa city program "I am Odesa" Web portal of open data Online broadcasts of meetings of the Public Budget Commission, City Council, etc.
Mykolaiv	In June, the Concept of the development of digital competencies of residents of Mykolaiv region was adopted	Video surveillance system, Data center, implementation of e-document circulation and provision of information security of the city—information system "Register of Territorial Community of the City of Mykolaiv", call center, public budget system
	In November, the first digital education and media literacy center in Ukraine was opened on the basis of the Mykolaiv Regional Institute of Postgraduate Pedagogical Education	
Zaporizhzhia	In September, the Ministry of Digital Transformation of Ukraine handed over 150 units of computer equipment to rural libraries with the aim of reducing the digital divide in the regions	Submission of meter indicators to online offices of communal institutions; Installation of smart plates with QR codes Electronic medical system Helsi
	The forum "Digital transformation of Zaporizhzhia region. The best practice and next steps" was organized in November. It initiated the creation Zaporizhzhia region community called the "Digital community!"	Contactless payment in municipal transport, tracking it through information boards installed at bus stops or using mobile applications; Webcam service Electronic petition service Safe City program for tracking speed violators on highways
Kherson	Provision of administrative services at the Central Administrative Office for the Crimeans, located on the borders	Smart metering—monitoring of energy consumption, notifications of infrastructure failures, control of water consumption, leaks in water pipes, optimization of resource costs, "smart water meters"

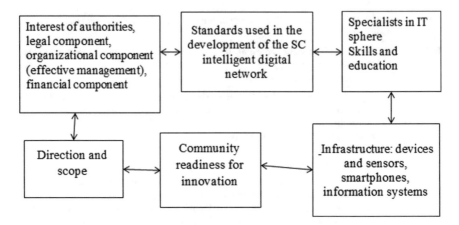

Fig. 1 Functional components of the SC mechanism in coastal regions (developed by the author)

On the basis of standardized processes in the EU, the future functional European Smart Grid reference architecture is now conceptually modeled. It is provided for the interconnection and interaction of standard interfaces in accordance with the European electricity market and the energy system as a whole. Moreover, it can be scaled to support the provision of livelihoods of territorial communities from the smallest administrative and territorial formations to megacities, taking into account their specificity [17].

Specialists, who configure, launch, and maintain the system, must have appropriate education and skills. At the same time, the residents of the community must have the appropriate digital skills in order to have interest and approval of such innovations. Areas of use of the intelligent network: intelligent housing and communal services system and resource accounting, street lighting management, transport system, environment, "green energy", health care system, tourism, public safety, waste management, social services, etc. [18].

In the digital city infrastructures, it is necessary to separate applications, Viber communities, chat-bots, and Telegram channels, which form separate digital platforms for communication.

Applications, used in the seaside regions of Ukraine (see Table 4), were selected for analysis, which simplify the functioning and communication of certain areas of the region.

It should be noted that, according to the data in the Table 4, the largest number of applications and covered areas of application are created and operated on the basis of the Odesa region, which is the largest among the seaside regions in terms of number, which confirms the effectiveness of the course taken by the local authorities on digitization, a sufficient number of IT specialists and users of these applications. So demand and supply for IT services in the region is high.

Viber communities, chat-bots, and Telegram channels of the seaside regions create opportunities for communication, advertising, job search, and business development

Table 4 Additives used in coastal regions and their original names (authors research)

Region	Appendix (original title)	Scope of operation
Odesa	Public transport in Odesa CityBus Odesa Odesa Tram Map Obshchestvennyi transport—Odessa CityBus Odessa Odessa Tram Map	Transport
	Let's Odesa Odesa Guide Odesa Map and Walks Prohulky po Odesse #Odessa Lets Odesa Odesa Guide Odessa Map and Walks Odessa—Putevodytel y karta	Tourism (about the city, walks, guides)
	Poster of Odessa GO-OD—Where to go in Odesa? Odesa news Afysha h. Odessy GO-OD—Kuda poity v Odesse? Odesskie novosti	News and events of the region
	Odesa.TV for smartphones Radio online "Odesa-mother" Odesa radio-stations-Ukraine Odesa.TV OTT for play-stations and TV Odesa radios online Radio Odesa QRZ-Odesa (social network for radio fans) Odessa.TV dlia telefona Radyo onlain Odessa mama Odesskye Radyostantsyy—Ukrayna Odessa.TV OTT dlia prystavok y TV Odessa radios online Odesskye radyostantsyy—Ukrayina Radio Odessa QRZ-Odessa (sotsialna merezha dlia radioliubyteliv	Radio and TV
	Aquapark Odesa Odesa Theater of Musical Comedy Ibiza Odesa The best restaurants in Odesa Akvapark Odessa Odesskyi teatr muzkomedyy Ibiza Odessa Luchshye restorany Odessy	Entertainment facilities and restaurants

(continued)

Table 4 (continued)

Region	Appendix (original title)	Scope of operation
	Odesagas supply Upravdom Odesa FFW Odesa (application for quick search of fire-fighting water supply) Odeshaz-postachannia Upravdom Odessa PPV Odessa (dodatok dlia shvydkoho poshuku protypozhezhne vodopostachannia)	Housing department
	Odesa	Guide
	Massage Odesa	Services
Mykolaiv	Mykolayiv Electric Supply Company Mykolaiv contact-center Personal cabinet of Nikolaivoblenergo	Housing department
	Mykolayiv Observer MykolayivNews	News
	CityBus Mykolayiv	Public transport
Zaporizhzhia	Zaporizhzhia GPS Inclusive	Map and Walks City navigation
	Zaporizhzhia stops "Easy Go" Dozor City	Public Transport
Kherson	Kherson	Guide
	Kherson News	News
	Kherson FM	Radio

[*] Dated: February 8, 2022
In addition to taxis and delivery from restaurants

in the seaside regions. Among the seaside regions, channels for communication, advertising, and news are leading in terms of number.

Communication is the key to building the community's social capital [19]. At the same time, A.P. Lelechenko, O.I. Vasylieva, V.S. Kuybida, and A.F. Tkachuk point to the decisive role of local means of information dissemination, whether it is social network, YouTube, messengers, local TV channels, newspapers. These channels of communication are the most trusted, they cover the news, problems, and achievements of the community, which is important for the dialogue between local authorities and residents. The world's most famous applications can be adapted and applied in the seaside regions of Ukraine, according to the disclosure of elements of competitiveness (see Table 5).

According to the results in Table 6, it should be noted that the applications used in the world in the littoral areas according to the elements of competitiveness of the littoral region do not reflect the demand for applications related to the element "Processes". The development of applications for smartphones that would help users in coastal regions in satellite fishing, tourism, tracking ship routes, summer sports,

Table 5 Examples of global applications that can be adapted to the needs of the coastal regions of Ukraine in accordance with elements of competitiveness (authors' research)

Examples of world-famous applications	Element of competitiveness
Windy Wind sea POSEIDON system weather Sea conditions	Territory (wind forecast, forecast maps of the ocean ecosystem and sea states, pressure, visibility, tides)
My sea	Processes (satellite fishing, tourism, ship itineraries, summer sports, booking tickets for sea vessels)
Sea proof Imray NavigatorWin GPS marine Marine radar Boat Navi 2 FishTrack—fishing charts Sea sector—maritime courses for sailors Seabook Maritime dictionary offline Seabook maritime Maritime knowledge The marine education Sensea maritime academy Middeterian sea GPS Nautical ang fishing charts	People (route planning, navigator, fishing chat, applications for sailors, communication between fellow sailors, training of sea rules and maneuvers, dictionary of marine terms)

and booking tickets for sea vessels would increase competitiveness and simplify the lives of residents of coastal areas. Since digital culture is now an independent resource and work tool, the improvement and development of applications is a component of the successful development of the region.

Digitalization in the region covers not only the supply side of services, including ecosystem services but also the demand side. In order to form a demand for ecosystem services from the consumer side, the use of the Big Date data storage system in the region is relevant, which allows the use of various digital tools for systematization and searching among large databases about the seaside region as a complete and competitive ecosystem.

Turning to analytics, in September 2020 Research And Markets published a report on the global Big Data analytics market. According to published information, the global market for big data analytics is estimated at $41.85 billion at the end of 2019. According to analysts' forecasts, it will grow to $115.13 billion, at an average growth rate of 11.9% during the forecast period from 2020 to 2028. According to the report, big data analytics can be called the heart of the digital world based on the analysis and transformation of data into information that provides valuable business insights. Cloud platforms are the main applications of Big Data analysis. At the same time, for the analysis of such data, large organizations mainly use a hybrid cloud platform, while public clouds prevail among small and medium-sized organizations. According to IDC experts, in 2021 the volume of the global Big Data and business analytics

Table 6 Digital marketing of Zaporizhzhia travel agencies (authors' development)

Criteria	"Drive tour" https://www.instagram.com/drive_tour_/	"Liuks" http://luxeua.com/
Advertising message and budget	Daily advertisements about countries, tourist attractions, entry rules, the most attractive value-for-money tours are promoted most actively and at the same time diplomatically in the Viber community (7 thousand members), which arouses the interest of potential customers and further communications in personal correspondence. Advertising monthly budget in social networks is 1–20 dollars	They specialize in international outbound tourism, sale of air tickets, processing of foreign documents. Excursion tours, vacations at sea, tours to Europe, specialized tourism are offered on the site in a tactful manner. Search optimization is used in Google Ads, thanks to which the travel agency site was in the third position in November 2021
Target audience	Audience interested in fast remote service, clients active in social networks Instagram, Facebook, Viber	Client data base since 2007. Consumers of senior age and their children have also become clients of the travel agency. Consumers are not very mobile on the Internet, they want to see the faces of travel agents and buy tours in the office. New customers come by personal recommendation of regular customers
A need that is closed	They cover the need for quick service, individual selection of tours depending on the wishes of customers, conclusion of a contract, payment remotely, and signing and presentation of documents—in the office, discounts for regular and to attract new customers	Close the need for security (for seniors, important information about insurance policies and reviews from 2011 on the site, common values in times of scarcity—stocking up on food, essential items, saving on accommodation or transport services during travel, closing the need for "living" offline communication

(continued)

Table 6 (continued)

Criteria	"Drive tour" https://www.instagram.com/drive_tour_/	"Liuks" http://luxeua.com/
Functional and emotional benefits	Instagram, Facebook are supported: the visual is beautiful, unobtrusive to attract the attention of potential customers	Emotional connections are built with clients in the tradition of the Facebook social network, the owner of a travel agency keeps a blog on her personal page, where the experience of traveling abroad is associated with the tradition of ordering tours in this agency. The site is simple, the design and navigation of the first sites, the transition to an agency in social networks is complicated, although it is represented on Facebook by business pages and a group with the appropriate name. It is not possible to contact the company on the website
Visual images, intonation	The logo with a bus on a background of palm trees and the sea is associated with a youth inexpensive tour of exotic countries. Instagram has good visual content, the latest customer reviews are up-to-date, information about quarantine restrictions, new entry rules. Customer care is manifested through the selection of tours and hotels for tourists with different budgets—from economy to luxury, unobtrusive service in the Viber community and communications. Facebook group with 1.3 thousand members at the end of September 2021. On Instagram—updates on tourism in the countries of the world, attractive tourist objects every 2–3 days	"Liuks" has a banal logo with a globe, under the title it says: "Advanced Tourism Technologies", although the navigation is 20 years out of date. Slogan: "If the stars light up, then someone needs it." On the website—photos of ancient Roman coins, a collection of teaspoons from different countries of the world, as of 13.12.2021—amateurishly outdated photos of the owners of the travel agency Yulia Volodymyrivna and Oleksandr Yosypovich with the use of photoshop—adding a hat to cover the baldness. This meets the expectations of those who are far behind... [16]
Communication channels	Mobile phone, Instagram (6.5 thousand followers), e-mail, Viber—community communication and personal correspondence in Viber	Mobile phone, e-mail, Viber, Facebook messenger. Customer reviews are outdated: on YouTube 2017, on the website—from 2011, on Facebook for 2019, on the website of the tour operator "Akordtur"—for August 2021

market was $215.7 billion, an increase of 10.1% compared to 2020 [20]. The use of the Big Data system and digital tools when forming a request for ecosystem services of the seaside region can be displayed in the form of the Big Data model in Fig. 2. This model is a Big Data architecture, that focuses on the competitive advantages of the seaside region and allows the consumer to form a digital query about his own preferences for ecosystem services of the region.

Accumulation, processing and systematization of large volumes of information about the seaside region, which starts with information about the region, "pull up" data on the weather forecast and water temperature, wind, tides and tides, the schedule of events in certain places, requires the operation of significant infrastructure such as analytical digital centers, dense Internet coverage throughout the region, regardless of the type of terrain, relief, size of the territorial unit, etc. Moreover, in the best case, it means accessing this information not only from portable smartphones and laptops. Ensuring the operation of this Big Data infrastructure will create a certain image of a digital modern city, which is based not only on the principles of Smart specialization, but also on the use of modern data processing technologies, which is currently implemented to a greater extent only in the banking sector, the manufacturing sector, and telecommunications services.

The use of a data array that is constantly updated, replenished, and structured provides prerequisites for Big Data analysis with the aim of identifying more competitive directions for ensuring the region's livelihoods on an ecosystem basis, directions of increased or decreased demand, drawing up a "portrait" of both a consumer of ecosystem services and a "profile" the ecosystem service itself due to the systematic analysis of consumer requests, forecasting and identification of trends, etc. That is, the functioning of the Big Data system and infrastructure opens up a wide range of opportunities for a thorough analysis of both the demand and supply of ecosystem

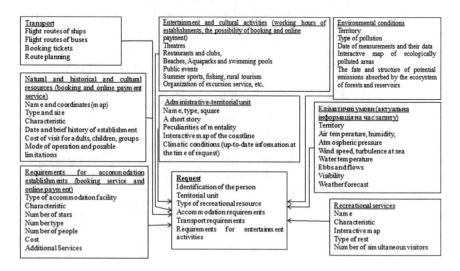

Fig. 2 Homeostatic loop in value co-creation (based on [19])

services in the coastal region. This tool becomes a powerful engine for increasing the competitiveness of the seaside region on an ecosystem basis in the digital space.

This analysis requires providing the information center with significant capacities in terms of digital infrastructure and a "brain center", which would consist of leading specialists in regional spatial planning, ecosystem development, business segment, and local management. Such a symbiosis of scientists, programmers-analysts, subjects of economic activity, and representatives of local authorities can function effectively, having as a "platform" for solving regional problems of development in the area of digital services and digital analytics, a basis, namely, that is formed on the basis of knowledge, skills, and competencies in the digital field.

3 The Record Management in Tourism in Coastal Regions: Digital Marketing, Electronic Documents, and Apps During the COVID-19 Quarantine Period

Despite existing works on the issue, research is constantly updated in connection with the active development of the information and digital environment. First of all, digital marketing in hospitality and tourism was studied in the context of research mainly devoted to the social behavior of travelers and the promotion of tourist services in social networks. So, travel agencies use digital marketing to attract customers and increase sales, and customers use it to get information about products, services, or topics related to the travel business. Separately, we will dwell on research devoted to the organization of office management in tourism in the context of enterprise management, which was supplemented by modern articles on documentation support for the activities of tourist institutions such as travel agencies and hotels [19]. Therefore, digital marketing and travel agency management have become inextricably linked, because the documentation of the activities of these enterprises and communications with clients mainly take place in electronic Form 5.

A modern challenge for the organization of travel agency activities in the conditions of a global pandemic is the monitoring of the rules of departure, entry to other countries, and return to Ukraine, the use of the most modern communication channels to increase the efficiency of business activities becomes especially relevant. Practical issues of digital marketing and electronic documents in tourism, related to the management of specific travel agencies, are not sufficiently studied.

The digital transformation of society has affected the business administration of travel agencies, in particular, the organization of their activities. Although online booking was used in travel agencies for several decades before that, digital technologies were even more integrated into the provision of travel services as a result of remote work during the conditions of the pandemic. In 2021, the share of online sales of tours increased compared to 2020, the year of cancellations and postponements, when tourists postponed the tour even by a year, for example, with the tour operator "Coral Travel". The money, paid for the early booking promotion at the tour operator

"TUI", was not deposited with the subsequent purchase of the tour at a later date. Travel agencies have increased the use of successful cases of IT solutions that help consumers of travel services establish contact and remotely receive consultations via Skype, voice assistants, video chats, and correspondence in Viber, Telegram, and other channels. Consequently, interactive tools become one of the main sales tools of travel agencies.

Digitalization of the tourism business has provided new opportunities to respond to modern challenges and has become a tool for increasing sales. Digital marketing of travel agencies is the use of various forms of promotion using digital communication channels in order to increase sales and profits. In modern conditions, travel agencies use the Internet as follows: create a website, and business page on social networks, post information about themselves on various sites, addresses, contacts, and social networks; regularly, at least 2–3 times a week, publish posts on social networks as Instagram, Facebook; run ads on Google and social networks.

There are 3.3 billion smartphones in the world, which is 50% of the world's population. In 2019, there were 2.77 billion. of social network users. According to Semantrum, a professional media content analysis agency, 21.6 million users are constantly using the Internet in Ukraine today, Internet penetration is 64.8% [20], among which there are three categories of Ukrainians: schoolchildren and students; owners or directors of large and medium-sized businesses; military. Social networks are used by the Zaporizhzhia researched travel agencies to promote their own brands, to set up loyal relations with consumers, but adding "buy" buttons on Facebook/Instagram is not justified, since the purchase of tours is multi-stage and consists of selecting a tour, checking the availability of relevant documents and removing copies from them, concluding a contract, paying for the tour, and issuing documents from the tour package.

Currently, digital marketing continues to develop new types of interaction with the audience thanks to the list of tools used in the digital environment: These are the most cognitively active categories of potential clients of travel agencies, who use social networks:

1. SEO promotion.
2. Content marketing.
3. SMM promotion.
4. Crowd marketing.
5. E-mail mailings.
6. Contextual advertising.
7. Viber/Telegram communication, etc.

Interaction with the audience is the key point in this direction. The use of a full range of digital marketing tools by travel agencies allows you to use all channels of communication with the consumer in order to promote and increase sales of tours and, accordingly, profits.

The methodological basis of the analysis of Digital-strategy 4 "W" (Digital is translated as "digital") is a combination of answers to 4 questions:

1. Why: why and why?;
2. Who: who is your target audience?;
3. When: how and when is the purchase made?;
4. What: what and how do we say?

Firstly, the question "Why?" offers a study of competitors on the Internet and shows what opportunities travel agencies have not yet used, which competitors have, to create analogs and ensure parity points, to determine their uniqueness. For example, travel agencies underestimate the registration in Google Maps, which is free and provides the opportunity to place the address of the office, information about services, contacts, and high-quality photos. Secondly, the analysis of the presence of competitors in all digital channels allows us to establish the size of the segment, how much money is allocated to it by others, and how much can be allocated by our travel agency. Thirdly, we study information about top competitors and find out their strategies. Fourth, we establish everything about the potential consumer, his behavior, and lifestyle (see Table 6).

Travel agencies study their regular and potential customers, and this relates to the "Who" element: who is your target audience? The answers to the questions "When" and "What" are provided, respectively, by the communication channels used to search for customers and arrange tours, and the quality and service of the tourist product.

We analyze the digital marketing of travel agencies in Zaporizhzhia. We identify competitors of the young travel agency "Drive Tour" on the Internet; we analyze the websites and social networks of competitors, specify the style of advertisements and visual images, the use of search engine optimization by competitors, the target audience, and the needs that they close; communication channels, we study the advertising activity of competitors.

The pandemic in the conditions of the information society became the challenge that stimulated the active use of digital marketing tools, which provide travel agencies with data about consumers and are used for their segmentation. The travel agency "Drive tour" has been operating in the travel market for less than 10 years, therefore it compensates for the lack of a website with daily offers of the most relevant tours in the Viber community, the travel agency office is temporarily located, so the owners have not posted information on Google Maps, although upon request to Google, a link is issued in Instagram [21], where there is an address, a link to the Viber community. Reviews, services, and contacts are placed in the actual. Instagram users are offered beautiful visual content in the form of regular videos and travel tips that suggest different countries for holidays in different seasons. "Drive tour" emphasizes hot prices, air tickets, insurance, and the fact that "We work with all of Ukraine" in the description of the business page.

Advertisements of the researched travel agencies are unobtrusive, the user independently decides which advertising message "hits". Correctly configured contextual advertising is shown only for specific requests and to the right target audience. Paired with an advertisement, the maximum number of targeted conversions to the site is obtained for the minimum price. Contextual advertising is a type of advertising on the Internet, according to the terms of which ads corresponding to the content attract users to the site [22]. At the same time, payment will be made for the "click" of users, interested in your service. The service is also known as PPC—"pay per click". Context advertising is also called context advertising or contextual advertising. Its main goal is to attract targeted traffic to the site, which will convert into customers.

The main indicator that can be used to determine interest in the ads of a particular travel agency is CTR. The click-through rate is the ratio of the number of clicks to the number of impressions. The higher this indicator, the more relevant the ad is to the audience, which means that the ad is optimized.

You can place contextual advertising to the owners yourself through Google through the special Google Awards service. The advantages of contextual advertising in what is perceived by the user as an answer to the question posed, the travel agency "Liuks" appeared third in the question about Zaporizhzhia travel agencies in November 2021, and already in December 2021, the travel agency stopped advertising on Google. SEO (Search Engine Optimization) includes search optimization of the site, the process of text filling (content), site structure, and control of external factors to meet the requirements of the algorithm of search engines, with the aim of raising the position of the site in the search results of these systems according to certain user requests. The higher a site's position in search results, the more likely a visitor will click on it since people usually follow links on the first page of the search results.

Contextual advertising can be managed online or pausing, decreasing, or increasing the number of advertisements, and adjusting the cost of going to the site. The principle of the auction is the most impressions and clicks received by the ad that most accurately corresponds to the requests or interests of users; payment for results and budget control; flexibility of setting: time of day and date of display, geographical location, type of user's device and its operating system; it is easy to measure the effectiveness with the help of Google Awards web analytics [22].

The digital marketing of the researched travel agencies of Zaporizhzhia is characterized by the use of contextual advertising in order to increase sales. SEO (Search Engine Optimization) is used by the travel agency "Liuks", and is the process of text filling (content), site structure, and control of external factors to meet the requirements of the algorithm of search engines, with the aim of raising the position of the site in the search results in these systems at certain user requests. The higher the position of the site in the search results (rating), the more likely the visitor will go to it, because people usually follow the links on the first page of the output. Ordinary users do not always pay attention to the difference between (1) an ad is an advertisement they see in a search result, and (2) a feature blog, when Google offers an answer to a search query and pulls up the best answers, for example, "travel agencies", "tour operators". Higher ratings attract better leads when potential clients search for services related to Zaporizhzhia travel agencies. Therefore, optimizing the web pages on the site with the help of relevant keywords helps a lot. Keyword selection makes a huge difference in on-page SEO. Off-page SEO and technical SEO techniques can be used to improve rankings and increase visibility in search engines such as Google.

Targeting in social networks is a special advertising mechanism that allows you to single out only the target audience from the entire audience and show advertising to it, this Digital-marketing tool is most actively used by the "Drive tour" travel agency. "Drive tour" adjusts advertising on Facebook, and Instagram, taking into account the age category of 18–40 years, the region is Zaporizhzhia region, and the interests of users are traveling, photography, etc. The travel agency offers high-quality content and customizes it according to the target audience. "Drive Tour" uses social networks to interest its audience with travel ideas and tips, as well as to attract the Viber community, the link to which is on the business page of the travel agency on Instagram, it is in the community that the most current tours are offered with a good price-quality ratio, further communications take place in personal correspondence. Unfortunately, it is understood from the experience of the travel agency "Drive Tour", that paid advertising on social networks does not turn potential customers into buyers of tours. Therefore, its effectiveness is limited today, but it attracts subscribers, inspires them to plan future trips, makes them loyal, and increases the trust of users in the travel agency. This travel agency strategy is justified because 3 out of 5 travelers watch online videos to narrow down travel choices. 73% of travelers' travel itineraries and plans depend on other travelers' photos.

99% of documents are processed by travel agencies in electronic form. The documentary process of interaction between travel agencies and clients takes place online: application processing, tour selection, with the exception of signing contracts for the provision of travel services and a mandate contract, issuing a tour package, even paying for a tour is possible online, with a tour operator-booking a tour, confirmation, receiving a tour package [19].

The analysis of the document flow of travel agencies showed that the largest volume is the original document flow. The volume of document flow depends on the number of clients served. The smallest one is the document flow of internal documents. Most of the documentary processes of interaction take place in the conditions of a global pandemic, for example, by phone and online correspondence by e-mail, various messengers: with client applications processing, tour selection, and even tour payment is possible online, with the exception of signing contracts for the provision of tour services and contract-assignment, issuing a tour package, but they can be sent remotely via courier delivery.

To solve the problem of organizing paperwork in tourism, the authors characterized the paperwork requisites of electronic documents for organizing departure and return to Ukraine, which appeared due to quarantine restrictions brought by the global pandemic, which led to an increase in the interest of clients in remote online interaction with travel agencies. The problem was solved by studying the details of electronic documents that tourists need to have for entering other countries and returning to Ukraine.

In modern realities, travel agents have to show miracles of dexterity and flexibility, and often steel endurance, in order to make money. Firstly, managers of travel agencies had to monitor entry rules and necessary documents for sending tourists several times a day. It was a real challenge for travel agencies, which we want to illustrate with a story from the experience of the owner of the "Liuks" travel agency, Yu.V. Khlebnikova, who she called on Facebook "How I spent the first day of summer!": "It rained all night long. The working day promised to be calm. In such weather, tourists have a passive mood, rarely does anyone want to come to the office, or want to plan something. 8:00 a.m.—I calmly read the news in a specialized travel group. Until I get to the topic: Urgent! Turkey from tomorrow with tests!" As a sane person, at first, he is sure that it is a fake. But soon the comments heat things up. We are counting the number of our departing tourists on June 2 and 3, there are 16 of them! 9.00 Valerian appears on the table. We are waiting for official information. From 10.00 to 12.00 the day stops being difficult. We are looking for solutions. At 13.00 we find different ways, including for those flying out of Kyiv. As soon as we announce the plan of action to our tourists... One of the tour operators received news that tests are required from June 4. We urgently cancel all planned actions. A glass of wine appears on the table 15.00. The latest news disappears from the operator's website. Again the date is given as June 2. We are looking for new solutions. Because according to the previous one, time has already been lost. We find new ways. We seem to have time for everything. 16.00... Flight Zaporizhzhia—Antalia departing on 01.06 is delayed. If they arrive on June 2, tests will be required. I drink Valiarianka wine".

Secondly, tour operators change prices. If a tourist withdraws cash from each ATM for 4 thousand and the amount of a tour to the Maldives is 100–120 thousand UAH for 2 people. Moreover, until the full amount is withdrawn or received on credit, the cost may increase, as well as the exchange rate set by the tour operator. That is why travel agents work even more flexibly and nimbly in the conditions of a global pandemic: they accept a 30% advance payment for early bookings on a bank card, for example, to minimize the risks of such a situation and earn an agency fee by selling a tour to a client. If the cost of the tour increases by more than 5% or more, the tourist can refuse the contract for travel services, in which case the travel agent is forced to return the agency fee in full. If the client buys a tour from a new travel agency for the first time, there may be concerns about the electronic transfer of funds to the bank card of an unknown person. Therefore, loyalty and trust in the travel agency must be one hundred percent. This is facilitated by personal recommendations of acquaintances and friends, firstly. Secondly, this is facilitated by current reviews on the pages of travel agencies on social networks. On the day of booking and payment, it is necessary to specify the exchange rate and the resulting amount at the travel agency, and this data is updated on the tour operator's website. For an ordinary person, these nuances are difficult and quickly forgotten.

For new agencies that have been working for about five years in the market of tourist services and are most interested in expanding the client base, regular customers appear, the costs of maintaining a website are not always justified for an agency that has been working in the market of tourist services for the first years. That is why travel agencies, which consist of 1–2 people, create Viber communities with an unlimited number of participants, where they post daily advertisements with profitable tour offers. Interested clients are asked to clarify the amount for a certain number of adults and children with an indication of age in the chat, and personal correspondence with the potential client continues. Then the tourist pays online for the tour by bank card transfer, the travel agent sends the application after receiving confirmation and a payment order from the operator, the cost of the tour is paid, and the travel agent receives a 6–12% agency fee, depending on the tour and the tour operator.

The document flow of outgoing documents of travel agencies, which is directed to clients, namely an agreement for the provision of travel services, a receipt for the payment of services, an assignment agreement, a voucher, an insurance policy, airline tickets, an information letter, was supplemented with additional electronic documents during the COVID-19 pandemic. COVID-19 certificates are already available in electronic form for those who have completed a full course of vaccination, you can generate your COVID-19 certificate in the "Diia" relationship. "Diia" is a mobile application on a smartphone [23]. By installing the "Diia", the user uses digital documents and shares copies of digital documents. To receive digital versions of your documents, you need to download "Diya" and log in. The following digital documents appear in the application automatically, if there is data of the relevant person in the registers. As of March 2021, up to 18.2 million Ukrainians could use a foreign passport in a smartphone [24].

Digital documents in "Diya" that tourists will need when traveling abroad with children are the following: biometric international passport and child's birth certificate. Before the child reaches 16, parents must present, both, an ID and a child's birth certificate. The lack of a document, even if the child's ID card is available, maybe the reason that the border guards will not allow your child to boarding the plane, because only this document contains not only the surname, and first name, but also the patronymic. If a tourist plans to rent a car, he needs a driver's license. If he buys a car, he may need a Certificate of registration of traffic pain and an insurance polis for traffic damage. Student International Choice may be useful for those returning to study abroad. Unfortunately, if one of the parents is traveling with the child, a notarized permission from the other parent is required. This permit is currently only in paper form. It is issued for a period of one year, according to the authors. This document also needs digitization. The procedure for signing with an electronic signature after updating the person in "Diya" should be simplified. The absence of the patronymic in ID cards, which are issued from the age of 14 to today, is ill-conceived in the tourism administration, because the "Birth Certificate" of A4 size, made of soft paper, needs a folder for transportation, and although the child has a foreign passport. It is necessary to carry a "Birth Certificate" until the age of 16. Separate attention should be paid to the COVID-19 certificates, the certificates of recovery from COVID-19, and the negative PCR test. Their validity can be checked by QR codes. The image with black squares on a white background can be seen in many places: this is a QR code (quick response code), which was developed for automation and technical progress to replace bar codes, which have a small amount of information: up to 30 characters. A QR code is a two-dimensional image into which certain text or numbers are inserted. To read a QR code on a mobile device, you need a camera and a special application for reading the code. For example, QR Droid for Android, Bakodo, or QR Reader for iPhone for iOS. In order for the application to decode the code correctly, good lighting is necessary, the code must completely fall into the camera, and the device itself must be as still as possible. The advantages of the QR code are ease of scanning and a large amount of data, which can be read by any mobile device with a camera and special software, and thanks to the error correction system, the information can be deciphered even if 30% of the code is damaged [25].

The EU Digital COVID Certificate is a document in the "Diya" mobile application that confirms the status of a vaccinated person. The certificate can be obtained in a few clicks. In the application, you need to click "Services", select the item "COVID-certificates" and click "Get certificate". Then "Diya". It is necessary to sign the request for the transfer of data from the Electronic Health Care System to "Diya" with a signature. Thus, Ukraine has joined the international infrastructure of EU COVID certificates. It is a common environment for all countries that ensures their mutual recognition. There were difficulties with the fact that Ukrainians began to be vaccinated, but not all vaccines are recognized in the countries of the European Union, for example, the Chinese vaccine CoronaVac. The EU Digital Directorate has provided access to the gateway, a tool that allows you to check certificates of other countries while traveling. COVID certificates will work like other documents in Action. You can check the validity of the certificate using the QR code [20]. At the same time, there were cases when tourists from Zaporizhzhia visited abroad with forged certificates, and with the growing competence of special services, the same tourists were detained at the airport after returning to Ukraine.

Tourists should be careful when choosing a vaccine. CoronaVac is not allowed entry into most countries, for example, Bulgaria, while until November 26, 2021, vaccination was not required for entry into this European country. At the same time, as of December 2021, Albania and Aruba continued to accept tourists without restrictions regarding certain vaccines. It is necessary to take into account the period of its validity indicated in the certificate so that it is not overdue, it must be a full course of vaccination, provided that at least 14 days have passed since the completion of the course. Therefore, it is necessary to clarify this on the Internet on the relevant websites of the travel countries or in a travel agency at the stage of planning the trip.

The details of the paper certificate and electronic certificate are different. A two-page paper certificate in Ukrainian and English consists of: (1) the title "International Certificate of Vaccination/Prophylaxis"; (2) text with full name, date, month, year of birth, gender, citizenship, passport number and its type, a table with the name of the vaccine, a requirement for the validity of the certificate; (3) signature of the supervising clinician; (4) doctor's seal; (5) official seal of the institution where the procedure was carried out; (6) two vaccination dates; (7) validity dates of the certificate. Making any changes, additions, or corrections to this certificate, the presence of blank columns may cause it to lose its validity.

The electronic "Vaccination certificate" is written in English in the "Diya" application, provided that you have a foreign passport. Without it, it cannot be downloaded.

Details of the certificate's record keeping: (1) name; (2) effective date; (3) text from the name of the vaccine, first and last name in transliteration, foreign passport number.

The document flow of outgoing documents, which is directed to customers, with the COVID-19 pandemic was supplemented, in particular, with insurance against no-shows or trip cancellation—travel cancellations. The peculiarity of this type of insurance is that, in addition to the standard set of insurance risks, it includes risks related to the deterioration of the health of children under the age of 14, who require inpatient or outpatient treatment under the supervision of parents, and also covers risks related to related to the impossibility of making a trip due to the illness of the insured person with COVID-19, in the presence of a laboratory-confirmed test result conducted by a certified state laboratory.

The details of the insurance policy are (1) the name of the document; (2) the insurance company logo; dates with the insurance period "from...to..."; (3) epy text; (4) the seal of the insurance company. The text indicates the insurer, the insured, the insured persons (surname and first name in English, passport series and number), insured events such as medical expenses, accidents, additional insurance conditions, including COVID-19, trip cancellation, and insured amount per person against insurance cases.

The procedure for case in the event of an insured event is "to contact the assistance company of the country where the tourist is temporarily located" and indicate the phone number. The seal impression is placed so that it captures the surnames and first names of tourists in the place "I agree with the terms of the insurance".

Tests for COVID-19, "Results Report for COVID-19 detection by RT-PCR" are a mandatory condition for entry into most countries, are completed in English, details: (1) PCR laboratory logo, QR code; (2) the date of the study; (3) text; (4) signature and date; (5) seal. The text contains the laboratory's license number, surname and first name, gender, date of birth, number, and series of the foreign passport, type of test, sample information, and medical report.

Analysis of the documentary process of tourist service in travel agencies and interaction with tour operators is presented by a set of documents, the composition of which depends on the tour and the country. As of December 2021, a certificate confirming the status of a vaccinated person was required by the rules of entry to most countries [26]. As of December 2021, countries that allowed tourists to enter without a vaccination certificate were the Dominican Republic and Egypt, the Maldives, and the UAE.

In Greece, the following documents could be equivalent to the entry document for a vaccination certificate:

- a negative PCR test made no more than 72 h before arrival in the country;
- a negative express antigen test, made no more than 48 h before arrival in the country;
- a certificate of a previous illness, recovery, and the presence of antibodies OR a certificate of the presence of antibodies, from the date of issue of which has passed from 2 to 9 months. In Ukraine, it was not possible to find out what the appropriate sample of the certificate for traveling abroad should look like.

In Greece, in addition, upon arrival, tourists are selectively given express tests at the airport at state expense. In case of detection of the disease COVID-19 during vacation, Greece assumes all expenses for the treatment of the tourist. It is clear that this is being done to control the situation regarding the spread of COVID-19 in the country.

For arrival in Madeira with an analog of the entry document according to the Certificate of Vaccination such as AstraZeneca, Pfizer, Moderna, Johnson & Johnson/Janssen, Sinovac/InstitutoBuntantan, Instituto Gambaleya (SputnikV) and Sinopharm (VeroCell) 2 injections 14 days before driving there could be the following documents:

– PCR test is allowed not later than 72 h before boarding, or rapid antigen test is allowed not later than 48 h before boarding;

Declaration of immunity (for those who have already been sick with COVID-19 in the previous 90 days);

– Valid EU digital COVID certificate.

After arriving at Madeira must undergo retesting (free of charge) or remain in self-isolation for 10 days at the place of planned residence.

For arrival in Turkey, instead of a vaccination document, which certifies that a person has been fully vaccinated at least 14 days before entering the country, the tourist can choose:

– a negative PCR test made no more than 72 h before entering the country;
– a document certifying that a person has contracted COVID-19 within the last 6 months, starting from the 28th day after receiving the first positive test;
– a negative express antigen test, made no more than 48 h before entering the country.

In Turkey and Greece, due to the restrictions due to the pandemic, electronic forms for travelers are required, while in Egypt this form must be printed and carried with you. It is the Declaration Health Form.

HES-code is an electronic tourist document for traveling to Turkey, starting with two-year-old children, which is issued in English. This document was introduced on March 15, 2021, due to quarantine restrictions, the document reads: "Form of Entrance due to Pandemic Threat". To do this, it is necessary to fill out an electronic questionnaire on the government website 72 h before departure.

Since the tourist will specify to the travel agency how to issue this document, many travel agencies issue this electronic document themselves and send it to clients, for example, via Viber/Telegram. The document indicates the hotel and the time of stay in Turkey, its purpose is to protect the health of the tourist during his stay in Turkey, to provide accurate information about the COVID-19 pandemic so that it is possible to find people who have been in contact with infected tourists with COVID-19. Temporary registration for the duration of the trip takes place at the hotel.

The HES code is required to enter the territory of Turkey, it is checked at the airport upon departure, upon landing, it can be checked when entering shops, banks, museums, cafes, restaurants, etc.; it is needed to fly within Turkey; ride intercity buses; to use public transport; for visiting hospitals. It is enough to present it in electronic form from a mobile phone. The details of the HES code are as follows:

(1) logo of the Ministry of Health of Turkey,
(2) text: HES code number and QR code, surname and first name, gender, passport number, date of arrival in the country, nationality, mobile phone number, e-mail address, country of residence; accommodation address as follows: city, region, street, house, apartment, hotel, list of countries, where the tourist was in the last 10 days. If it is a document of a minor child, the mobile phone number and email address of one of the parents are indicated.

Passenger Locator Form (PLF) is an electronic tourist questionnaire that must be filled out before arriving in Greece. The document is printed in red: "A negative PCR or antigen test is required to be allowed to enter the country. Certificates of negative test results for COVID-19 must be issued in English and contain the personal information of the traveler.

You may be tested again upon entering Greece." Details include (1) surname and first name, (2) PLF QR code, (3) text: personal information: surname, first name, mobile, business, home phone, gender, age, email address, passport number, date of submission documents; information about transportation/air transportation or airlines, plane number, date of arrival, place of entry into the country; permanent address and country, state, province, city, street, house; temporary address and country, state, province, city, street, house, name of the hotel, cruise ship; contact information in the event of an emergency: last name, first name of the contact person to be contacted, their country, city, state, e-mail address, mobile phone; companions traveling with family or non-family, other groups, household: tourist group, team, business group, other), age, seat number, last name, first name, passport, ID.

"Vdoma" is an application to prevent the spread of COVID-19 in Ukraine. If a person has returned from countries where there has been an outbreak of COVID-19, has been in contact with a sick person, or has symptoms of the virus, they must observe self-isolation or observation. The application must be downloaded and installed when passing through border control after returning to Ukraine, without this the person will not be passed, technical difficulties arise for those who have an iPhone, the application was not activated after filling out the form and pressing the "Send" button, an "Error" was issued and the program did not go into working mode, we note from the experience of close acquaintances and friends that there have been repeated cases of such force majeure working moments. The application provides the possibility of an emergency call to the hotline of the Ministry of Health of Ukraine, but when a person with an iPhone asked for help in activation, she was told that she was doing something wrong. Only receiving the results of a negative test made it possible to solve this problem, but all this time the tourist was in a state of stress after the trip. The functionality of the application is registration at the place of self-isolation or observation, for which the tourist is given 24 h; photo confirmation of stay at the place of self-isolation or observation: after arrival at the place of self-isolation/observations must include geo-location and take a photo when prompted by the application; another shooting may be requested at any time.

Difficulties in taking pictures arise because the application requires you to turn your head in different directions, and when a tired person does it, it can take up to 10 min, from the authors' own experience, those who tilted or turned their heads a little wrong. If a person is not in a place of self-isolation, he will be fined. 120 min per day are allowed to visit shops, and pharmacies, and walk the dog. If the person does not take the test, then he has 14 days to be in self-isolation, when he does and receives a notification that the negative test results and self-isolation are completed, the relationship can be deleted from the phone. The document with the test results is sent to e-mail by the medical institution where the test was performed at the beginning of the next day. As of June 2021, the cost of the test was about UAH 800–900, at Zaporizhzhia airport it was possible to do it on a first-come, first-served basis for UAH 1,100, but the office of the St. Nicholas Clinic did not have a signboard or any identifying marks, so its location few people knew at the time. In addition to the fact that the application is installed by travelers after returning from another country, the application is installed in the smartphones of persons who may be potential carriers of the COVID-19 virus and who are registered in medical facilities as those who need self-isolation or observation.

Therefore, in connection with the restrictions of the pandemic, the documents for tourist departure and return were supplemented with electronic documents, these are tourist questionnaires with personal information, information about their state of health in relation to COVID-19, location in the country of entry, a feature of the questionnaire for entry to travel to Greece is to have a contact person in Ukraine with a phone number and an address that can be contacted in case of an emergency. In addition, the "Diya" and "Vdoma" software applications have appeared for downloading documents from the central document database to control the situation regarding the incidence of COVID-19 among tourists.

4   Conclusion

According to the analysis of the state of development of the concept of Smart Cities and digital services in the seaside regions, it has been confirmed that this is happening most effectively in the territory of Zaporizhzhia and Odesa regions. The functional components of the SC mechanism in the coastal regions are defined, namely: infrastructure, the interest of the authorities (normative and legal component), standards of the system, specialists (their skills and abilities), the readiness of the community to accept changes.

The largest number of applications, chatbots, and Telegram channels in terms of their scope of application operates on the basis of the Odesa region, which confirms the effectiveness of the digitalization course taken by local authorities, a sufficient number of IT specialists, and users of these tools. Social networks, YouTube, messengers, local TV channels, and newspapers are important for the dialogue between local authorities and residents because they help build the social capital of the community.

Big Date system and digital tools when forming a request for ecosystem services is reflected in the form of a Big Date architecture model, which focuses on the competitive advantages of the coastal region and allows the consumer to form a digital request regarding their own preferences for the region's ecosystem services. The functioning of the Big Data infrastructure creates the image of a digital modern city, which is based on the use of modern data processing technologies, which are constantly updated and replenished. Drawing up a "portrait" of a consumer of ecosystem services, a "profile" of an ecosystem service is the ultimate goal of using this system. Using the results of digital analytics data is the basis of timely response and flexibility of the management system of a competitive region.

Thus, the tourism business is one of those most affected by global transformations as a result of the pandemic in 2020–2021. A large number of bookings that were "frozen" and shifted for a long time, refusals of tourists, the closing of offices, their transition to remote work, closing the borders of the countries, and then some constant changes in the rules for crossing the border caused a certain burden on the owners of travel agencies. The principles and methods of work, the organization of communication with clients, and the management of document management have undergone changes as the appearance of electronic documents and the mandatory use of smartphone applications when organizing tourist trips.

The implementation of digital technologies in the activities of an increasing number of enterprises has become possible due to the large audience of the Internet, which is constantly increasing, relatively cheap promotion as classical marketing methods and the possibility of setting up paid advertising campaigns, using the capabilities of Google, Facebook, Instagram, Viber, etc. through segmentation and creating a comfortable digital business environment for business. The analyzed digital marketing strategies of travel agencies in Zaporizhzhia according to the selected criteria made it possible to single out competitive strategies and study the communication channels of travel agencies.

Digitization processes affected not only the communication channels of travel agencies but also the introduction of new electronic documents and applications during the registration of the tour, related to the crossing of borders by tourists and the tracking of tourists after the trip in order to prevent the spread of the Coronavirus, which meet modern world standards. The issue of further improvement and implementation of electronic documents and special applications developed by state authorities for smartphones, on the one hand, simplifies the mechanism of tracking patients, and on the other hand, requires constant monitoring by managers of travel agencies.

References

1. Obolentseva, L.V.: Metody formuvannia stratehii upravlinnia konkurentospromozhnistiu promyslovykh kompleksiv rehioniv (The Methods of forming of strategy for managing the competitiveness of regional industrial complexes). Bus. Inform **12**, 413–418 (2017)

2. Tulchynska, S.O., Pohrebniak, A.Ju., Miedviedieva, A.D.: Upravlinnia konkurentospro-mozhnistiu promyslovykh kompleksiv rehioniv (The management of the competitiveness of regional industrial complexes). https://ela.kpi.ua/bitstream/123456789/29496/1/2019-13_4-01.pdf (2019). Accessed 9 Jan 2023

3. Pushkarchuk, I.M.: Mekhanizmy pidvyshchennia konkurentospromozhnosti promyslovosti yak osnova finansovoi stabilizatsii rehionalnoi ekonomiky (The mechanisms of increasing of industrial competitiveness as a basis for financial stabilization of the regional economy) Efficient economy 3. http://nbuv.gov.ua/UJRN/efek_2016_3_40 (2016). 10 Accessed Jan 2023

4. Gudz, P., Gudz, M., Vdovichena, O., Tkalenko, O.: Scientific approaches for planning the architecture for urban economic space. In: Onyshchenko, V., Mammadova, G., Sivitska, S., Gasimov, A. (eds.) Proceedings of the 2nd International Conference on Building Innovations. ICBI 2019. Lecture Notes in Civil Engineering, vol. 73, pp. 581–589 (2020)

5. Gudz, P., Oliinyk, Y., Shkurupska, I., Ivanchenkov, V., Petrenko, O., Vlasenko, Y.: (2020) Formation of foreign economic potential of the region as a factor of competitive development of the territory. Int. J. Manag. **11**(Issue 5), Maj 590–601

6. Costanza, R.: The ecological, economic, and social importance of the oceans. Ecol. Econ. **31**(2), 199–213 (1999)

7. Costello, M., Ballantine, W.: Biodiversity conservation should focus on no-take Marine Reserves. Trends Ecol. Evol. **30**, 507–509 (2015)

8. Morrissey, M.K.: The role of the marine sector in the Irish national economy: an input-output analysis. Mar. Policy **37**(1), 1–21 (2013)

9. Pontecorvo, G.: Contribution of the ocean sector to the United States economy. Mar. Technol. Soc. J. **23**(2), 1000–1006 (1989)

10. Witte, P., Slack, B., Keesman, M., Jugie, J.-H., Wiegmans, B.: Facilitating start-ups in port-city innovation ecosystems: a case study of Montreal and Rotterdam. J. Transp. Geogr. **71**, 224–234 (2018)

11. Stoshikj, M., Kryvinska, N., Strauss, C.: Service systems and service innovation: two pillars of service science. Procedia Comput. Sci. **83**, 212–220 (2016). https://doi.org/10.1016/j.procs.2016.04.118

12. Engelhardt-Nowitzki, C., Kryvinska, N., Strauss, C.: Strategic demands on information services in uncertain businesses: a layer-based framework from a value network perspective. In: 2011 International Conference on Emerging Intelligent Data and Web Technologies (2011). https://doi.org/10.1109/EIDWT.2011.28

13. Fauska, P., Kryvinska, N., Strauss, C.: The role of e-commerce in B2B markets of goods and services. IJSEM **5**, 41 (2013). https://doi.org/10.1504/IJSEM.2013.051872

14. Smart cities need to be heading in a human-centric, digitally inclusive direction or the tech won't pay off, panelists agree (2021) International Institute For Management Development. https://www.imd.org/research-knowledge/videos/smart-cities-need-to-be-human-centric-and-digitally-inclusive/. Accessed 20 Jan 2023

15. Smart City segodnya—eto obychnyj gorod zavtra (Smart City today is an ordinary city tomorrow). http://sib.com.ua/sib-01-116-2021/smart.html?fbclid=IwAR2IBXfuWJStfEXMDXmI2FO7MJemfqW4Y0tB4LnUHX52BKVYWoDgTLz_bVA. Accessed 10 Jan 2023

16. Ministerstvo tsyfrovoi transformatsii Ukrainy (The Ministry of Digital Transformation of Ukraine). https://thedigital.gov.ua/. Accessed 10 Jan 2023

17. Smart_cities: The European Commission. https://commission.europa.eu/eu-regional-and-urban-development/topics/cities-and-urban-development/city-initiatives/smart-cities_en (2022). 10 Jan 2023

18. Lelechenko, A.P., Vasylieva, O.I., Kuibida, V.S., Tkachuk, A.F.: Mistseve samovriaduvannia v umovakh detsentralizatsii povnovazhen: navchalnyi posibnyk (The local self-government in the conditions of powers' decentralization: a study guide), p. 89. Kyiv (2017)

19. Byelikova, M.V.: Dilovodstvo v turyzmi: analiz dokumentalnoho protsesu vzaiemodii turysty-chnoho ahentstva (The records management in the tourism: analysis of the documentary process

of tour agency's intercommunication). In: Innovatsiini tekhnolohii v turystychnomu ta hotelno-restorannomu biznesi (The Innovative Technologies in the Tourist and Hotel-Restaurant Business), pp. 60–66. Cherkasy (2020)

20. Factu Group Ukraine Prezentaciya «Proniknovenie Interneta v Ukraine» (Factu Group Ukraine Presentation "Internet penetration in Ukraine"). https://promo.semantrum.net/uk/2017/04/21/v-ukrayini-na-pochatok-2017-roku-narahovano-21-6-mln-koristuvachiv-internetu/ (2017). 12 Jan 2023

21. Turisticheskoe agenstvo «Lyuks» (The "Lux" tour agency). http://luxeua.com/main/photo_my.html. Accessed 12 Jan 2023

22. Contextual Targeting: The Next Big Thing in 2022 UPQODE. https://upqode.com/contextual-targeting/ (2022). Accessed 12 Jan 2023

23. Diia: Application in Google Play. https://play.google.com/store/apps/details?id=ua.gov.diia.app&hl=uk&gl=US (2020). Accessed 12 Jan 2023

24. Українські COVID-сертифікати в Дії офіційно визнані ЄС. Урядовий портал : веб-сайт. https://www.kmu.gov.ua/news/ukrayinski-covid-sertifikati-v-diyi-oficijno-viznani-yes. Accessed 12 Dec 2021

25. Fedorov, M.: Tsyfrovi dokumenty v Dii – tse shliakh do borotby z koruptsiieiu (The Digital documents in Dii are a way to fight corruption). https://www.kmu.gov.ua/news/mihajlo-fedorov-cifrovi-dokumenti-v-diyi-ce-shlyah-do-borotbi-z-korupciyeyu (2021). Accessed 12 Jan 2023

26. Turahentsiia Drive tour (The "Drive tour" tour agency). https://www.instagram.com/drive_tour_/. Accessed 12 Jan 2023

Economic Assessment of Outsourcing of Intellectual and Information Technologies

Petro Pererva⑩, **Mariya Maslak**⑩, **Szabolcs Nagy**⑩, **Oleksandra Kosenko**⑩, **and Tetiana Kobielieva**⑩

Abstract The efficient operation of industrial enterprises is impossible without digitalization of production processes. The methodological essence of intellectual and information technologies, the definition of which is proposed to be introduced into scientific circulation, is substantiated. The authors include such information technologies, which include an intellectual component that requires legal protection and protection of exclusive rights to use it. Methodological bases of economic assessment of outsourcing of intellectual and information technologies in industrial enterprises have been developed. It is proposed to determine the effectiveness of the use of IT outsourcing using the graph-analytical method, the essence of which is to provide a graphical comparative analysis and analytical justification of the effectiveness of decisions on IT outsourcing and the implementation of IT functions by the company's own forces. The proposed methods of economic evaluation of IT outsourcing were tested on the example of enterprises in the Kharkiv industrial region.

Keywords Information technologies · Outsourcing · Industrial enterprises · Digitalization

P. Pererva (✉) · M. Maslak · O. Kosenko · T. Kobielieva
National Technical University "Kharkiv Polytechnic Institute", 2 Kyrpychova Str., Kharkiv 61002, Ukraine
e-mail: pgpererva@gmail.com

S. Nagy
Miskolc University, Miskolc-Egyetemváros, Miskolc 3515, Hungary

© The Author(s), under exclusive license to Springer Nature Switzerland AG 2024
A. Semenov et al. (eds.), *Data-Centric Business and Applications*, Lecture Notes on Data Engineering and Communications Technologies 195,
https://doi.org/10.1007/978-3-031-54012-7_7

147

1   Introduction

1.1   Relevance

It is difficult to imagine the modern world without information technology. They are widely used in all spheres of society [1–3]. The use of information technology helps to accelerate the development of industrial enterprises, which helps to strengthen their position in the market and increase the competitiveness of products [4–6]. Currently, society is surrounded by various information and information technologies. Every year the amount of information needed by each person increases and, in this regard, there are new methods and means designed for its processing and storage—more relevant and improved.

The methodology of the concepts of technology, innovative technology, information technology, and technological process today has already been worked out in some detail, but separately from each other [7–9]. There is still no analysis of intelligence factors in the composition of the information technologies, their places, and roles in the system of economic relations. In most cases, technology is understood as a category of production (technical, informational, environmental, etc.), although its consideration in the economic context, in the context of information and intellectual development of the production and entrepreneurial system today is extremely necessary [10–12]. The relevance and importance of this task involve the definition of the methodological essence of intellectual information technology (IT), its place in the system of economic relations, as well as the justification of its connection with the basic, close economic content economic categories.

If the developed technology is of commercial importance, then in this case it is logical to determine the future consumer [13–15]. Most often, the buyers of new technologies are technologists in industrial enterprises. Usually, they are interested in new technologies as the amount of knowledge necessary to start producing new or increasing the production efficiency of existing products. As a result, there is a situation when there is a buyer, there is a seller, however, is no product. The author proposes to buy or license technology from him as a set of ideas about the sequence of transformations of raw materials, while consumers are ready to use a set of technological documentation, as well as a set of appropriate equipment [16–18]. That is, buyers are interested in formalized knowledge that can be transferred, protected, and copied. So, intermediaries are needed between the author and the potential buyer who can Unformalized individual knowledge is turned into the product of a transaction for the transfer of technology. The ideal case would be from the very beginning of the development of the technology to focus on a predetermined format for presenting the results. In other words, the author-developer will get a better chance that his idea will be perceived with enthusiasm if this idea is initially focused on certain market aspects [19]. All this determines the specific feature of information technology as an object of intellectual property, which is characterized by all intangible factors of determination, evaluation, transfer, use and legal protection and protection. All this determines the intelligence of both the origin and the use of information technology as a product.

A number of researchers note that IT services urgently need their own economic justification since their number, scope, and terms of use for each enterprise differ significantly [20–25]. If for one enterprise this IT service is effective, then for another it can be unprofitable. All this implies the need and importance of developing new and improving existing methods of economic evaluation of outsourcing of intellectual information technologies.

1.2 Goals and Objectives

Studies [10, 16, 21, 24, 26, 27] suggest that the manager, manager, or entrepreneur is not at all interested in technological ways of converting raw materials, materials or technological methods of organizing and managing production as such. They are interested in the economic effect of applying these methods in production and entrepreneurial activity. Although such a statement is not entirely unambiguous and may be perceived differently, there is a certain line in it, which is aimed at the presence in the technology of elements of research of existing knowledge and methods, as well as the generation of new ones, which according to existing practice [28–30] relate to scientific activities. Within the same framework, it is necessary to consider the information, innovation, and intellectual activity of the enterprise, which is focused on the intellectualization of various kinds of information technologies and obtaining an economic effect from their use, that is, the generation and consumption of knowledge and methods to obtain a qualitatively new (effective, progressive) result from the production and business activities of the enterprise.

This work aims to promote a better understanding of the economic methods and mechanisms that contribute to the joint creation of value in production systems by introducing an IT outsourcing system. Focusing on systematic approaches to determining the effectiveness of IT outsourcing can help both researchers of the process of digitalization of production processes, in the disciplines of service science, and managers of industrial enterprises to achieve a better understanding of the specific possibilities of IT outsourcing in order to increase the efficiency of joint creation of production values.

This chapter is structured as follows. First, we provide an understanding of the conceptual basis of intelligent information technologies (IT), which form the basis of IT outsourcing. The definition of "intellectual and information technology" at this time is not yet actively used and the authors propose to introduce it into scientific circulation.

The current state and prospects for the development of IT outsourcing are considered in detail. At the same time, the main attention is paid to the consideration of existing types and forms of outsourcing, the methodological essence of IT outsourcing, and the study of its features and factors that form the effectiveness of its use. A detailed analysis of the existing methods of economic evaluation of IT outsourcing was carried out, their main advantages and disadvantages were determined, and the direction of development of the existing methodological base for the economic assessment of IT outsourcing was formed.

Finally, the graph-analytical model of economic evaluation of IT outsourcing proposed by the authors is considered in detail, which is based on comparative analysis and effectiveness of IT outsourcing solutions from a third-party organization and the practical implementation of a similar IT function by the enterprise's own forces.

The section ends with the results and prospects for further research.

2 Methodological Essences of Intellectual and Information Technologies

2.1 Structural and Characteristic Components of the Modern Concept of "Technology"

The concept of "information technology" was introduced into scientific circulation relatively recently—in the XX century in the process of formation of computer science as a science. The peculiarity of information technology is that the subject and product of labor in it is information, and the tools of labor are the means of computer technology and communication. In a broad sense, information technology can be defined as the use of computers, software (operating system/tools and applications), communications, and networks to ensure that the information needs of an organization are met. They represent a wide class of disciplines and fields of activity related to the technologies of formation and management of processes of working with data and information, using computing, computer, and communication technology.

Based on the generalization of the definitions of the terms "technology" and "information technology" Tsareva T.O. and Zozulyov O.V. provide mainly traditional, albeit with an application for modernity, understanding of technology, attributing to its characteristics the following features [31]:

- knowledge, skills, and experience;
- research in the field, which is associated with the use and development of a certain technology, the development of a scientific component that ensures the existence and improvement of this technology;
- a set of material support, including personnel with the necessary skills, and documentation;
- a process of any nature (production, management, etc.).

In our opinion, this list is not exhaustive, since the authors do not consider such important characteristics of the technology as market-commodity nature; opportunities for sale, purchase, licensing; the need for legal protection and legal protection of the intellectual content of technology, etc. Insufficient attention to the intellectual orientation of modern technologies prevented the authors from developing a complete description of the components of the technology, limiting themselves only to scientific, methodological, process, and infrastructure components. Without objecting to these components, we consider it necessary to supplement them with distributive, intellectual, innovative, and information components, as the most characteristic of technologies and technological processes at the present stage of development of scientific and technological progress both in Ukraine and around the world (Table 1).

Our research shows that technological processes in each of the activities or industrial sectors differ significantly, have a different level of efficiency, and, accordingly,

Table 1 Structural and characteristic components of the modern concept of "technology" [11, 26, 31]

Components	The content of the component	Nature of functioning
Scientific and methodical	A set of knowledge (methods, principles, techniques, methods, experience) allows one to bring the object of using technology to the desired state in a rational way (including the definition of rationality criteria)	Providing research with a more rational, more efficient, optimal way to achieve the goals and objectives
Process-technical	A set of processes (management, research, engineering, etc.) containing a certain order of operations and allowing to obtain a product of a given quantity and quality	Direct transformation of the subject of labor. The nature of the functioning is set by the developer
Infrastructural and provisioning	A set of conditions for ensuring the scientific, methodological, and process components, which allow you to effectively use the technology in practice	The nature of the functioning is determined by the place in the technological process and the composition of the work performed/provided by the infrastructure element
Innovative	Reproduction in technology of modern requirements aimed at improving production processes in order to increase the competitiveness of both the production itself and its results (products or services)	It has a proactive character, aimed at the constant improvement of both the technology itself and the object of its use
Intellectual	The totality of creative and intellectual knowledge regarding the implementation of research activities and, in fact, such activities to create objects of intellectual property for technological purposes	Obtaining, accumulating, generalizing, systematizing, and storing, technological knowledge for the development of technology
Distributive (transfer)	Providing opportunities for progressive development of enterprises both in this industry and related industries	Transfer and commercialization opportunities, legal protection and legal protection of the intellectual component

have different methodological content. For example, in machine-building enterprises, technological processes form the basis of production and entrepreneurial activity, providing the necessary level of competitiveness of both the enterprise itself and its products. The task of creating a market, transfer, and commercialization of existing technologies by almost top management is not set, nor is the task of creating new technologies for the purpose of their commercial sale a priority.

Much more often, production personnel are tasked with developing a new or improving existing technology for their own consumption. Almost every type of engineering product sold in the domestic and especially in the foreign market must have a new level of property and meet the ever-increasing requirements imposed by a potential consumer for functional, environmental, and aesthetic properties.

2.2 Economic Content and Definition of Intellectual and Information Technology

In world practice, in the scientific and practical aspect, it is increasingly about the need to possess critical information technologies—technologies that provide or can provide in the future a significant increase in the technical characteristics of systems and give them qualitatively new properties, *basic information technologies*—elementary parts of defense technology, oriented to improve samples of defense systems, means of their subsystems only according to one technical parameter, *high information technologies*, which are characterized by a low level of operational capabilities, in-line production; progressive information technologies that keep ahead of the growth rate of labor productivity, reducing the burden on what surrounds the environment [32].

These names of new information technologies are associated with one or another feature of the technological process, which is accepted by the authors as defining. In this case, the precision of the production process is most often taken into account. These terms are not exhaustive, since they do not reflect the entire complexity and capacity of new Information Technologies. In this regard, it becomes difficult to compare their characteristics. It can be concluded that, regardless of the terminology used by us, the above types of information technologies are objectively the main elements of a single independent direction within the framework of general information technology, the essence of which, in our opinion, is more fully reflected in the concept of *intellectual* information *technology* (IIT), which is proposed to be introduced into scientific circulation.

The proposed approach implies the fact that the composition of information technology may also include the results of intellectual activity that do not provide for legal protection, including technical data. The composition of information technology may also include the results of intellectual activity, which have legal protection, but the author is the person who organized the creation of information technology [33].

This definition directly connects the terms "information technology" and "object of intellectual property", which is the subject of this study.

A certain sequence of development of this definition can be found in the interpretation of the term "technology" by the World Intellectual Property Organization (WIPO), which defines technology as "... systematic knowledge of the production of products, the application of the process or the provision of a service, regardless of whether this knowledge is reflected in the invention, industrial design, utility model, new technological installation, technical information or services provided by specialists in the design, installation, management of production or its activities" [34].

The above analysis of definitions that confirm the intellectuality of the origin of the term "information technology" at the present stage of development of science and technology, shows a certain evolution in the formation of this concept, which the legislative framework considers as an intangible object in which a number of intellectual property objects find their place.

The definition and methodological content of the concept of "intellectual and information technologies", in our opinion, can be reduced to the following.

Intellectual information technologies (IT) are such technologies that are based on intellectual property objects and which have a set of main features: science intensity; consistency; physical and mathematical modeling for structural and parametric optimization of highly efficient technological process of dimensional processing; computer technological environment and automation of all stages of development and implementation; stability and reliability; ecological purity—at appropriate technical and personnel support (precision equipment, equipment and tools, working technological environment, diagnostic system, computer management network and specialized personnel training), which guarantees the receipt of products that have a new level of functional, aesthetic and environmental properties.

With appropriate scientific, technical, legal and personnel support (precision equipment, equipment and tools, a certain nature of the working technological environment, a diagnostic system, a computer management network and specialized personnel training, the presence of legal protection), the use of IIT guarantees the receipt of products with a new level of properties that ensure intelligence and innovation. IIT in compliance with economic feasibility, which is exactly what interests the consumer.

The study of the achievements of theoretical scientists who devoted their works to the study of the essence of the concept of "technology" and most widely and in detail studied this phenomenon allows us to draw the following conclusion: *intellectual-in-formation* technology is that part of the culture of society that combines the intellectual, scientific, technical and commercial achievements of innovators with the *socio-economic side of human life, transformations or movement of materials, information and people, as a result of what is achieved by the goal and something new is formed at a new qualitative level.* This definition, in our opinion, most accurately defines at the essential level the concept of intellectual and formative technologies, laying in it already at this level a certain intellectual and innovative novelty, rationality and efficiency in achieving the goal.

3 Current State and Prospects of Development of Outsourcing of Intellectual and Information Technologies

3.1 The Essence of IT Outsourcing and Its Place in the Global Services Market

Significant achievements of scientific and technological progress in recent decades, manifested, among other things, in the rapid development of IIT, caused noticeable changes in the nature of modern economic processes [1, 7, 16, 23, 27]. The key role of IIT in the activities of industrial enterprises necessitates the continuous development and actualization of the intellectual and information components of the business. The overall effectiveness of the organization depends on the availability of progressive tools for managing information processes. In the context of rising costs for the introduction of modern IIT, the practice of IT outsourcing acquires a special role [3, 7, 14, 21, 26, 28, 30]. Outsourcing of services in the field of intellectual and information technologies (IT outsourcing) provides a high degree of optimization, and high-quality management, taking into account the specifics of the production and commercial activities of a particular enterprise, allows you to get access to the latest developments and contributes to the introduction of innovations, as well as the growth of flexibility and the formation of competitive advantages.

The practice of outsourcing services in the field of IIT is widespread in European practice, but in many countries, outsourcing relations are at the stage of formation [34, 35]. However, the innovative nature of the transformations of national economies, the current trend of informatization of society, as well as the growing presence of national industrial enterprises in the international market, prove the importance of studying the processes of IT outsourcing development. Determining the place of IT outsourcing in the national and international IT outsourcing market will allow us to assess the degree of development of this modern business practice, as well as reflect the aspects of its adaptation to the peculiarities of economic relations in our country [36].

If we study in more detail the evolution of the concept of "outsourcing", we can emphasize the fact that in one form or another, it was used in the 20s of the twentieth century. Thus, Ford G. noted: "If there is something that we cannot do better than our competitors, then there is no point in doing it; we must pass these works on to those who perform them with a known best result" [37]. It is these words that an outstanding American businessman can describe in detail, the fact how long ago the relationship was born, which later became known as "outsourcing" and the main essence of outsourcing, which manifested itself in the form of cooperation between enterprises that tried to cooperate together in the production of certain goods or in the provision of services, where the key advantages of each of the participants were used and the areas of responsibility of each of the participants to achieve the best result.

Outsourcing as a result of interaction between enterprises and the transfer of certain functions from the main enterprise to counterparties makes it possible to gain a competitive advantage, although given the past, the first official attempts to outsource the implementation of certain tasks were destructive for certain companies, for many reasons. The main ones were insufficiently qualified management, weak regulatory framework, and business processes that were outdated and did not correspond to a certain type of interaction.

Outsourcing is often called the "phenomenon of the XX century", as well as "the greatest business discovery of recent decades", because only from the end of the 80s of the XX century did this concept become widely used and used to establish business processes in progressive Western companies. Outsourcing is widely used by leading companies in countries with developed market economies, for example, General Motors, Apple, Hewlett Packard, Blackberry, etc.

The main effect of outsourcing is achieved due to the fact that a specialized enterprise provides more efficient and high-quality performance of the processes or functions transferred to it. The indisputable advantage of outsourcing is the lack of large long-term investments, which is extremely important for the current state of Ukrainian enterprises.

It should be noted that there are a certain number of types of outsourcing, the main of which are presented in Table 2.

IT outsourcing is the leader of the current services market. IT outsourcing is the practice of hiring resources from another organization to handle certain functions of the IIT. The most commonly used IT services include transfer to a specialized enterprise in whole or in part of the functions related to IIT, that is, maintenance of network infrastructure; design and planning of business systems with further constant

Table 2 Types of outsourcing [38, 39]

Types of outsourcing				
Outsourcing of management functions	Production outsourcing	Outsourcing information technology	Outsourcing in service sector	Outsourcing knowledge management
Subspecies of outsourcing				
– Accounting Accounting and finance; – Personnel management; – Marketing; – Advertisement; – Logistics; – Recruitment	– Outsourcing Main production; – Outsourcing Auxiliary production	– Offshore programming; – Testing software ensure; – Electronic business	– Business services (leasing, construction, architectural); – Trading services, hotel services; – Transport and communication services; – Public sector services, including social	– Study and analytical data processing; – Formation and management knowledge bases; – Media monitoring; – Archiving and indexing of data

development and support; system integration; placement of corporate databases on servers specialized companies; creation and support of public web-servers; management of information systems; leasing of computer equipment; web development; hosting; software and application development; website maintenance or management; IIT technical support; database development and management, communication and infrastructure, etc. Industrial enterprises often carry out IT outsourcing because it is cheaper to enter into a contract with a third party than to buy and maintain their own storage devices.

The transfer of IIT development and support to outsourcing has recently worried more and more Ukrainian enterprises. For the vast majority of enterprises are non-core assets, so the opportunity to transfer them to a professional external partner and free up resources for priority areas of business looks quite attractive.

In the global structure of the outsourcing services market, business process outsourcing (about 35%) occupies the largest share, which is associated with a large development of this sector of the outsourcing services market and minimal risk compared to production outsourcing. The second place is occupied by IT outsourcing (30%), and the third—is production outsourcing (14%) [2, 40]. Ukrainian realities have a slightly different look. ANCOR personnel conducted a study of the demand for various types of outsourcing services in the country and also assessed the quality of their work [41]. The study was conducted among the top management of Ukrainian and foreign enterprises that carry out their production and commercial activities in our country. As a result, it was found that about 40% of the studied enterprises testified to the fact that they use different types of outsourcing. At the same time, almost every second of these enterprises use IT outsourcing services (Fig. 1).

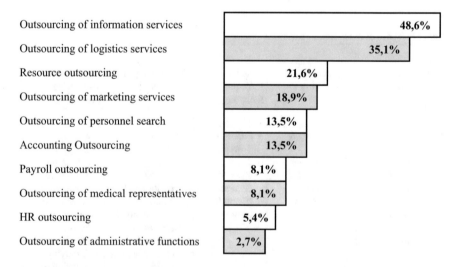

Fig. 1 Rating of needs for services of various types of outsourcing in Ukraine [41]

In the first place are IT outsourcing services—48.6%, in the second—are logistics—35.1%, and the least important is personnel accounting services—5.4% and administrative functions—2.7%.

Outsourcing IIT is quite widely represented in business. A large number of enterprises for themselves have already decided to switch to outsourcing work models. Such schemes for providing individual functions turned out to be convenient in all respects. The customer receives qualified support and at the same time gets rid of problems associated with the work of non-core or inefficient processes. He gets the opportunity to concentrate on the main business area, which ultimately increases the competitiveness of the company. In this regard, the question arises, what functions can the customer be ready to give to IT outsourcing, and where are the boundaries of IT outsourcing?

This question can be answered only after studying the risks associated with refusing to perform certain actions on their own. For example, if an organization uses so-called self-writing IT products, then you should not outsource full IT support. In this case, it would be appropriate to think only about the transfer to the side of the functions of maintenance of computer equipment [19, 29]. All functions that are not on the list of risky theoretically can be outsourced. There are many examples when enterprises, for the sake of one-time, rarely repeated work, contain internal services only because it has happened historically. However, in conditions when all enterprises are forced to save maximize, and optimize their business processes, having extra units and services in the staff is simply unprofitable.

At the present stage of development of the world economy, IIT is one of the effective tools for doing business, and the world market for services in the field of IIT is characterized by a high degree of dynamism.

Despite the widespread applicability of the concept of IT outsourcing, its definition in the modern scientific literature is not sustainable. The study of the definition of IT services by competent international institutions, including the International Organization for Standardization (ISO—International Organization for Standardization) and the Library of Information Technology Infrastructure (ITIL—Information Technology Infrastructure Library). On this basis, it was concluded that the common point of all authoritative definitions of IT outsourcing. There is a mandatory interaction between the supplier and the customer regarding the business processes of the organization based on the use of information technology. At the same time, the role of this interaction in the process of providing services is considered by competent institutions in different ways.

As a working definition of an IT service, it is proposed to apply the ITIL approach, expanded by a subset of ITSM (Information Technology Service Management—IT Service Management). Based on this, it is proposed to define the concept of IT outsourcing as follows.

IT outsourcing (from English IT-outsourcing) is the provision (transfer) of the function of servicing the information infrastructure of an industrial enterprise (ensuring the stable operation of computers, information network, etc.) to another enterprise that specializes in information services for enterprises and organizations and has a staff of narrow highly qualified specialists in the field of IIT.

The inclusion of IT outsourcing in the activities of an organization is a process of using knowledge, skills, abilities, tools and methods of managing business processes for the formation, and development of ensuring the functioning of the information infrastructure of organization management in the context of organizational goals and objectives in order to create additional value. In fact, IT outsourcing is a form of joint management of an organization, both by its managers and by specialized information (digital) outsourcing enterprises that have no competencies in the customer organization. For small and medium-sized businesses, as well as for other small organizations, IT support is important and necessary, but it is carried out irregularly, so IT outsourcing is the best solution to this problem for them and can ensure the efficient operation of such enterprises.

3.2   Forms of IT Outsourcing Implementation and Factors of Their Effectiveness

Today, there are many different ways to implement outsourcing services in the field of IIT. Most organizations outsource some activities in the field of IIT, and many work with more than one supplier. The trend of outsourcing some or all of the organization's activities in the field of information technology greatly affects the careers of employees and managers of the information technology department. Outsourcing decisions can be assigned to departments IIT is top management, but the results have to be provided to professionals in the field of IIT.

In the practice of production and commercial activities of industrial enterprises, five main forms of outsourcing are used: full and partial outsourcing, joint outsourcing, intermediate outsourcing, and transformational outsourcing. A brief description of these forms of IT outsourcing is provided in Fig. 2.

Based on the above, it is possible to identify a number of features and general characteristics of outsourcing:

- outsourcing is carried out on a contractual basis: outsourcing is defined as the practice of planning, managing, and implementing certain types of work by a third-party organization in accordance with the terms of a bilateral agreement;
- one-time purchase of a certain product (service) will not be the object of outsourcing; outsourcing is always a permanent cooperation on the basis of contractual (contractual) relations;
- granting the right to perform certain types of work to a third-party organization is always carried out for a sufficiently long period of time;
- outsourcing takes place only when it comes to transferring to another (external) enterprise such works that could, under certain conditions, be carried out within the enterprise itself;
- each existing situation using outsourcing involves an individual solution reflected in the contract between the parties;

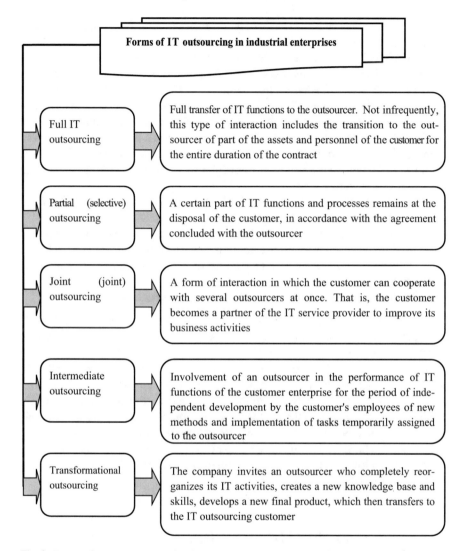

Fig. 2 Forms of outsourcing in industrial enterprises and their characteristics [8, 15, 23]

- outsourcing is carried out in order to more fully meet the requirements of the external environment: outsourcing acts as a tool that allows you to distribute internal and external resources to achieve the goals and objectives of the enterprise.

The results of the study indicate that outsourcing certain functions of the enterprise is strategically expedient if:

- independent business partners will perform these functions better and cheaper. For example, a significant number of manufacturers of computer equipment abandoned the function of final assembly, entrusting this work to outsourcers, which gave them the opportunity to receive significant savings on the purchase of components for computer equipment, as well as on the organization of the assembly process itself. Cisco has transferred almost all the production and assembly of routers and switching equipment to a partner company that owns 37 enterprises whose work is coordinated via the Internet;

- this activity is not competitively significant and its outsourcing does not threaten changes in key competencies, production capabilities, and intellectual know-how of the enterprise. The practice of outsourcing services for the maintenance of industrial equipment, information data processing, management and accounting, and a number of auxiliary administrative functions to enterprises specializing in this activity is widely used;
- it reduces the level of commercial risk associated with changes in business processes, technologies, and/or product benefits for potential consumers;
- it significantly increases the flexibility and efficiency of economic decision-making, reduces the time for the development and launch of innovative goods on the target market, and reduces the company's costs for coordinating all these actions;
- it allows the enterprise to focus its efforts on its core business.

The conducted studies allow us to conclude that the consideration of IT outsourcing as one of the components of the portfolio of IT projects of an industrial enterprise allows, based on the implementation of the process approach, to quickly create new competitive advantages. This becomes possible due to strict adherence to all the main stages of portfolio project management, a built-in system of interconnection of the project strategy, portfolio strategy, and overall strategy, transparent requirements for all responsible for the implementation of each IT project, compliance with the established requirements for the results of IT projects and understanding their impact on the overall result of the entire enterprise. At the same time, the technological basis for the implementation of these approaches is the integrated digital infrastructure of enterprise management.

4   Analysis of Existing Methods of Economic Evaluation of IT Outsourcing

4.1   Problems of Basic IT Outsourcing Performance Evaluation Models

The main source of economic effect for recipients of IT services when using IT outsourcing is to increase the level of efficiency of the industrial enterprise as a whole, as well as the possibility of freeing up appropriate organizational, financial and human resources, in order to develop new directions or concentrate existing efforts on existing areas that require increased attention from this enterprise.

A significant contribution to solving the problems of IT outsourcing, justification of the advantages and disadvantages that it provides, was made by various scientists [6, 8, 12, 15, 20, 24, 27, 29, 38, 40], etc.

Despite the significant contribution of various authors to the development of the concept of IT outsourcing, scientists have not investigated this topic deeply enough. Thus, in the scientific literature, there are isolated articles that only partially reveal the concept of IT outsourcing and its essence, which creates the problem of adapting the existing concept to the conditions of a particular national market.

The study of available publications has revealed a large number of tools aimed at effective relationship management in the field of IT outsourcing. At the same time, a specialist who wishes to put these tools into practice will face some problems, among which we propose to highlight the following:

- to date, there are no generalizing publications in which these tools would be systematized and described in detail (the only exception is the work of Bravar Jean-Louis and Morgan Robert "Effective Outsourcing: Understanding, Planning, and Leveraging Successful Outsourcing Relationships" [37], but its small volume did not allow to cover all the available algorithms), the relevant information is scattered over a large number of articles and monographs. As a result, outsourcing specialists are deprived of the opportunity to compare different tools and choose the most suitable one and are often forced to independently develop appropriate tools;
- usually, the proposed methods are general in nature and are not adapted to the specifics of a particular industry or a particular enterprise. Because of this, an enterprise or organization that wishes to apply IT outsourcing in its activities will face the need to independently adapt existing methods to a specific situation.

Despite the in-depth theoretical study of certain aspects of outsourcing in industrial enterprises, the issues of economic efficiency of its application have not yet found sufficient scientifically based reproduction in the scientific literature. In particular, the problem area is methodological approaches to deciding on the use of IT outsourcing in industrial enterprises of various forms of ownership and size and determining its effectiveness.

4.2   Analysis of Single-Criteria and Multi-criteria Models for Evaluating the Effectiveness of IT Outsourcing

At the same time, there is no single method for calculating the economic effect of outsourcing. Our studies allow us to conclude that at this time there are and are actively used mainly two criterion approaches to assessing the effectiveness of outsourcing: single-criteria and multi-criteria [42].

In the case of a decision using one efficiency criterion, a certain financial indicator (indicator) is most often used, which indicates possible savings due to the use of IT outsourcing. One of the possible options for applying a single-criteria approach is Boltava I.D. [43]. At the same time, to determine the value of the effect of the use of IT services outsourcing, Boltava I.D. recommends the use of the following dependence (1):

$$E_{ITa} = \left[(B_1 - B_2)/D_\Sigma \right],$$ (1)

where E_{ITa} is the value of economic efficiency of IT outsourcing in relative units, %; B_1 is the cost of an industrial enterprise for IT functions before outsourcing, UAH; B_2—expenses of the enterprise after using IT outsourcing, UAH; D_Σ is the total revenue of the enterprise from its production and commercial activities, UAH.

As can be seen from the formula (1), the indicator E_{IT}, the use of which is recommended by Boltava I.D., has a rather limited value, as it shows the savings due to the use of IT outsourcing services as a certain share in the total income of the enterprise [43]. At the same time, a somewhat strange pattern can be traced: the lower the income of the enterprise, the larger the size of the indicator of the economic efficiency of outsourcing. Note that the amount of income of the enterprise is almost directly related to specific results of IT outsourcing.

The presence of this deficiency in model (1) is also indicated in his study by Ignatiev A.V. [44], who recommends the use of dependence (2), which to some extent eliminates the lack of model (1).

$$E_{ITa} = \sum_{i=1}^{i=n} \left[(B_{ITs} - B_{ITa})/1 + (\alpha/100\%)^i \right],$$ (2)

where E_{ITa} is the economic effect of the practical use of outsourcing in monetary units, UAH; n is the duration of outsourcing, years; B_{ITs} is the estimated costs for the independent performance by the enterprise of the IT function in the first-time interval, UAH; B_{ITa} is the cost of performing an IT function by an outsourcer in the i-th time interval, UAH; α is the discount rate valid at the time of the calculations, %.

Obviously, the formula (2), although it makes it possible to more accurately (taking into account the change in the value of money) to calculate the savings obtained through the use of IT outsourcing, nevertheless, does not introduce anything meaningfully new in understanding the structure of the economic effect of the use of outsourcing, that is, a step forward compared to formula (1) is not. The correctness of this conclusion is also drawn in his study by Salman [24].

Subsequently, Ignatiev [44] offers an improved model for assessing the economic efficiency of using IT outsourcing in an industrial enterprise (3), which to some extent is the development of the model (1).

$$E_{ITa} = \sum_{i=1}^{i=n} \left[(B_{ITs} - B_{ITa})/1 + (\alpha/100\%)^i \right] - OT_c + OT_i,$$ (3)

where OT_c is the one-time costs associated with the transition of the enterprise to IT outsourcing; OT_i is the one-time income of the enterprise after the transition to obtaining IT services from outsourcing.

Model (3) in comparison with formula (2) has several advantages, since it takes into account in its composition the one-time costs of OT_c associated with the transition to outsourcing (for example, the introduction of control over the activities of an outsourcer, the restructuring of activities in the field of IT technologies, etc.), as well as one-time income OT_i associated with the transition to outsourcing (for example, the elimination of some services or departments in the field of IT technologies, etc.). These components allow you to reflect the fact that the use of outsourcing entails a change in the structure of the company—in connection with the transfer of the process to an external operator, the preservation of that internal IT unit that previously performed information functions is impractical.

In our opinion, single-criteria methods for evaluating the effectiveness of IT outsourcing approach the process of making a final decision unilaterally and do not take into account many objective and subjective factors (for example, the dissemination of confidential information or becoming dependent on an outsourcer, etc.).

In scientific studies on the application of a single-criteria approach, a wide range of formulas is given that can be used for this purpose. However, there are no recommendations for choosing the optimal formula for a particular situation, as well as an analysis of the advantages and disadvantages of these formulas. Apparently, the authors of these models propose them, without performing a preliminary substantive analysis of the structure of the economic effect of the application of outsourcing, easier, without identifying the components that should be included in the relevant formulas. As a result, the composition of key effect-creating factors is more or less random in nature and reflects not the content of the effect of outsourcing, but rather the intuitive ideas of the author of a particular model.

A significant drawback of the formulas (1–3) is that they are limited to considering only the financial component of the IT outsourcing effect, which greatly simplifies the situation and does not give the customer a complete picture of all the positive and negative consequences of its transition to IT outsourcing [24].

As a result of the presence of significant shortcomings in the use of single-criteria methods of adoption in the practice of industrial enterprises, they began to use multi-criteria methods for assessing the effectiveness of IT outsourcing. This group of methods allows us to assess the integral effect of the use of IT outsourcing. The multicriteria methodical approach is based on the convolution of a certain set of partial indicators that evaluate different types of economic effects from the use of IT outsourcing into a single complex indicator.

The general view of the approach to evaluating the effectiveness of IT outsourcing E_{IT} on the basis of a multi-criteria model can be reduced to the following dependence (see Eq. 4):

$$E_{IT} = \sum \gamma_i (F_{2i} - F_{1i}), \qquad (4)$$

where γ is the importance of the i-th factor influencing the effectiveness of IT outsourcing in the overall assessment; F_{2i} is the value of the i-th factor in the overall assessment after the transition of an industrial enterprise to the use of IT outsourcing services; F_{1i} is the value of the i-th factor in the overall assessment before the transition of an industrial enterprise to the use of IT outsourcing services.

Since disparate indicators are compared in formula (4), they are preliminarily reduced to a single dimensionless scale. Accordingly, the effect calculated by this method is also a dimensionless value, which does not allow to evaluate the economic component of the effect quantitatively.

A single list of partial indicators, as well as a single convolution procedure, has not yet been proposed by the researchers. The list of criteria and the economic and managerial meaning of each of them are determined by the enterprise independently. In existing scientific studies, different sets of partial indicators are given (as a rule, they do not fully coincide with each other, although, of course, they partially intersect), as well as various options for convolution procedures (arithmetic mean weighted, geometric mean weighted, etc.). However, if the existence of different sets of indicators can be explained by the specifics of the manifestation of the effect of outsourcing in different industries and for different processes [19, 37, 43], then for convolution procedures such a justification is usually absent.

Based on these considerations, it should be argued that multicriterial methods of this type also have disadvantages. In particular, the deterioration of some criteria can be compensated by the best values of others, that is, in such a generalizing indicator of the effect (4), negative effects in some areas can be compensated by positive results in other areas. However in the final generalizing indicator, information on the magnitude of partial positive and negative effects is absent.

As a result, the final value may indicate a positive effect, despite the fact that the most important criteria for the customer (the reasons why he wants to outsource the function) will be worse than with internal performance [33, 44]. For this reason, in our opinion, it would be fairer, along with the assessment of the integral effect, to also separately evaluate the manifestations of the effect of IT outsourcing for all significant for the person, decision-making, and directions (factors).

The researcher of the effect of the use of outsourcing Kotlyarov I.D. generally supports the multi-criteria method but recommends getting rid of its shortcomings by determining not the overall economic effect, but to calculate the effect for each effect-creating factor [32, 33]. This can be either a direct economic effect (directly related to the performance of a specific function or business process) or indirect (not directly related to this function) [33]. This statement can be formulated somewhat differently: the direct effect refers directly to the process or function (or their combination) that is outsourced, while the indirect effect applies to the entire enterprise as a whole. Of course, this division cannot always be carried out clearly, but it is important to keep it in mind. In turn, direct and indirect effects can be decomposed into separate components, which is presented by us in Table 3.

The economic effect of IT outsourcing using this technique is calculated by summing up the effect for all factors (the effect for each factor is calculated by subtracting from column 3 of column 2 if the corresponding factor is a source of

Table 3 Types of economic effect from the use of IT outsourcing [32, 45]

Factors of economic effect	The composition of expenses (income) when performing the IT function on the customer's own forces	The composition of expenses (income) when performing an IT function by an outsourcer
Study of factors of direct economic effect		
Transaction costs	The cost of preparing, organizing, managing and completing the process on its own	– The cost of finding and selecting an outsourcer, concluding a contract, managing and controlling relations with an outsourcer; – The release of excess personnel, the organization of the sale of surplus assets; – Possible losses associated with the inability of the outsourcer to perform his functions
Transformation costs	The current costs of the process on their own	Outsourcing service fee
Investment streams	Acquisition of assets necessary to complete the process on your own	– The cost of forming specific acts-leads necessary for interaction with the outsourcer; – Income from the sale of assets that are not needed in connection with outsourcing
Income from the increase in quality and/or sales of products	Proceeds from sales of the final product when performing the process on their own	Income from the sale of additional products and/or from the increase in product quality, obtained through outsourcing
Investigation of factors of indirect economic effect		
Taxes	The amount of taxes (on property, on the wage fund, etc.) and other mandatory payments	The amount of taxes paid by the customer when performing the process on their own
Increase in asset utilization efficiency	Proceeds from assets when performing the process on their own	Increase in revenue from the same amount of assets when transferring a function to outsourcing (for example, due to the lease of vacant premises)
Change in indirect transaction costs	The cost of managing a company when performing a process on its own	The cost of managing the company when performing the process by the outsourcer

income, and deduction from column 2 of column 3, if the factor is a source of costs; then the results of each line are summed up). The direct economic effect is an effect associated with the direct implementation of the IT function. By adding an indirect economic effect to it, you can calculate the total economic effect of using IT outsourcing for the customer. Of course, depending on the timing of IT outsourcing, the corresponding values can be discounted. When carrying out calculations, it is necessary to take into account the probability of achieving the predicted results (in no case should they be taken as reliable).

The authors of the methodical approach that is analyzed deliberately do not offer a detailed methodology for calculating the economic effect of IT outsourcing. In our opinion, this contradicts the statement made by the authors of the methodology that an important problem in calculating the effect of IT outsourcing using a single-criteria methodology is the lack of a single list of components of the economic effect. Details of the effect are given in Table 3 is actually such an enlarged list. It shows only those areas where the economic effect of using IT outsourcing is manifested. Detailing this list, identifying specific indicators for assessing the economic effect listed in Table 3 areas, in the future should be carried out by specialists from a specific sector of the economy. Obviously, such indicators will differ for an oil production, engineering, or energy enterprise. Moreover, the list of specific indicators may differ in the study of different enterprises within the same industry and even for one enterprise when assessing the feasibility of outsourcing different functions. All this presents quite significant difficulties in the practical use of these proposals.

In addition, the analyzed methodology, in our opinion, can be recommended with certain precautions and additions only for use in large enterprises and in the case of analyzing the feasibility of implementing large outsourcing projects that require large expenditures of resources.

One of the researchers of economic efficiency of outsourcing Lipatnikov V.S., in our opinion, rightly points out the need to take into account the uncertainty factor, which takes place in the relationship between the customer and the outsourcer [45]. To do this, the researcher considers it appropriate to use the mathematical expectation of the economic effect of $EV_{EE,}$ the value of which is calculated by the following formula (see Eq. 5):

$$EV_{EE} = \sum_{i=1}^{n} W_i \left(\Delta B_{EE}^i + \Delta I_{EE} - \Delta C_{EE} \right), \tag{5}$$

where n is the number of predictive options for cooperation with the outsourcer; W_i is the probability of implementation of the i-th option (determined by expert means); ΔB_{EE}^i is the reducing the cost of performing a business process or IT function when implementing the i-th option; ΔI_{EE}—additional income from the use of outsourcing in the implementation of the first option; ΔC_{EE}—cost increase associated with cooperation with the outsourcer, in the implementation of the i-th option.

Model (5), according to V.S. Lipatnikov, allows us to take into account the uncertainty of the external environment. For example, as one of the options for cooperation with a foreign outsourcer, a situation of a significant drop in the cost of IT services can be considered in combination with a decrease in the national currency exchange rate of the country in which the IT enterprise operates [45]. Obviously, in this case we can hardly talk about additional income, while the cost of IT outsourcing services (in national currency) will increase significantly. Another option for implementing cooperation may be early termination of the contract, due to the introduction of sanctions against the country, and in this case, outsourcing will not bring any additional benefits, and at the same time the customer suffers losses associated with already made investments in projects that are planned for practical implementation.

At the same time, the model (5) does not allow to take into account the existing or foreseeable risks of cooperation with the outsourcer associated with unfair, poor-quality or unauthorized behavior of the outsourcer, which, in our opinion, must be included in it.

The amount of losses of the customer enterprise L_{ent}^i due to the unscrupulous behavior of the outsourcer in the i-th version of cooperation can be calculated using the following formula (see Eq. 6):

$$L_{ent}^i = \sum_{j=1}^m W_j L_{ent}^j,$$ (6)

where m is the number of situations (within the framework of the i-th option of cooperation with the outsourcer), which can lead to losses for the customer enterprise.

Note that these situations can be superimposed on each other. For example, an oil and gas company consider the environmental risk from an oil spill during drilling operations and the risk of a decrease in the price of oil. Obviously, these situations can be implemented both separately and in aggregate, and, thus, we are not talking about two, but about three specific situations of economic losses for the customer of IT services, each of which is characterized by its own value of possible losses and probability of occurrence; L_{ent}^j is the amount of economic losses at the onset of the j-th situation. It is important to note that this is not about the absolute amount of losses, but about how much these losses exceed the losses that could occur in the case of performing a business process (or IT function) on the own forces of the customer enterprise. This is due to the fact that the methods of analyzing the economic effect of the use of outsourcing involve the calculation of not a complete, but an additional (compared to the independent implementation of the IT process by the customer) economic effect. Thus, it is possible that in some cases the value of L_{ent}^j will be negative (which can be explained by the higher level of technological competencies of the outsourcer); W_j is the probability of the occurrence of the j-th situation.

At the same time, the use of the model (5, 6) requires significant costs from the customer enterprise to develop predictive scenarios for the cooperation of the consumer of IT services with the outsourcer, assess the probability of practical implementation of these scenarios, as well as determine the integrity and professionalism of the outsourcer. All these problems are quite difficult to solve, the accuracy and reliability of their results, in our opinion, is small, which ultimately gives us grounds to determine the low level of utility of the model determination of the effectiveness of outsourcing, which is analyzed.

Another conditional differentiation of outsourcing decision-making methods existing at this time in industrial enterprises: economic methods and graphic methods.

The group of economic (numerical) methods is based on the calculation of costs for the internal performance of the function and their comparison with the costs using IT outsourcing. Almost all the methodological approaches we have considered to the economic justification of the effectiveness (expediency of use) of IT outsourcing can be attributed to the group of economic methods.

Graphic methods, in turn, can be divided into three groups: graphs, matrices, algorithms. As an example of graphical methods for shaping the effectiveness of IT outsourcing, consider the McKinsey model [46]. Each enterprise is a set of business units (or business processes) that are practically ready for outsourcing, so one of the most rational models for deciding the feasibility of performing these operations and implementing an outsourcing scheme is the McKinsey model. The methodical essence of the model is reduced to two-stage graphical modeling.

At the first stage, the McKinsey outsourcing efficiency field is formed (Fig. 3), on the second—the McKinsey efficiency matrix (Fig. 4).

The essence of the formation of the efficiency field (Fig. 4) consists in the distribution of business units or divisions in the form of points according to the coordinate

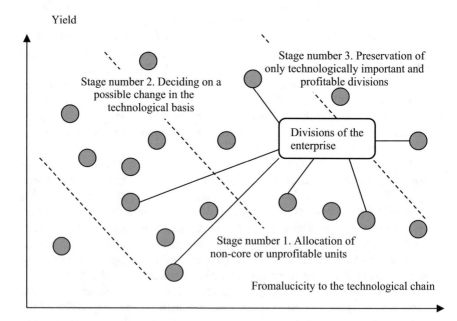

Fig. 3 Formation of the McKinsey outsourcing efficiency field [46]

The power of business

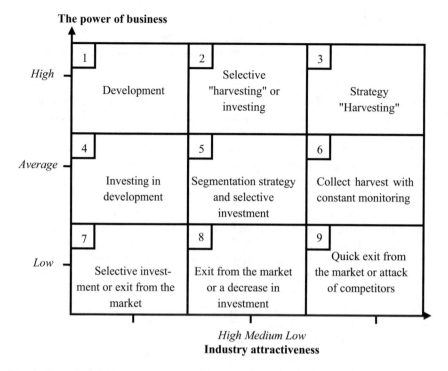

Fig. 4 Formation of the McKinsey outsourcing efficiency matrix [46]

system, where the X axis is involvement in the technological chain (the degree of involvement is estimated by experts), and the Y axis is profitability (estimated in terms of net costs and contribution to the total profit of an industrial enterprise). On the basis of the efficiency field, a certain ranking of divisions and functions of the enterprise is carried out and as a result of these actions, a matrix is built the effectiveness of outsourcing McKinsey (Fig. 4).

The McKinsey matrix is based on two parameters: the strength of the business and the attractiveness of the industry. These parameters are generalizing values of the following factors:

- business strength: relative size, growth, market share, position, comparative profitability, net income, technological condition, image of the enterprise, management and employees (personnel);
- industry attractiveness: absolute size, market growth, market breadth, pricing, competition structure, industry rate of return, social role, environmental impact and legal constraints.

In the coordinate system "the strength of business—the attractiveness of the industry" there are nine quadrants (Fig. 4). The processes trapped in the three quadrants on the upper left side (quadrants 1–3) are promising and profitable. Three medium quadrants (4–6) are less attractive and require careful investment. One of the solutions for three quadrants in the lower right corner (7–9) is to sell them (outsourcing).

If there are significant advantages when using the McKinsey IT outsourcing effectiveness model, which takes into account important aspects of the functioning and evaluation of the effectiveness of the enterprise, its disadvantages should be identified. Firstly, the model has a purely expert idea of the effectiveness of the functioning of individual functions and divisions of the enterprise, which indicates the presence of a significant impact on the adoption of managerial subjective factors. Secondly, the model reproduces the current state of production and commercial activity of the enterprise and does not take into account the prospects for the development of both the enterprise as a whole and its individual divisions. Thirdly, the lack of numerical (quantitative) characteristics may cast doubt on the objectivity of the distribution of IT functions and IT departments of an industrial enterprise according to the quadrants of the McKinsey matrix (Fig. 4). Fourth, in addition to two evaluation criteria, such as "involvement in the technological chain" and "profitability" (Fig. 3), there are other aspects that force an industrial enterprise to leave some divisions, functions and processes in its structure.

Practically on similar principles, graphic models for determining the effectiveness of IT outsourcing by Boston Consulting Group are built (the model is built in a coordinate system: strategic business goals—the effectiveness of business processes) [44], Price water house coopers (coordinate system: competitiveness—strategic significance) [46], IBS model (coordinate system: quality—cost) [47], model Khlebnikov D.V. (strategic importance—comparison with the market) [48], model of Kuryanovich V.R. (strategic importance—level of business processes) [49], model of Kurbanov A.Kh. (level of system efficiency—index of outsourcing expediency) [50], model of Moiseeva N.K., Maliutina O.N. and Moskvina I.A. (strategic importance—level of competence in business processes) [51], etc.

Existing graphic methods are distinguished by clarity and the presence of both economic and non-economic components. But they often do not allow to take into account the specifics of the industry, to obtain clear results and give an unequivocal answer to the question of the feasibility of outsourcing in a particular situation.

To overcome the imperfection of the approaches under consideration, it is necessary to include a strategic aspect in the economic group. This is due to the fact that economic methods only allow you to compare the costs before and after the implementation of outsourcing, but do not take into account the importance of the work/competence/business process.

Well-known methods for assessing the effectiveness of outsourcing provide for standard indicators: reducing costs, improving product quality, increasing the rhythm of the organization, etc. In most cases, the calculations take into account the interests of the parent company-customer, which transfers individual functions to a third-party performer-outsourcer. Taking into account the impact of IT outsourcing risks can lead to the formation of a multidirectional effect among participants. Therefore, it is recommended to evaluate the effectiveness of outsourcing taking into account the effect that is formed in all participants in the outsourcing process. Possible negative consequences of IT outsourcing will be covered by the positive results of others Participants. Although the specific economic results from the use of IT outsourcing will differ for enterprises in different industries and for different functions outsourced, however, those areas in which these results will be detected will be similar for different enterprises. This allows you to compile a general list of such areas in order to create a basis for them to search for specific forms of manifestation of the beneficial effect of IT outsourcing, taking into account the peculiarities of individual industries, enterprises and functions (processes) transferred to IT outsourcing.

When deciding on the use of IT outsourcing, it is necessary to take into account not only the magnitude of the integral effect, but also the value of partial types of effects (that is, in fact, to combine single-criteria and multi-criteria approaches). To do this, it is advisable to use not only quantitative, but also a qualitative convolution of partial indicators, since it stores information about the partial values of the effect.

The conducted research allows us to conclude that the use of both numerical and graphical methods for evaluating the effectiveness of IT outsourcing provides certain positive results for the top management of the enterprise and increases the accuracy and objectivity of management decisions. At the same time, these solutions would be more justified when using integrated methodology to evaluate the effectiveness of IT outsourcing, as if combining both quantitative and graphic elements.

Thus, the need to optimize the management strategy as the main direction of improving the efficiency of the organization's functioning determines the high degree of relevance of IT outsourcing and necessitates further comprehensive scientific substantiation of the methods of its implementation.

5   Development of Methods of Economic Evaluation of IT Outsourcing in Industrial Enterprises

5.1   Justification of Directions for the Effective Use of IT Outsourcing by Industrial Enterprises

Domestic industrial enterprises at this time in a certain way lack an idea of the essence of IT outsourcing and the possibilities of its use in their production and commercial activities. In our opinion, in order for an enterprise to understand the need to create its own IT department or to provide its functions to an outsourcer, it is necessary to investigate the tasks and functions of the IT department in each of the four following areas:

- strategic expediency;
- quality of operational capability (ability);
- economic (primarily financial) benefits;
- available opportunities to improve IT operations in your enterprise.

Studies have shown that outsourcing has one very important feature: it is not identical to the sub-assignment, since it makes a completely free choice to fulfill its obligations (in our case, the provision of IT services) to the enterprise-customer of its services. In contrast to sub-assignments, outsourcer services are an element of strategic partnership with the customer of IT services, while each of the parties tries to achieve maximum commercial success by adapting their business processes to the production needs of their partner [52].

An industrial enterprise that solves the problem of transferring IIT or their parts to IT outsourcing, one way or another, must find the right solutions to the following problems:

- will IT outsourcing help solve the tasks?
- how and using what methodological approaches to assess the benefits that will then be obtained?
- are the benefits comparable to the costs of IT outsourcing?

ANCOR personnel holding conducted a study of the effectiveness of outsourcing services in various areas in Ukraine. The study was conducted among the top management of Ukrainian and foreign enterprises that carry out their production and commercial activities in our country [41]. As a result, it was found that 41.2% of surveyed enterprises consider saving resources of the customer enterprise to be the main source of efficiency from the use of outsourcing services, 37.6% of respondents prefer optimization and acceleration of business processes, 32.9% are sure that their business activities have become more efficient, 14.1% noted improvement of control processes and improving the quality of outsourcing processes, 14.3% are sure that thanks to outsourcing they have the opportunity to make decisions based on the latest innovative achievements of their field of production activities (Fig. 5).

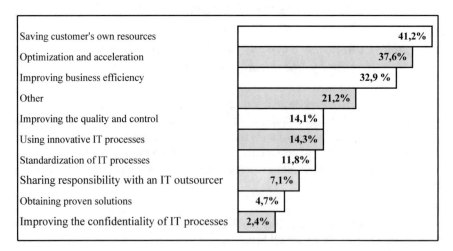

Fig. 5 Rating of directions of efficiency of outsourcing by industrial enterprises [41]

For Ukrainian enterprises, obtaining certain efficiency factors (competitive advantages) through the use of outsourcing is extremely important, based on a sufficiently high level of competition, a significant exaggeration of supply over demand in the target market, variability of economic and social conditions of production and commercial activity, high speed of innovative changes in various fields of activity and practically no opportunities to receive state support.

5.2 Graph-Analytical Model for Evaluating the Effectiveness of IT Outsourcing

It should be noted that, despite the rather optimistic ideas of many researchers and theorists in the field of bringing information services to IT outsourcing, its effectiveness and rationality of use is not always unambiguous. There are a number of reasons why enterprises do not dare to outsource part of their business processes. This implies a number of disadvantages associated with the risks of outsourcing [53]:

- possible increase in costs in case of insufficient study of the capabilities of your own enterprise and outsourcing too many functions and processes;
- the possibility of bankruptcy of an outsourcing company;
- outsourcing may limit the flexibility of company management;
- lack of legislative definition and regulation of outsourcing;
- outflow of confidential information;
- it is possible to reduce the liability of persons to whom business processes are transferred on the terms of outsourcing;
- the difficulty of controlling technological business processes;
- outsourcing may limit the technological flexibility of the enterprise.

When deciding on outsourcing, it is necessary to analyze financial and organizational costs, without letting go of the analysis zone the risks arising from the organization of IT services outsourcing. The lack of a full-fledged economic analysis in making this decision can, instead of benefiting, cause significant losses to the user of outsourcing services. Outsourcing service providers (outsourcers) usually use techniques to calculate the effectiveness of outsourcing, allowing to show financial advantages and very often deriving from the field of analysis possible losses from outsourcing risks that often arise in their commercial activities [54].

The preliminary economic justification for the use of IT outsourcing is associated with the definition of criteria for choosing the effectiveness of outsourcing—assessing the costs of implementing an IT function on the company's own forces or outsourcing it. The decision to outsource IT functions should be preceded by an economic analysis of the effectiveness of such a solution, which, in our opinion, can be carried out by comparing the costs of implementing this function when outsourcing it with the costs of its implementation by the customer enterprise, that is, on its own without involving the services of an outsourcer [54].

Investigating the effectiveness of outsourcing, N.I. Chukhrai focuses the attention of production top management on the fact that the function, process, work or service transferred to the outsourcer should bring really tangible profit [52]. Therefore, the decision to outsource IT functions urgently requires a comparison of costs when carrying out these works on the company's own and by an outsourcer. At the same time, it is necessary to take into account the thresholds of profitability and sales volumes of products, which will form the appropriate boundary when making a management decision by an enterprise to perform IT operations by an external enterprise (outsourcer). It is proposed to determine the critical limit (the upper break-even threshold of IT outsourcing) of making a management decision using the concept of the volume of outsourcing sales of an IT function—the $N_{out_{max}}^{crit}$ quantitative value of which can be found by comparing the total costs of an enterprise to perform an IT function (IT operations, IT services) on its own TC_{ent} and similar costs when performing the same work by a TC_{out} outsource using dependency (7). At the same time, the total costs of the enterprise and the full costs of the outsourcer include fixed costs—SF and variable costs V (see Eq. 7):

$$SF_{out} + V_{out} N_{out_{max}}^{crit} = SF_{ent} + V_{ent} N_{out_{max}}^{crit}, \qquad (7)$$

where SF_{ent}, SF_{out} is the fixed costs of the enterprise to perform the IT function (IT operations, IT services), respectively, on its own and using the services of an outsourcer; V_{ent}, V_{out} is the variable costs of the enterprise to perform IT functions (IT operations, IT services), respectively, on their own and using outsourced services.

From the formula (7) it is possible to determine the quantitative value of the critical limit (break-even threshold of IT outsourcing) of making a management decision using the concept of the volume of outsourcing sales of the IT function—$N_{out_{max}}^{crit}$ (8):

$$N_{out_{max}}^{crit} = (SF_{ent} - SF_{out})/(V_{out} - V_{ent}). \qquad (8)$$

With the volume of work N (for example, the number of months of support for the corporate website of the enterprise), exceeding the value of the critical limit for the outsourcer ($N > N_{out_{max}}^{crit}$), the outsourcing services are inefficient for the enterprise, so its own costs for creation and maintenance will be less.

At the same time, not during the entire interval of the volume of work from 0 to $N_{out_{max}}^{crit}$ the outsourcing service will be effective for the enterprise. There is a certain critical limit (lower break-even threshold of IT outsourcing) $N_{out_{max}}^{crit}$, less than which outsourcer services will be unprofitable for the $N_{out_{max}}^{crit}$ customer enterprise. The value $N_{out_{max}}^{crit}$ is recommended to be found from the following equation (see Eq. 9):

$$SF_{out} + V_{out} N_{out_{min}}^{crit} = I_1 N_{out_{min}}^{crit}, \qquad (9)$$

where I_1 is the level of income received by the enterprise when performing a unit of work on the use of this IIT (for example, income for one month of using the corporate website of the customer enterprise).

From the formula (9) we find the value $N^{crit}_{out_{max}}$ (see Eq. 10):

$$N^{crit}_{out_{min}} = SF_{ent}/(I_1 - V_{out})$$ (10)

To determine the effectiveness of the use of IT outsourcing, a number of researchers recommend to determine the use of the graphical method [52]. Taking into account these recommendations, we propose to combine analytical calculations—models (7) and (8) with their graphical interpretation. This proposal is implemented in Fig. 6.

A graphic comparative analysis of the effectiveness of solutions for outsourcing and practical implementation of IT functions by the company's own forces, presented in Fig. 6, allows us to draw a number of, in our opinion, sufficiently important conclusions and generalizations.

Firstly, the ordering company should bear in mind that not always every management decision in the field of IT services is effective. To achieve at least a break-even zone for the use of a particular IT service, it is necessary to assess the amount of need for it. With a small amount of need, it is very difficult to ensure the effective operation of information equipment, the premises where it is located, to load the maintenance personnel with the corresponding tasks. From these Fig. 6 it follows that there is a certain zone of unprofitability, within which the creation and use on its own is unprofitable. At the same time, the total costs of own production and use of the TC_{ent} information service exceed the company's income from its use IIT, that is, $TC_{ent} > I_{IT}$. At the same time, one should not abandon this IIT at the same time. On the graph Fig. 6 clearly shows the relevant zone in which the creation and use of this IIT can be effective in outsourcing it to an external enterprise (outsourcer). This zone is located between the end of the unprofitability zone and the beginning of the zone of efficiency of its own production for the implementation of this procedure, a number of conditions must be met, the most important of which is to achieve such a level of income from the consumption of this technology, which would ensure its entry into the zone of IT outsourcing efficiency, which is characterized by two inequalities:

- the total costs of outsourcing do not exceed the total costs of the enterprise for the creation and use of this IIT, i.e. ($TC_{out} < TC_{ent}$);
- the total costs of outsourcing not given IIT do not exceed the level of income from its use in the outsourcing customer enterprise, i.e. ($TC_{out} < I_{IT}$).

Secondly, in the presence of unprofitability in the field of independent creation and use by an industrial enterprise of IIT, it should be borne in mind that the decision to abandon its own provision of this service and transfer it to outsourcing will not always be effective. We again pay attention to the amount of need for this IIT. In Fig. 6 clearly shows the zone of the size of the needs under which the creation and use of this IIT is unprofitable not only for the company's own capabilities but also when transferring this technology to outsourcing to an external enterprise. We can state the fact that the size of the unprofitability zone of IT outsourcing is not constant. Its size depends on the economic capabilities of the outsourcing enterprise. But when using

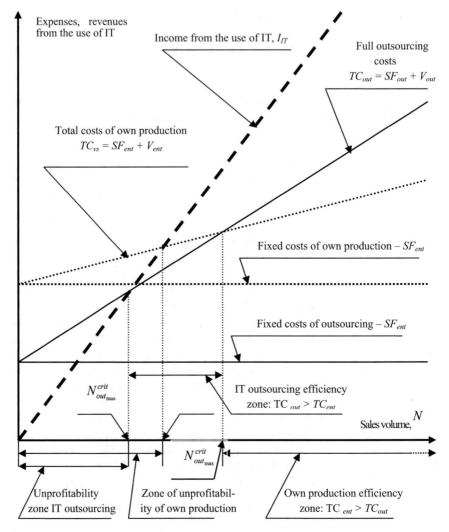

Fig. 6 Graphical comparative analysis of the effectiveness of IT outsourcing solutions and implementation of IT function by the enterprise's own forces

any outsourcer with free IT services, which he will provide, will not. Therefore, it is inevitable on the graph of Fig. 6 there will be such a zone within which the total costs of the IT outsourcer will exceed the company's income from its use of I_{IT}, that is, $TC_{out} > I_{IT}$. In the presence of such conditions, the enterprise should simply abandon this kind of IIT due to the presence of its absolute unprofitability. Note that for resuscitation (creating opportunities for effective use) of such technology, larger volumes of production needs should be created, where its use will be appropriate. Accordingly, the volume of income from the consumption of technology in this enterprise will also increase, which will change the inequality of $TC_{out} > I_{IT}$ for a more acceptable case for this case: $TC_{out} < I_{IT}$.

Thirdly, a graphical representation of the effectiveness of solutions for outsourcing or the implementation of the IT function by the enterprise's own forces allows us to identify a zone in which the transfer of IIT to outsourcing is ineffective for this enterprise. This zone is characterized by the presence of such amounts of the total costs of the enterprise for the creation and use of IIT and the outsourcer for similar needs, in which the following inequality is performed: $TC_{out} > TC_{ent}$, that is, the total costs of a TC ent enterprise are less than its similar costs to outsourcing this *TC* out technology. At the same time, there are certain limitations: both the amount of total costs of the enterprise TC_{ent}, and the number of costs for outsourcing this IT TC_{out} should be less than the amount of income from technology consumption at this enterprise I_{IT}, that is: $I_{IT} > TC_{out}$ and $I_{IT} > TC_{ent}$. Note that in this situation for an industrial enterprise outsourcing of this IIT is not unprofitable, just the level of economic effect in its creation and use on its own is more effective. In principle, an enterprise can leave the use of outsourcing this IIT if it has more profitable and more efficient areas of investment that can be released when accepting the option of outsourcing IIT. In this case, it is necessary to carry out additional economic analysis and make appropriate calculations that are outside our study.

The proposed methodological approach allows us to evaluate the economic efficiency of IT outsourcing to a greater extent from the point of view of the economic justification of the management decision to accept the outsourcing services or perform them on their own. Such a task is faced by the enterprise at the very first stage of outsourcing relations and is a sufficiently important and rather complex economic task. At the same time, with the development of outsourcing relations, the customer company will be interested in increasing their economic efficiency. To do this, it is necessary to search for weaknesses and develop guidelines for their elimination. Graph-analytical method for evaluating the effectiveness of outsourcing (Fig. 6) does not allow this. Its algorithm involves enlarged economic indicators (total costs, fixed costs, variable costs), the structure of which is not provided for in the mechanism of using this method. Based on this, we consider it appropriate to develop methodological recommendations for detailing these indicators in order to identify their components, to which the customer company needs to pay special attention.

5.3 The Practice of Using the Model of Economic Evaluation of IT Outsourcing in Industrial Enterprises

Our research at industrial enterprises of the Kharkiv industrial region allowed us to identify certain initial data for a visual presentation of the practice of using the developed proposals, the essence of which is reproduced in Fig. 6.

As an object for studying the effectiveness of IT outsourcing services, a service was chosen to create and maintain a corporate website of an industrial enterprise.

Creating a corporate website is the most important stage in the marketing of any enterprise. The Internet has long overtaken such advertising channels as radio, print media, and outdoor advertising in terms of audience coverage and quality. A properly developed corporate website will help strengthen the positive image of the enterprise, establish communication with the consumer, and become a confident channel for attracting consumers.

The cost of creating a site is any cost associated with the preparation of site content both on its own and by third-party organizations, with registration and payment of a domain name, payment for hosting, updating a computer park, training employees to work with the site, etc., as well as with paying for the direct production of the site. When assessing the cost of the site, it is necessary to take into account a sufficiently large number of indicators, including indicators of its future work. The more objective criteria amenable to numerical expression are taken into account when assessing the cost of creating a site, the more accurate the results of the assessment. The calculation of the number of costs for creating a website for a particular business should be carried out taking into account many factors of the micro and macro environment: the size of this market segment, the degree of its saturation with specific goods and services, the level of competition among sites on this topic, the degree of competitiveness of the goods and services themselves, etc.

If you know the average number of site visitors per day or per year and the average maximum allowable cost of attracting a site buyer, then you can set the estimated range of the real cost of the site. Figure 7 presented the cost of developing sites for a specific given project by various web studios in Ukraine [55, 56].

The cost of the site is a large item that must be calculated in advance. Conventionally, they can be divided into two parts: the creation (development) of the site (fixed costs) and the cost of its maintenance (variable costs, depending on the time of use of the site). The study of the practice of creating sites of industrial enterprises [56]

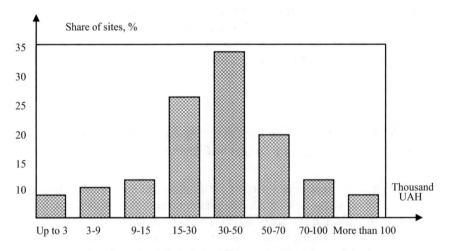

Fig. 7 The cost of creating enterprise sites by outsourcers in Ukraine [55, 56]

allows you to form models for determining costs, which have the following form (see Eq. 11):

$$B_{site} = \sum_{i=1}^{n} B_i^{prep} + \sum_{j=1}^{m} B_j^{dev} + \sum_{k=1}^{p} B_k^{exp} + \sum_{f=1}^{s} B_f^{othn}, \tag{11}$$

where B_{site} is the total costs (preparation, creation, operation, etc.) on the site of the enterprise; B_i^{prep} is the preliminary expenses of the i-th type for discussion and consultations on the idea of the site, development of a business plan for creating a site, feasibility study of the feasibility of its creation, registration and renewal of ownership of a second-level domain, preparation of site content, etc.; n is the number of types of work *of* the i-th type; B_j^{dev} is the costs of the j-th type on the creation of the site (employee salaries, design, content, etc.); m is the number of types of work of the j-th type; B_k^{exp} is the cost of the k-th type of site operation (hosting, site support, redesign, optimization, site promotion, etc.); p is the number of types of work of the k-th type; B_f^{othn} is the other expenses of the fth type for site maintenance (overhead costs, rent, payment for consultations, stationery, payment for transport, telephone, Internet, etc.); s is the number of types of work of the f-th type.

Taking into account the model (11) and the actual state of costs at industrial enterprises of the Kharkiv industrial region, a study of the costs of these enterprises for the creation and maintenance of their corporate website on the Internet was conducted. Table 4 contains a fragment of this study, which reproduces the costs of creating and maintaining the site PJSC "Ukrelectromash" on its own.

With the use of the developed recommendations, actual cost data on the independent use of IIT at industrial enterprises was obtained and similar data were determined when the same technologies were handed over to IT outsourcing to various outsourcers. The analysis of the calculation results received from industrial enterprises makes it possible to draw a number of important conclusions for each of the studied enterprises.

1. It is usually significantly more difficult for an industrial enterprise to carry out all the preparatory work for the implementation of a specific IIT than an enterprise specialized in these works. This conclusion is confirmed by the fact that the level of fixed costs for IT outsourcers is significantly lower than that of enterprise-consumers of IT services.
2. Maintenance work at a specialized enterprise is usually of higher quality due to the high qualification of outsourcer specialists. While the same work at the enterprise-consumer of IT services is usually performed by existing specialists by expanding their area of work and adding new types of work to their official duties. This to a certain extent explains the fact that the level of variable costs of outsourcers is to a certain extent greater than when performing the same work by specialists of the enterprise-consumer of IT services.
3. The calculations made indicate that with short periods of consumption of IT services (up to 1… 1.5 years) it is advisable to perform them using IT outsourcing. With longer periods (more than 2 years)—it is more efficient for enterprises

Table 4 Expenses for the creation and maintenance of the site PJSC "Ukrelectromash" on its own (according to the enterprise)

№	Item of expenditure	Expenses, UAH
Fixed costs, UAH		
1	Discussion and consultations as conceived by the site	2500
2	Registration of the right to own a second-level domain	600
3	Preparation of site content	12,000
4	Site creation	55,000
5	Hosting	1200
6	Other expenses	8800
Total, SF_{ent}		**80,100**
Variable costs, UAH./moon		
1	Website promotion	2500
2	Adjusting the content of the site	800
3	Site support	1500
4	Site redesign	800
5	Search engine optimization of the site	750
5	Overhead costs, office equipment, consumables, payment for consultations, stationery, payment for the Internet, transport, telephone and so on	1100
6	Other expenses	500
Total, V_{ent1}		**7950**

to perform these works independently. But, in our opinion, at the same time, the company should agree that the quality of these works will not always be sufficiently high

4. Conducted research and calculations on the example of creating and maintaining a corporate website allowing us to form specific recommendations for each of the studied enterprises for the implementation of this type of IIT on their own or using one or another outsourcer. Moreover, extremely important in these recommendations is the period of consumption of IT services.

For example, for PJSC "Ukrelectromash" when performing this IT service, the most effective will be the conclusion of an outsourcing agreement with the company "Ukrfast IT-outsourcing". With an annual term for the provision of IT services, PJSC " Ukrelectromash" will receive an effect in the amount of 14.5 thousand UAH, and with a two-year period—3.1 thousand UAH. The maximum term of the transaction, according to the calculations made, is 27.26 months (with a longer period, it is more efficient for the enterprise to perform this IT service itself), and the minimum is 8.52 months (a shorter period does not ensure the payback of the funds invested by the outsourcer). The outsider "Pirise software company" is not recommended to include PJSC "Ukrelectromash" in the tender list, so its services at any time of the transaction will be unprofitable for this enterprise.

This methodical approach to assessing the effectiveness of outsourcing the management information system at real business enterprises is sufficiently universal. It takes into account the conceptual foundations of IT outsourcing on the one hand and allows you to adapt the very structure of performance assessment depending on the complexity and amount of costs for the technologies used in this business, the complexity of the organizational structure of the enterprise itself, the goals and scale of IT outsourcing.

6 Conclusion

6.1 Synopsis

Studies of the historical development of the definition of the economic category "technology" have shown that this term is important to consider taking into account the theoretical and methodological, scientific-technological, economic and intellectual component. The analysis emphasizes the presence of such a specific feature of the technology as its belonging to intellectual property, which are characterized by all intangible factors for the definition, evaluation, transfer, use and legal protection and protection.

The concept of "intellectual and information technologies" is introduced into scientific circulation, which means such technologies that combine the intellectual, scientific, technical, commercial and informational achievements of innovators, which are created on the basis of intellectual property, and their use provides the consumer with the innovation of its production activities.

A detailed scient metric analysis of the development of information technologies in the activities of Ukrainian industrial enterprises and the possibility of their outsourcing was carried out. The list, sequence, and content of stages that consistently "alienate" information technologies and their infrastructure support from the production processes of the main production and commercial activities of an industrial enterprise and contribute to the formation and development of IT outsourcing are substantiated.

It is determined that the main source of economic effect for the customer enterprise when using IT outsourcing is to increase the level of efficiency of the industrial enterprise as a whole, as well as the possibility of freeing up appropriate organizational, financial and human resources, in order to develop new directions or concentrate existing efforts on existing areas that require increased attention from this enterprise. The analysis of existing methods of economic IT outsourcing assessments. The main advantages and disadvantages of modern approaches to substantiating the economic advantages of IT outsourcing over the independent implementation of services in the field of IT are identified. It was concluded that the use of both numerical and graphical methods for evaluating the effectiveness of IT outsourcing provides certain positive results for the top management of the enterprise and increases the accuracy and objectivity of management decisions.

A methodical approach to the economic assessment of IT outsourcing has been developed by dividing all effect-forming factors into direct factors (related directly to the IT function that is outsourced) and indirect (related to the entire enterprise as a whole) economic effect, which eliminates the negative impact on the calculations of the effectiveness of IT outsourcing of factors of multidirectional action. Methodical recommendations for determining the economic efficiency of IT outsourcing using the graph-analytical method have been formed, which allows for a direct comparison of the effectiveness of IT outsourcing solutions and the practical implementation of the IT function by the customer's own forces.

6.2   Further Research

For developing countries at this time, outsourcing IIT would open up many opportunities that could be the impetus for the effective development of the national economy. Outsourcing of IIT provides the national economy with income in foreign currencies and contributes to the creation of additional jobs, and the formation of the middle class. In Ukraine, there is a positive dynamic in the development of IT outsourcing over the past 10 years. This country occupies the first position in terms of IIT outsourcing among the countries of Central and Eastern Europe. But it is too early to talk about competition with such world outsourcing leaders as India, China, or Malaysia because Ukraine does not fully realize its potential in this area. IT outsourcing should become a strategic direction for the development of the state and requires certain encouraging actions on the part of the domestic leadership. The directions of further research should be associated with the definition of trends and priorities for the development of IT outsourcing in the national economy during the political and economic crisis. The practical significance of such research is to determine the guidelines for the development of state policy in the field of outsourcing IIT. In addition, the obtained results can be used in further research in the field of IT outsourcing in the national economy and determining its competitive advantages compared to other countries that have a strong IT outsourcing sector. Such studies can become the basis for Western enterprises' decision-making on the choice of the state for the transfer of their intellectual and information technologies. by license or on the basis of outsourcing.

The proposed methodological approach allows to evaluation of the economic efficiency of IT outsourcing to a greater extent from the point of view of the economic justification of the management decision to accept services for IT outsourcers or perform them on their own. Such a task is faced by the enterprise at the very first stage of outsourcing relations and is a sufficiently important and rather complex economic task. At the same time, with the development of outsourcing relations, the customer company will be interested in increasing their economic efficiency. To do this, it is necessary to search for weaknesses and develop guidelines for their elimination. Graph-analytical method for evaluating the effectiveness of outsourcing (Fig. 6) does not allow this. Its algorithm involves enlarged economic indicators (total costs, fixed costs, variable costs), the structure of which is not provided for in the mechanism of using this method. Based on this, we consider as a direction of further development and research to develop methodological recommendations for detailing these indicators in order to identify their components, to which the customer company needs to pay special attention.

Being built on the basis of consideration of outsourcing as a process, this approach has limitations that determine the possibility of its application to evaluate the effectiveness of IT outsourcing only when outsourcing is not yet implemented in practice. The last remark determines the direction of further research on the effectiveness of IT outsourcing, the results of which should include not only initial work in the field of IT outsourcing, but also an economic evaluation in the process of its use at an industrial enterprise.

References

1. Zuev, D., Kalistratov, A., Zuev, A.: Machine learning in IT service management. Procedia Comput. Sci. **145**, 675–679 (2018). https://doi.org/10.1016/j.procs.2018.11.063
2. Pererva, P., Nazarenko, S., Maistro, R., Danko, T., Doronina, M., Sokolova, L.: The formation of economic and marketing prospects for the development of the market of information services. East.-Eur. J. Enterp. Technol. **6**(13 (114)), 6–16 (2021). https://doi.org/10.15587/1729-4061. 2021.245251
3. Rajaeian, M.M., Cater-Steel, A., Lane, M.: A systematic literature review and critical assessment of model-driven decision support for IT outsourcing, viewed 20 June 2018. https://www. sciencedirect.com/science/article/abs/pii/S0167923617301240. (2017)
4. Oleg, B., Viktoriia, T., Viktoriia, K., Olga, K., Oleksandr, D.: Improvement of the method for selecting innovation projects on the platform of innovative supermarket. TEM J. **8**(2), 454–461 (2019). https://doi.org/10.18421/TEM82-19
5. Grishko, N.Y., Hlazunova, O.O., Vorobiova, K.O.: Approaches to the management of the costs of innovation activity of mining enterprises: aspects of economic security Naukovyi Visnyk Natsionalnoho Hirnychoho Universytetu, No. 5, pp. 137–145. http://nduv.gov.ua/ UJRN/Nvngu_5_22 (2017)
6. Maziarczyk, A.: Impact of outsourcing on the productivity of Polish industrial enterprises. The Małopolska School of Economics in Tarnów Research Papers Collection, vol. 1445, no. 1, pp. 41–52 (2020). https://doi.org/10.25944/znmwse.2020.01.4152
7. Kosenko, O., Cherepanova, V., Dolyna, I., Matrosova, V., Kolotiuk, O.: Evaluation of innovative technology market potential on the basis of technology audit. Innov. Mark. **15**(2), 30–41 (2019). https://doi.org/10.21511/im.15(2).2019.03
8. Chang, Y.B., Gurbaxani, V., Ravindran, K.: Information technology outsourcing: asset transfer and the role of contract. MIS Q. **41**(3), 959–973 (2017). https://doi.org/10.25300/MISQ/2017/ 41.3.13
9. Chou, Y., Shao, B.B.M.: Total factor productivity growth in information technology services industries: a multi theoretical perspective, viewed 14 December 2017. https://www.sciencedi rect.com/science/article/pii/S0167923614001195 (2017)
10. Hutsan, O., Kobieliev, V., Kosenko, A., Kuchynskyi, V.: Evaluating elasticity of costs for employee motivation at the industrial enterprises. Probl. Perspect. Manag. **16**(1), 124–132 (2018). https://doi.org/10.21511/ppm.16(1).2018.12

11. Maksym, U., Svitlana, C., Ludmila, L., Viktor, R.: Methods for assessing the investment attractiveness of innovative projects. Stud. Appl. Econ. **39**(6) (2021). https://doi.org/10.25115/eea.v39i6.5167

12. Khalilian, A.T., Ibrahim, O., Nilashi, M.: Integrated feedback control reporting for improving quality of telematics and informatics, viewed 20 June 2018. https://0-www-sciencedirect-com.tkplib01.tut.ac.za/science/article/pii/S0736585317304082 (2017)

13. Kobielieva, T., Tkachova, N., Tkachov, M., Pererova, P., Diachenko, T.: Management of relations with enterprise stakeholders on the basis of value approach. Probl. Perspect. Manag. **19**(1), 24–38 (2021). https://doi.org/10.21511/ppm.19(1).2021.03

14. Delen, G.P.A.J., Peters, R.J., Verhoef, C., Van Vlijmen, S.F.M.: Lessons from Dutch IT-outsourcing success and failure. Sci. Comput. Program. **130**, 37–681 (2016). https://doi.org/10.1016/j.scico.2016.04.001

15. Hamzah, A.K., Sulaiman, R., Hussein, W.N.: A review on IT outsourcing approach and a proposed IT outsourcing model for Malaysian SMEs in e-Business adoption. In: International Conference on Research and Innovation in Information Systems (ICRIIS), pp. 521–526, IEEE, viewed 12 August 2019. https://ieeexplore.ieee.org/abstract/document/6716763 (2019)

16. Kobielieva, T., Kuchynskyi, V., Garmash, S., Danko, T.: Ensuring the sustainable development of an industrial enterprise on the principle of compliance-safety. Stud. Appl. Econ. **39**(5) (2021). https://doi.org/10.25115/eea.v39i5.5111

17. Maslak, O.I., Maslak, M.V., Grishko, N.Y., Pererva, P.G., Hlazunova, O.O., Yakovenko, Y.Y.: Artificial intelligence as a key driver of business operations transformation in the conditions of the digital economy. In: Proceedings of the 20th IEEE International Conference on Modern Electrical and Energy Systems, MEES 2021 (2021). https://doi.org/10.1109/MEES52427.2021.9598744

18. Salman, R.H.: Exploring capability maturity models and relevant practices as solutions addressing IT service offshoring project issues, viewed 14 December 2018. https://pdxscholar.library.pdx.edu/cgi/viewcontent.cgi?article=2843&context=open_access_etds (2018)

19. Ashmyanskaya, I.: Outsourcing as a new way of production in the global economy [Electronic resource]. http://mirec.ru/2008-02/autsorsing-kak-novyj-sposob-proizvodstva-vmirovoj-ekonomike (2019)

20. Varajãoa, J., Cruz-Cunha, M.M., Da Glória Fraga, M.: IT/IS outsourcing in large companies—motivations and risks, viewed 18 June 2018. https://0-wwwsciencedirect-com.tkplib01.tut.ac.za/science/article/pii/S1877050917323372 (2017)

21. Pererva, P., Maslak, M.: Commercialization of intellectual property objects in industrial enterprises. Probl. Perspect. Manag. **20**(3), 465–477 (2022).https://doi.org/10.21511/ppm.20(3).2022.37

22. Liu, Y., Tyagi, R.K.: Outsourcing to convert fixed costs into variable costs: a competitive analysis. Int. J. Res. Mark. **34**(1), 252–264 (2017). https://doi.org/10.1016/j.ijresmar.2016.08.002

23. Rhodes, J., Lok, P., Loh, W., Cheng, V.: Critical success factors in relationship management for services outsourcing. Serv. Bus. **10**(1), 59–86, viewed 06 August 2019. https://link.springer.com/article/10.1007%2Fs11628-014-0256-8 (2016)

24. Salman, R.H.: Exploring capability maturity models and relevant practices as solutions addressing IT service offshoring project issues, viewed 14 December 2018. https://pdxscholar.library.pdx.edu/cgi/viewcontent.cgi?article=2843&context=open_access_etds (2014)

25. Varajãoa, J., Cruz-Cunha, M.M., Da Glória Fraga, M.: IT/IS outsourcing in large companies—motivations and risks, viewed 18 June 2018. https://0-wwwsciencedirect-com.tkplib01.tut.ac.za/science/article/pii/S1877050917323372 (2018)

26. Kuchynskyi, V., Kobielieva, T., Kosenko, A., Maslak, O.: Economic substantiation of outsourcing the information technologies and logistic services in the intellectual and innovative activities of an enterprise. East.-Eur. J. Enterp. Technol. **4**(13 (112)), 6–14 (2021). https://doi.org/10.15587/1729-4061.2021.239164

27. Wuyts, S., Rindfleisch, A., Citrin, A.: Outsourcing customer support: the role of provider customer focus, viewed 20 June 2018. https://doi.org/10.1016/j.jom.2014.10.004

28. Lee, G.R., Lee, S., Malatesta, D., Fernandez, S.: Outsourcing and organizational performance: the employee perspective. Am. Rev. Public Admin. **49**(8), 973–986 (2019). https://doi.org/10.1177/0275074019855469

29. Maziarczyk, A.: Global financial crises, profitability and outsourcing in industrial companies in Poland. Org. Manag. Sci. Q. **49**(1), 87–101 (2020). https://doi.org/10.29119/1899-6116.2020.49.6

30. Sheehan, C., Cooper, B.K.: HRM outsourcing: the impact of organisational size and HRM strategic involvement. Pers. Rev. **40**(6), 742–760 (2011). https://doi.org/10.1108/00483481111169661

31. Tsaryova, T.O., Zozulyov, O.V.: Technology as an economic category. Econ. Bull. NTUU "KPI". **6**, 345–351. https://ela.kpi.ua/handle/123456789/6351 (2019)

32. Kotlyarov, I.D.: Outsourcing: experience of theoretical description [Electronic resource]. http://economics.openmechanics.com/articles/190.pdf (2018)

33. Kotlyarov, I.D.: Making a decision on the use of outsourcing based on the assessment of its effect for the enterprise. Innovations **9**, 88–92. https://cyberleninka.ru/article/n/prinyatie-resheniya-ob-ispolzovanii-autsorsinga-na-osnove-otsenki-ego-effekta-dlya-predpriyatiya-1 (2017)

34. WIPO: The official website of the World Intellectual Property Organization World Intellectual Property Organization [electronic resource]. http://www.wipo.int/portal/en/index.html

35. Podlevsky, A.A., Matviychuk, M.M.: Outsourcing as a factor in increasing the competitiveness of enterprises. Bull. Natl. Univ. Water Manag. Nat. Manag. Series "Economics" **1**(49), 162–168 (2018)

36. Zozulyov, O.V., Mykalo, O.I.: Outsourcing as a tool for increasing the competitiveness of domestic enterprises in the conditions of globalization. Econ. Ukraine Sci. J. **8**(573), 16–24 (2019)

37. Jean-Louis, B., Robert, M.: Effective Outsourcing: Understanding, Planning, and Leveraging Successful Outsourcing Relationships, p. 288. Balance Business Books, Dnipropetrovsk (2017)

38. Andrienko, V.N., Belopolskaya, T.V., Kirilishen, Ya.V., Levchuk, E.A., Plakhotnik, E.A.: Effective mechanisms of industrial enterprise outsourcing: monograph. Dneprodzerzhinsk: State Technical University. 343. chrome-extension://efaidnbmnnnibpcajpcglclefindmkaj/https://www.dstu.dp.ua/Portal/Data/5/10/5-10-b4.pdf (2016)

39. Partyn, G.O., Didukh, O.V.: The main types of outsourcing and their application in the management of the company's activities. http://ena.lp.edu.ua:8080/bitstream/ntb/16661/1/371-637-638.pdf (2018)

40. Repin, N.V.: Universal system of choosing an IT outsourcing supplier by a customer. Econ. Internet J. **6**. http://www.e-rcj.ru/ArticIes/2018/Repin.pdf (2018)

41. Demand analysis and quality assessment of outsourcing services in Ukraine [Electronic resource]. http://pravotoday.in.ua/ua/press-centre/market-viewing/view-38.

42. Kotlyarov, I.D.: Algorithm for making a decision on the use of outsourcing in the oil and gas industry. Probl. Econ. Manag. Oil Gas Complex **11**, 33–38. https://publications.hse.ru/articles/64030217 (2016)

43. Boltava, A.L.: Outsourcing as an effective way of doing business. Sci. Thought Caucasus J. **4**, part 2 (48). https://www.dissercat.com/content/autsorsing-kak-instrument-diversifikatsii-regionalnoi-ekonomicheskoi-sistemy-rossii (2014)

44. Ignatiev, A.V.: Algorithm for making a decision on the transfer to outsourcing of functions in the field of ICT in small and medium-sized industrial enterprises. Mod. Stud. Soc. Probl. (Electron. J.) **7**(15). https://cyberleninka.ru/article/n/algoritm-prinyatiya-resheniya-o-perevode-na-autsorsing-funktsiy-v-sfere-ikt-v-malyh-i-srednih-promyshlennyh-predpriyatiyah (2012)

45. Lipatnikov, V.A.: Evaluation of the economic effect from the use of external oil service, taking into account the risk factor. Probl. Econ. Manag. Oil Gas Complex **12**, 7–13. https://rucont.ru/efd/444621 (2015)

46. Makhmutov, I.I., Murtazin, I.A., Karpova, N.V.: Methods and models of outsourcing. In the world of scientific discoveries. **1**(61), 80–104. https://naukarus.com/methody-i-modeli-autsorsinga (2015)

47. Mikhailov, D.M.: Outsourcing: a new business organization system. M.: KnoRus. 256. https://search.rsl.ru/ru/record/01002872102 (2016)
48. Khlebnikov, D.: Outsourcing as a Tool to Reduce Costs and Optimize a Business System. https://www.studmed.ru/hlebnikov-d-autsorsing-kak-instrument-snizheniya-zatrat-i-optimizacii-biznes-sistemy_454fbc49044.html
49. Kuryanovich, V.: Firm restructuring and transition to outsourcing. Sales Bus. **4**, 34–31. http://www.md-management.ru/articles/html/article32685.html (2015)
50. Kurbanov, A.Kh.: Methodology for assessing the feasibility of using outsourcing. Mod. Probl. Sci. Educ. **1**. www.science-education.ru/101-5437 (2016)
51. Moiseeva, N.K., Malyutina, O.N., Moskvina, I.A.: Outsourcing in the development of business partnership. In: Moiseeva, N.K. (ed.) Finance and Statistics, vol. 240. https://znanium.com/catalog/document?id=376558 (2014)
52. Chukhrai, N.I.: Logistic solutions for outsourcing. Logistics **6**, 37–39. https://galicianvisnyk.tntu.edu.ua/?art=480 (2017)
53. Agburu J.I., Anza N.C., Iyortsuun, A.S.: Effect of outsourcing strategies on the performance of small and medium scale enterprises (SMEs). J. Glob. Entrep. Res. **7**(26) (2017). https://doi.org/10.1186/s40497-017-0084-0
54. Fauska, P., Kryvinska, N., Strauss, C.: The role of e-commerce in B2B markets of goods and services. IJSEM **5**, 41 (2013). https://doi.org/10.1504/IJSEM.2013.051872
55. Website development services market research: prices, terms, guarantees, tools. Andrey Konovalov, Center for Marketing Solutions, Moscow, 39 p. https://marketing.rbc.ru/research/30164/ (2019)
56. Petrik, V.L., Golovanov, I.S., Golovanova, M.A.: Approaches to estimating the cost of creating a website. No. 4. S. 88–102. http://nbuv.gov.ua/UJRN/eupmg_2011_4_10 (2019)

Digitalization of Consumers' Behavior Model in the Dairy Market

Tetiana Kulish, **Yana Sokil**, **Darya Legeza**, **Oleh Sokil**, **Iryna Budnikevich**, and **Bahriddinova Diyora**

Abstract In the quickly changing digital economics, a producer have an opportunity to expend sales if he knows consumer attitudes in the online marketing. In spite of the elastic demand on milk products, a produce may get low benefits to compare with a conventional market. The reasons are different approaches to price formation, ignorance of consumer behavior, overload of information in the Internet. Do not to lose your consumer a company have to develop a digital brand strategy. The aim of this study is to devise a consumer behavior model in the milk market using digitalization tools. We use survey of consumers of milk products in Ukraine to describe consumer model. We interviewed 450 potential consumers who live in Ukraine and buy milk product of domestic Ukrainian brands. The questionnaire has 23 answers in communicative, general, research, and final types. We evaluated a brand strength of seven national companies in Ukraine with a help of Fichbein's model. The results show that most respondents pay attention to taste qualities (95%) and product naturalness (85%). The most popular places to buy dairy products are supermarkets (39%), farm markets (28%), and grocery stores (19%). Respondents

T. Kulish · Y. Sokil · D. Legeza (✉)
Dmytro Motornyi Tavria State Agrotechnological University, Zhukovsky Street, 66,
Zaporizhzhia 69600, Ukraine
e-mail: darya.legeza@tsatu.edu.ua

T. Kulish
e-mail: tetiana.kulish@tsatu.edu.ua

Y. Sokil
e-mail: yana.sokil@tsatue.du.ua

O. Sokil
Lviv Polytechnic National University, Stepan Bandera Street, 12, Lviv 79013, Ukraine
e-mail: oleh.h.sokil@lpnu.ua

I. Budnikevich
Yuriy Fedkovych Chernivtsi National University, Kotsyubynsky 2, Chernivtsi 58012, Ukraine

B. Diyora
Samarkand Institute of Economics and Services, 9 Amir Temur Street, Samarkand 140100, Uzbekistan

A. Semenov et al. (eds.), *Data-Centric Business and Applications*, Lecture
Notes on Data Engineering and Communications Technologies 195,
https://doi.org/10.1007/978-3-031-54012-7_8

187

(49%) assume products variety as wide, 31% of respondents notice that variety is insufficiently broad, and 16% of respondents consider the variety narrow. The brand should be historically well-known existing in the market for 25% of respondents. About 25% of respondents pay attention mainly to the constant availability of a milk brand on a store shelf. We concluded that the leading brands in the dairy market should defend their positions, focus their activities on strengthening their competitive advantages and consumer loyalty.

Keywords Purchasing decision · Consumer profile · Consumers' behavior model · Sociometric assessment method (SAM) · Fishbain model

1   Introduction

At the present stage of the dairy market development, consumer and competitor orientation, and flexible adaptation to an ever-changing market conjuncture are necessary conditions for effective enterprise performance. The most important influence on market competition in milk production has price adaptation [1]. There are unequal price trends in developing and developed countries. While in African countries farmers use bicycles to transport their products, producers from the EU have deep experience to distribute their products with a help of cold supply chains and a digital monitoring system. Pris et al. have studied that the implementation of cold supply chains extends the shelf life the food products [2]. The other issue is a lack of modern digital and AI technologies in milk processing. Such factors do not allow small companies to set sufficient prices for their products because they do not provide mass production. Only a few countries of the European Union form the current market and set prices on milk products such as Poland and Holland. One reason is the geographical position of such countries in the center of the EU where the main consumers locate. On the one side, mass production allows a well-known brand company to conquer the market and set an acceptable price. On the other hand, a consumer may evaluate such prices as highly unacceptable and select other products, for example, plant-based milk. The results obtained by Lindström in 2022 suggest that the milk market has elastic demand [3]. So, leading companies support the idea to increase prices gradually, offering more dairy products at that time. Today big milk factories influence on market prices through mergers and acquisitions [4]. However, the digitalization of sale processes leads to the development of new opportunities, for instance for Asian and African countries, in the world dairy market. Current digital business analytics plays a vital role in the market expansion because the digitalization of marketing is spreading quickly [5]. Under such conditions, digital marketing tools determine market success, not mass production. The more a producer will use digital tools to connect with a consumer, the closer brand will be to a consumer, and the more successful a producer will be.

Milk products are a part of the strategic consumer basket, and the major part of consumers buy dairy products, comparing them by safety, availability, and nutrient ingredients. Information overload presented in advertisement on TV, billboards, and ad booklets, lead to consumer disorientation. There are a lot of questions that consumer has when he or she studies a product. What makes a product safe on retail shelves? What marks may describe the nutrition of a product? Where can a consumer get information about a milk producer? How a consumer may connect directly with a producer? When a milk producer uses conventional types of Ads, he does not contact personally with potential consumers and does not know their 'consumption pain'. Contrarily, digital transformation leads to enterprising shifts in marketing strategies and tactics, particularly in social media platforms. There are opportunities to keep track of the consumer purchase history to personalize their services and communicate with customers through convenient network channels. Therefore, milk companies create social network pages, interactive sites, and mobile applications to broaden their presence in the digital market [6]. Today, digital marketing of food is extremely boosted by direct communication between producers, mostly SMEs, and consumers with help of the development of online ordering and interactive feedback [7].

The modern way of communication is the creation of forums with discussion, and evaluations of a company, products, and services based on buyers' communication with each other, to both carry out C2C operations and share shopping experiences.

2 Literature Review

Digitalization in the modern business environment has been developing a novel marketing philosophy. The marketing mix of Ps broadens through digital tools and interactive communication in various business models. Business promotion in social networks and the formation of Internet platforms for consumer communication caused a surge in a new consumer marketing movement. Digitalization has opened the borders for the development of a global market, where a buyer from one country can find a seller in another country without leaving home. Gradually, the opinion and needs of the consumer become the basis for the formation of the food market. We are witnessing a transformation from a competitive market to a consumer market. If earlier the main goal of a company was to study its competitive position, today the main task is the creation of a personal behavior model. Today, a buyer is not looking for products on the shelves, but the enterprise is finding a point of contact with its personal consumer. In words of one syllable, they strive to create a personalized brand that is understandable to the consumer. To be successful on the market, businesses offer broad or original product variety, individual prices, and interactive digital communication.

Milk companies develop marketing strategies based on the differentiation and expansion of milk diversity. Kotyza et al. examined that the main dairy products in the EU market are fresh milk (25.4%), whey (46.2%), and cheese (8.7%) [8]. This list is different in various countries according to consumer preferences. The majority of milk producers in the EU choose a list of products connected with natural and organic products [9]. The main reason why producers aim to process organic products is the ever-increasing demand for safe and natural foods. In 2018, Gassler et al. evaluated various factors in consumer behavior and their readiness to buy milk products [10]. They concluded that consumers prefer finding natural and fresh products to selecting milk by price and availability. Results of Knuck and Hess demonstrate that middle-aged consumers in Germany with high income are ready to select organic products among others [11]. The more variety of organic products presents in stores, the more demand for milk products increases because consumers evaluate milk products as a necessary part of healthy nutrition.

Price in the milk market directs the main trend of sales. Research by Olipra shows that milk prices influence significantly dairy market cycles among other economic and market environment factors [12]. Roman also found that price trend has a long-oriented balance in various regions in Poland [13]. Because the break-even point of the business is highly developed by milk costs, a dairy company should make an effort to increase its profit. The producer should increase productivity by 21% to get a profit of 5% more [14]. A lot of research analyses and compares various aspects of producers' prices. Nevertheless, there are still relevant problems of retail prices to be addressed. There was little discussion on retail price influence in the milk market. When Bor et al. have been studying price asymmetry, they established the market power of large retailers in the Turkish milk market [15]. These days, retailers have their market power during the negotiation process.

Chen and Liu conclude that retailers evaluate producers by brand size, market position, and milk industrial market share to get their benefits [16]. Packaging is not only product storage and transportation, but it is a modern tool of marketing communication. From this side producers post information on packaging to communicate with consumers and help them to make a decision in a supermarket during selecting a brand. The last researches by Woś et al. show that a producer in European countries places various communicative messages and has a different amount of information on a package to attract a consumer [17]. They studied, that the Netherlands' brand prints information on animals' welfare, environmental protection, and the quality of a product. Consumers in German may get knowledge about processing conditions and pleasure. The more a producer give ingredients' images on the packaging, the more often a consumer pay attention and select such milk product [18]. Packaging has become the first step of interactive communication with a consumer.

However, given the dynamics of changes in the external environment of the dairy market, the recent study of consumer behavior in the context of digitalization is a prerequisite for the effective functioning of the dairy industry. Social media is increasing its influence on consumers making decisions. Surfing the Internet, people read opinions about products and services, follow influencers, and live comments how to use such products. In 2021, Zhao et al. examined Weibo to set the network influence on the mothers' decision to breastfeed [19]. After studying the most prominent companies with more than 1 billion followers, they concluded that 12.3% of messages contented the celebrity endorsement, and 19.3% of them presented parenting experience. Parents as influencers play a vital role in digital social media. Schiano's group headlines that 38% of parents encourage children to drink natural milk [20]. Various social networks make a brand closer to clients and consumers, making interaction on the Internet and mobile application. They upgrade the model of market members (business-government-consumer) because social influencers affect consumers significantly in decision-making. The market subject triangle from producers, government and consumers is added by one more partner, such as influencers.

Based on the literature review, there are various factors, which influences a behavior model, such as variety, natural products, price, packaging, and digital tools. The more a brand knows a personalized consumer in a digital environment, the more clearly it foresees market changes.

3 The Aim and Tasks

The aim of this study is to devise a consumer behavior model in the dairy food market using digitalization tools. The main idea of the results is to illuminate the major instruments of digital marketing, which stimulate well-known brands and local companies to communicate with consumers directly. Recent studies have been done examining a series of practical questions in digital marketing such as SMM influencers and price formation. However, a few following practical questions related to brand development in the digital environment were left unanswered. What makes a brand value in digital marketing? What main narrative do milk consumers look for on the Internet? Which digital resources inspire confidence in consumers?

To answer these questions, we set the following research tasks:

1. To evaluate the level of dairy product demand.
2. To investigate consumer motivations in purchasing dairy products.
3. To determine the degree of consumers' perception and loyalty to dairy market brands.
4. To analyze the level of credibility in resources in the purchase process.
5. To reveal effective digital communication channels with consumers at the production promotion of the dairy company with a processing plant.

4 Methods

The object of the study is a model of consumer behavior in the digital dairy market. Concerning demand and consumer motivations, we define the scopes of the study:

– on the territory: the country's market;
– by time: short-term demand and consumer motivations;
– variety: dairy products.

In order to investigate consumer digital model in the milk market we have used following methodology:

The study was carried out based on official printed sources of information. The methods of economic analysis presents a combination of elements with mathematical statistics. The main analytical indicators are price for various milk products and variety depth. Furthermore, we applied statistical grouping and classification to analyze raw information.

The software application used to collect consumers' answers was Survio.com platform. The survey method used in our study is a online interview with closed-ended questions. The advantage of this method is the ease of questionnaire sharing among respondents from social media, direct mailing, and placement of QR-code surveys on the research companies' sites. Another convenience is the analysis automation of the collected data, the inexpensiveness, and the availability to fill out the questionnaire to any point without the interviewer's influence.

We used a random sampling technique. Respondents exhibited age, gender, family size, and consumer preferences. The planned sample was 450 respondents and the error margin is 92–95%. We interviewed potential consumers who live in Ukraine and buy milk product of domestic Ukrainian brands. The survey was done from September to December of 2021. We used a questionnaire with 23 answers in communicative (to establish contact with respondents), general (to reveal the main issues of marketing of milk companies), research (to examine the influence of digitalization on local brands), and final types (to get information about respondents). We used the SPSS software package to study the results of the questionnaire and the applied empirical method to group the determinant influence on consumer brand loyalty.

To evaluate the attitudes consumers to a product or a process many researches use the Fichbein's model [21]. We adapted the model to consumption peculiarities in the dairy marketing. We answered consumers about beliefs and attitudes about 7 Ukrainian brands with a help of 7 attributes (taste, quality, variety, price, packaging, advertising, relationships with consumers, and assessment of consumer attitudes to brand).

We use the sociometric assessment method to examine a group of social indicators, which may influence brand strength. Among other sociometric methods, which Avramidis et al. recommend, peer-rating methods are most applicable to examine the brand strength of milk companies [22]. The main tasks were to evaluate and compare the brand expansiveness of Ukrainian milk companies (Slovyanochka, Galychyna, President, Yagotynske, Prostokvashyno, Molokiya, Ferma).

5   Results

The main peculiarity of milk is its nutritional and organic value, which presents by calcium, animal protein, and vitamins (Krivošíková et al.) [23]. The results of consumer surveys in the dairy market show that respondents generally have a positive perception of dairy products. In response to question, 'Do you agree with the statement that human health is largely determined by the quality of nutrition?' majors of those who surveyed (72%) agreed with this statement. Only 10% disapproved, and 18% did not even think about this problem. Thus, when companies develop an advertising campaign for dairy products, they should emphasize the usefulness of these products for public health to attract potential consumers.

A total of 450 respondents were recruited for the survey, including 42% of employees, 16.2% of entrepreneurs, 15.5% of adults, 13.5% of schoolchildren and students, and 8% of temporarily unemployed. Most parts of respondents are women (64%) and men (36%). In Ukraine, women visit supermarkets and markets more often to compare with men because women usually buy products for their families. Moreover, milk is the product that women use for baby feeding, and they should select a product that matches breast milk. Women who have a one-year-old baby buy milk more often than other respondents do. Only 13.5% of respondents use dairy products daily. Such results report that mothers prefer breasting to feeding animal-based milk. Determining the frequency of product purchases shows that majority of consumers buy milk products less often then 2–3 times a week. Approximately two-thirds of respondents purchase once a week (35%) or twice a week (33.5%). A small number (8%) of young men suggested that they purchase milk products 2–3 times a month. Deep analysis shows that such men are single, do not have babies, and use milk products as an ingredient in dishes. In general, the frequency of dairy product purchases depends on the size of the family.

The majority of respondents are people from 14 to 22 years (26%) and from 23 to 40 years (31%). It means that the dairy market in Ukraine has young consumers which present approximately half of all shoppers. Such people purchase milk products daily and do not have a lot of time to select a milk brand. They usually compare milk products by price and available variety. It should be mentioned that young consumers use numerous websites, social networks, and mobile applications to get information during the day. To make a consumer's choice more effective, companies should target website content to the behavior of young people in finding the necessary information. For example, respondents propose companies display on a website a way of the usage of traditional products from other countries such as ayran, koumiss, and Greece cheese. The vast majority of respondents (86.7%) live in urban cities. Approximately 13 percent of the surveyed are rural people and produce milk products in their households. Rural consumers usually look on the shelf for extraordinary milk products that they can't process individually. Employment or owning a household plays a crucial role in consumer's making-decision what product to purchase. The survey includes answers from 42% of employees and 16.2% of employers. Employees usually visit markets daily and select a narrow list of milk products such as milk, kefir, sour cream, and fruit yogurts. When employers select a product he or she compare brands, test qualities, and availability. The temporarily unemployed people (8%) consume fewer dairy products due to the low-paying capacity of this population category.

Respondents' preferences parceled out the groups of milk products as follows. Most of them (23%) buy milk, one-fifth prefer fermented milk, about 16% give priority to sour cream, and 14% like yogurts. Although these differences are insignificant depending on demographic indicators, answers show the dependence on consumers' skills of digital-centered search of information. Those who select fermented milk spend more than eight hours a day on the Internet. Moreover, they use various mobile applications to select and order goods because they do not want to spend time in supermarket lines.

Only young people and retired persons prefer milk mostly because they lower income during studying at a University or on a pension. More deep research has detected that these two groups of consumers selected low-price milk and sour milk to use in the processing of further dishes. It was revealed that most people who selected yogurts evaluate a product by the presence of natural ingredients. It is very crucial for them to buy safe health products.

Ukrainian consumers prefer traditional dairy products (milk, sour cream, cultured yogurt) as basic food. At the same time, respondents consider as a luxury the new-day fermented milk category (yogurts, cottage cheese with fillings, and soft cheeses). The Ukrainian consumer choice is still conservative. Even during the pandemic and isolation, people have not changed their preferences in the milk product basket because they use to consume their conventional products. Moreover, the majority of those who responded noticed the specifics of the consumption culture in Ukraine. They underline that kefir is primarily a healthy product that normalizes the functioning of the digestive system. Furthermore, people like the specific taste of fermented baked milk and consume it for mood. On this side, a company may consider it during the workup of a media plan and include posts about the usefulness of fermented dairy products on a website. People eat fruits and berries in drinking yogurt due to additional vitamin proof of the product's usefulness. It would be effective to make an attractive visualization of the list of vitamins on the package. The popularity of sour cream is due to the culinary traditions of Ukrainian consumers. Therefore, almost ten percent of respondents buy it every week. This part of the survey shows that Ukrainian consumer has a distinctive model of behavior. People the more conservative in selecting of type of milk products but progressive in the usage of digital tools to search for information about milk companies and products.

In the study of consumer preferences, respondents identified the main factors that influence product choice. The study establishes that most respondents pay attention to taste qualities (95%) and product naturalness (85%). Respondents search the information about the milk naturalness of various brands on the packaging, on websites, and on market platforms where other people leave their feedback. A consumer becomes more intelligent in the selection of products by safe and natural peculiarities. They visit various blogs about proper nutrition, and websites for consumer feedback. They load mobile applications with interactive monitoring of their nutrition and experts' consulting. The pandemic and ecological issues make consumers think about what they eat and how it is safe for nutrition.

An important criterion is the producer brand (80%) and the level of confidence in it (72%). The more a brand is digitalized, the more attractive it became for readers, and the more often a company may communicate with a consumer. While consumer preferences in Ukraine are stable, increasing partnerships with EU countries change participators in the Ukrainian market. European brands have developed modern digital strategies, which allow them to contact directly with influencers and get feedback through websites. Using a personalized consumer profile and a journey map, exporters may manage Ukrainian consumers' making-decision. Table 1 demonstrates the main importers of milk products in Ukraine in 2021 [24].

Ukraine has negative trade balance, which is equal to $2.7 billion and has rank 64 in the total import market of milk and cream, not concentrated nor containing added sugar or other sweetening matter. The data shows, that market concentration has index 0.41 what means a high competition between importing countries. The main exporters of milk products to Ukraine are Poland, Lithuania and Germany that supply 10.6 billons tones in 2021 and have group A by Paretto principle. The average export price was $935 per ton. The highest price on milk was from Netherlands ($3544 per ton), France ($3349 per ton) and Denmark ($3333 per ton). It should be mentioned that Poland supply the greatest quantity of milk products by low prices ($740 per ton). Exporters from Poland cause completion with Ukrainian producers by prices.

The value of importing milk products from Poland increased extremely from 2019 to 2020 approximately in four times because quantity of products boosted from 2.7 to 9.6 billion tons.

The price level is significant for more than 60% of respondents, although over one-third of respondents pay attention to loyalty programs. Studies have shown that the ratio of the influence of price, quality, and naturalness when choosing dairy products changes significantly with age. The older a consumer, the more often he selects a milk product by price. People of these ages usually do not use smartphones

Table 1 Importers if milk products (HS 0401) to Ukraine, 2021

Importers	Value imported in 2021 (USD thousand)	Trade balance 2021 (USD thousand)	Quantity imported in 2021	Unit value (USD/ ton)	Cumulative share in Ukraine's imports (%)	ABC group
World	13,176	−2662	14,087	935	*	*
Poland	8134	−8134	10,997	740	61.73	A
Lithuania	1382	−1382	528	2617	72.22	A
Germany	1066	−1064	535	1993	80.31	A
Belgium	948	−948	367	2583	87.51	B
Belarus	799	−799	1310	610	93.57	B
Netherlands	241	−240	68	3544	95.40	B
France	211	−211	63	3349	97.00	C
Italy	177	−177	65	2723	98.35	C
Slovakia	154	−154	128	1203	99.51	C
Romania	26	−26	10	2600	99.71	C
Cyprus	24	−17	8	3000	99.89	C
Denmark	10	−9	3	3333	99.97	C
Estonia	4	−4	4	1000	100.00	C

*The world total value import is not included in the calculation of the cumulative share of a country's import

in their life, thus they could not be users of mobile applications and need extra loyalty programs on the websites of producers and supermarkets. According to the results of the survey, because retirees like to visit farmers' markets and local fairs, they could not be full clients in the digital milk market in the nearest future. At the same time, people from forty to fifty pay the minimum attention to the price. They are more digitally intelligent persons and meet current market changes with enthusiasm and opportunities in order to save time on searching the product. On the one hand, they react positively to the implementation of digital marketing tools by producers and retailers. On the other hand, the digital market of milk products in Ukraine is developing slowly and still does not use elements of augmented reality and artificial intellect to promote the program. Using the chips in supermarkets and cookies on the Internet to collect information about behavior consumers' model.

Retirees pay less attention to quality and even naturalness—the price comes to the fore. Older people use such an approach to selection because they pay more attention to their budget planning. The average monthly pension is $87.5 in Ukraine, while a bottle of milk costs $0.98 per liter. Besides, milk and milk products are an integral part of the nutrition of retirees. Therefore, the vital criterion of making-decision for such consumers is price affordability. Other target groups of consumers also pay attention to the retail price among other factors. Unemployed and students compare prices with the naturalness, freshness, and taste of milk products. At the same time, every Ukrainian brand competes on the market by product peculiarities and qualities rather than by price. Monitoring of leading milk producers in supermarkets, conducted in January 2023, shows that the price limits are not so big and fluctuate from $0.9 to $1.05 per liter. Prices of organic milk products are higher by almost 15–20%. During the pandemic, online sales of milk products were growing and changing the principles of price formation because perishable foods were in high demand. People were ready to pay extra money for the timely delivery of fresh milk. Consumers with high income or employers used to order milk at high prices and use online platforms from that time.

Research reveals that a current trend in the milk market is the consumption of natural products; therefore, a successful brand should ensure the maximum naturalness of the product. A consumer requires proof of product consistency and natural ingredients. One way to solve such a task is to put information on the packaging. In a digital environment, it would be more forward-looking to use QR codes on a pack with links to video files of milk processing in a dairy plant. If a company offers milk products for children, it should be practical to print a QR link on a women's forum. Augmented reality in billboards and other printed materials will allow explaining the real process of milk production. Because numerous consumers follow the trend of proper nutrition, they should get proof of safe and healthy nutrition. Such people reject the presence of flavors, preservatives, and chemical additives. Hence, the rejection of flavorings is the absence of exotic flavors, in yogurt desserts for instance, on behalf of traditional flavors.

According to the survey, almost half of the respondents (49%) assess dairy product variety as wide and sufficient. This segment of people usually limits its consumption to milk, kefir, and sour milk. Consumers prefer to visit retailers daily and buy fresh products. They buy products spontaneously and don't spend time searching for a certain type of milk product. Such people compare prices of various brands but do not pay attention to the presence and price of extraordinary milk products. Over one-third (31%) of those interviewed highlighted that variety is insufficiently broad. Most women, employees, and social workers want to see a more varied list of dairy products on supermarket shelves. Such a segment of consumers compares brands by product ingredient, nomenclature, and availability. Women usually need dietary products and milk baby food. They are not limited by popular milk products and are always ready to search for rare types of products on the Internet and even order them online. COVID-19 has influenced on raising of online purchases around the world. They like various natural yogurts, cheese deserts, and cottage cheeses. They prefer Ukrainian brands to imported ones and follow the promotion in stores. People who notice a variety as insufficiently broad like to take part in product testing and are willing to buy new types. Therefore, it should be effective to use interactive digital platforms and forums to get knowledge from consumers on how to innovate milk product varieties. The smallest group of respondents (16%) emphasize that range of products is narrow. This group presents the interests of employers and workers with high income who visit usually hypermarkets and order on the Internet. They do not like to spend a lot of time shopping and visiting retail stores once a week. They prefer ordering on the Internet to getting stores in traffic daily. Because they value time and information, they prefer to pick up and order in grocery delivery. Therefore, they take new ideas of presented products on the Internet with enthusiasm. Leading digital companies offer current technologies for the promotion of foods such as true-to-scale 3D visuals, online scanning of languages, and immersive neuromarketing screening.

Our study demonstrates that 37% of respondents are satisfied with the quality of dairy products. They evaluate milk food by tasty and chemical criteria such as the fat content of milk and acidity. Such consumers do not require the existence of additional types of products with vitamins, protein, or gluten-free. They are usually advocates of the one brand, which they select among others. Almost half of interviewed (46%) are not completely satisfied with the quality of the existing product. Analysis of answers shows that this segment requires more natural and healthy milk. They believe in local companies, which products they used to buy for a long time. The brands organize excursions and interactive events for schools and parents. Consumers notice that they like training in cuisine, which organizes by regional milk factories. The involvement of the population in the production process creates a positive opinion about the quality of the product, including milk. The response rate of unsatisfied consumers is 17%. They are bright representatives of people with whom brands can't establish direct beneficial communication. Analysis of big data of feedback in forums will help brands engage such a segment of people. Supplementary data on clients' negative opinions may give the idea for innovative decisions on what should be changed. Last time consumers leave their advice on such forums and propose their own criteria for nutrition.

Most surveyed when choosing products prefer local production (80%), but 8% consider the imported product to be of better quality for 12% of respondents it does not matter the manufacturer. In an investigation into the influence of Covid-19 on market development, Sokil et al. concluded that, in the future, globalization transmute into national localization [25]. It means that the domestic market should have a sufficient number of potential consumers of new products for dairy companies.

The type of retail stores is substantial for respondents, especially when they buy dairy products. Different factors influence consumers' preference to visit grocery, hypermarkets, farm markets, or store near the house. The most popular Ukrainian format for consumers is supermarkets where they may purchase various brands and varieties. Almost two-fifths of respondents (39%) visit supermarkets when they look for milk products. Consumers buy products not often than 2–3 times per week and fill a food basket with milk products of different brands. They spend time searching for required foods by peculiarities and qualities. In Ukraine, people still look for food on a farm market where rural households offer agricultural products that are produced by them. It is a family tradition to visit a farmers market on a weekend and select unbranded natural products from farmers or individual entrepreneurs. These consumers value naturalness and freshness mostly. They get advice from sellers about how to store and cook milk products. Consumers go to a farm market for mood mostly for purchases.

According to respondents' answers, the less popular types of stores are grocery (19%) and branded (16%) stores. While such retailers are food-oriented sellers, consumers visit them more rarely. Remote location, lack of a wide range of products, and limited opening hours are the key reasons why consumers pay them less attention. Although these stores always provide fresh milk products, the variety is still limited. Respondents, who do not visit branded store notice that they are not followers of the brand proposed in the store. Moreover, products of the brands are highly-price compared with other brands. According to respondents' answers, the factors that most influences the choice of retailer are the price level (66%), the variety (54%), and product quality (44%). Grocery and the branded store have their small net and may manage their assortment inside the net using principles of the Internet of things. It would be a competitive-based benefit in the struggle with national leading brands if each store inform others and use Kanban technology.

Besides, 80% of consumers pay attention to the brand, when they select products. Only a small number (6%) of respondents felt that a brand is not important to them when they purchase milk. The last like visiting a farmers market to purchase milk products. Therefore, companies should make significant efforts to form a positive image of their brands and develop convenient packaging and attractive design. For one-fourth of respondents, the brand must exist in the market for a long time. These people notice that constant availability on the shelves is the key requirement for milk brands. Moreover, 16% of respondents would like to see periodic updating of the range.

To study the value of national brands, we develop a behavior model of milk product consumption. The model is based on data grouping of respondents' answers. We divided respondents according to their answers to certain questions The model includes a part of the questionnaire and consists of answers connected with consumer behavior and making decisions. Table 2 demonstrate the main features of consumer behavior.

The definition of the most popular brands showed that the majority of respondents prefer Prostokvashyno (18%), Yagotinske (16%), in third—Galychyna and Slovyanochka 14% and 12% respectively.

The Fishbein model is used to determine consumer commitment to a particular brand, which is based on the Formula (1):

$$A_O = \sum_{i=1}^{n} b_i e_i \qquad (1)$$

where A_o is an attitude toward the object; b_i is personal belief about the probability of the object possessing the ith attribute; e_i is a personal evaluation of the ith attribute; and n is a the number of significant indicators (attributes).

Respondents rated the importance of the corresponding attribute on a 7-point scale of the type {"very good" $+3$ $+2$ $+1$ 0 -1 -2 -3 "very bad"}, and the value of the bi on a point scale of the type {"very likely" $+3$ $+2$ $+1$ 0 -1 -2 -3 "unlikely"}. The results of the multifactor model of the Fishbein study are shown in the Table 3.

The results are reduced to the general indicator of A_o. This criterion presumes the consumers' attitude to the corresponding brand of dairy products. The respondents' attitude to Prostokvashino is better, in second place—Yagotynske, although there is a small difference between the general indicators.

For many companies in the dairy market, the brand is a determining factor of success or failure of competition. The rivalry is increasingly conducted not only by price methods but also by non-price factors. At the same time, a well-developed marketing strategy for dairy brands is becoming increasingly important.

Sociometric assessment method allow the specification of the sociometric status of dairy product brands. The average purchase size was determined to calculate the variance of the average bill size. Towards this, an experiment was conducted—a pilot study, which included observation and oral experience of a control group of 150 buyers of milk and dairy products.

We determine the strength of brands by the next stages of sociometric analysis:

1. The list of competing brands of a parity class in the dairy market that were brought in a sociometric brand card is made.
2. A socio-survey was conducted, during which sociometric brand cards were filled out.
3. The sociometric matrix was completed based on brand cards.
4. The sociometric status of each brand in its class and the index of brand expansion were calculated. The sociometric status of the trademark was calculated by Formula (2).

$$S_m = \frac{V_m - W_m}{N} \qquad (2)$$

Table 2 A model of consumer behavior on the dairy market (450 respondents)

#	Feature	% of answers
1	*Gender*	
	Male	36.0
	Female	**64.0**
2	*Age*	
	14–22	26.0
	23–40	31.0
	41–54	22.0
	55 and over	21.0
3	*Social class*	
	Student	13.5
	Employee	42.0
	Unemployed	8.0
	Entrepreneur	16.2
	Retired people	15.5
	Others	4.8
4	*Variety*	
	Milk	23.0
	Fermented milk 9 (kefir)	20.0
	Sour cream	16.0
	Fermented milk (ryazhenka)	8.0
	Yogurts	14.0
	Cottage cheese	12.0
	Others	7.0
5	*Frequency of purchase*	
	Daily	13.5
	1 time a week	35.0
	2 times a week	33.5
	2–3 times a week	18.0
6	*Place of residents*	
	Urban	86.7
	Urban-type settlement	5.6
	Rural	7.7
7	*Key factors of selection milk products*	
	Price	64.0
	Taste	95.0
	Naturalness	85.0
	Expiration date	75.0

(continued)

Table 2 (continued)

#	Feature	% of answers
	Trademark, brand	72.0
	Packaging	45.0
	Loyalty programs	32.0
8	*Influences*	
	Trust only own opinion	12.0
	Recommendations of famous people, bloggers	28.0
	Advertising	14.0
	Recommendations of medical experts	31.0
	Recommendations of friends and relatives	10.0
	Recommendation of sellers	5.0

Table 3 Consumer attitudes to Fishbein model

Characteristics (attributes)	Personal evaluations (ei)	Personal beliefs (bi)			
		Prostokvashyno	Yagotinske	Galychyna	Slovyanochka
Taste	3	+2	+2	+1	0
Quality	3	+2	+2	1	1
Variety	2	+3	+3	+2	+2
Price	1	+3	+2	+1	+1
Packaging	2	+2	+1	0	0
Advertising	2	+3	+3	+2	+2
Relationships with consumers	3	+1	0	−2	−2
Assessment of consumer attitudes to brand		31	28	6	5

where S_m is a sociometric status of the trademark; V_m is a the volume of goods of the mth brand; W_m is a the amount of deviations of the mth brand; N is a number of respondents.

The index of brand expansiveness is calculated by Formula (3):

$$I_m^{exp} = \frac{V_m}{N} \tag{3}$$

where I_m^{exp} is the index of brand expansiveness.

5. The obtained individual sociometric indices were ranked based on the study results to identify the leading brands. The ranking was carried out in descending order of importance. Thus, the first rank is appropriated to the leading brand.

Maximum rank scores should correspond to the maximum values of the indices (Table 4).

6. As a result of the analysis and calculations, we can conclude that the leading brand in the dairy market is the trademark Prostokvashino, which received the maximum rank in sociometric status. In second place is the trademark Yagotynske, and closes the top three brands—Galychyna.

Based on the analysis, it can be concluded that the leaders in the dairy products market need to defend their positions, focus their activities on strengthening their competitive advantages and consumer loyalty. The possible changes in the macro and microenvironment can lead to enterprise deterioration. So, it is necessary to ever-evolve and improve the production process, use a high level of the innovation process, qualified personnel, develop and adopt a program for effective brand management, communication with consumers when promoting the products of dairy enterprises.

Almost everyone in the world has access to an endless flow of information on the Internet. With the development of information technology, it has become much easier to search for and share information. Therefore, a significant percentage of Ukrainians spend a lot of their free time on social media and Internet resources. Due to this, social media and Internet resources are becoming not only a social leisure activity but also a platform for networking with consumers and other business representatives.

Therefore, today social media have begun to be used as one of the most effective labor-intensive digital tools to promote products. The functionality of social networks provides a wide range of tools and opportunities for doing business.

According to the research results, most respondents (31%) listen to the recommendations of well-known medical experts when choosing dairy products, 28% trust the opinion of famous people, 14% of respondents are influenced by advertising, 12% of respondents trust only their choice, 10%—listen to the recommendations of friends and acquaintances and only 5% make choices with the guidance of sellers.

Table 4 Brand valuation

Brand	The sociometric status		Index of brand expansiveness	
	S_m	Ranking	Iexp	Ranking
Slovyanochka	−0.24	4	0.38	4
Galychyna	0.16	3	0.58	3
President	−0.32	5	0.34	5
Yagotynske	0.4	2	0.7	2
Prostokvashyno	0.56	1	0.78	1
Molokiya	−0.48	6	0.2	6
Ferma	−0.62	7	0.12	7

6 Approbation of the Research Results

The results of the study have been presented to representative of the local brand Molochna Rika. The authors studied the potential brand strength of the brand. The authors and representatives of the company have discussed the opportunity to enter national supermarket net Silpo. The results suggest that Molochna rika has had a brand attitudes to local consumers as a sustainable brand. To expansion brands' position in retail, Molochna Rika should pay attention on product taste, variety and relations with consumers. Following a four-month negotiations with Silpo, Molochna Rika has entered to Silpo in the one region. It could be beneficial to make packaging more interactive and print QR-code on video of milk production in the company, and to make 3-D imagines to attract children buy milk.

7 Conclusions

As a result of the conducted research, the model of the consumer behavior in the dairy products market was developed:

1. Dairy products in human nutrition come first, especially in childhood nutrition and the elderly diet. Therefore, we consider expanding the range of milk processing enterprises, increase their efficiency and competitiveness.
2. Overall, it was determined that the majority of respondents often buy dairy products, and the size of purchases depends on the number of family members.
3. Consumers prefer the following groups of dairy products milk—23% of respondents; in second place fermented milk products—20%; about 16% gave priority to sour cream, yogurt—14%, which is due to the specifics of the culture of consumption of dairy products.
4. When buying dairy products, consumers pay attention to the following characteristics: taste, naturalness, brand, producer's image, price level, loyalty programs. In the dairy industry, there is a strong tendency to choose natural products, so a successful brand must ensure maximum naturalness of the product.
5. Most consumers prefer national producers, and 68% of respondents supported the decision of milk processing plants to expand their range, which indicates a sufficient number of potential buyers of dairy products.
6. It was determined that the most popular places to buy dairy products are—a supermarket, a market, and a grocery store respectively, 39%, 28%, and 19% of respondents chose these months of purchase. The price level (66%), the variety of an assortment (54%), and product quality (44%) influenced the choice of sales point.
7. It was concluded that the gender and age of respondents, social class, as well as their place of residence, affect the demand for milk and dairy products: in cities, 70% of respondents buy products, in rural areas 30%, as in rural areas more often produce dairy products themselves in households.

8. To increase the level of consumers' loyalty in the dairy products market, there is a good reason to consider modern marketing tools for digitalization:

   - on the company's official website to provide detailed information about the company and its products with mandatory feedback;
   - preparation and posting on the Internet of news, releases, articles, interviews with an emphasis on the usefulness and naturalness of dairy products;
   - holding contests and raffles for consumers on the official pages of social networks of companies;
   - to organize client support in social networks, that is carrying out mass consultations specially created for this purpose network community (groups, on blogs);
   - conducting an image advertising campaign on social networks, which will increase brand awareness and loyalty to it among consumers;
   - involvement of culinary bloggers in the advertising campaign;
   - maintaining promotional calendars, which will optimize the communication activities of companies;
   - use CRM systems to establish effective communication with consumers and optimize the company's business processes.

Implementation of these measures will allow: to increase the volume of branded traffic, the number of visitors who visited the site of the company, entering the name of the company or brand in the request; increase the number of potential consumers of products (buyers) increase the level of consumer loyalty; increase sales.

Acknowledgements The study is related to the Ukrainian State scientific and technical programs No. 0116U002738 "Marketing strategy development of agrarian enterprises" of Dmytro Motornyi Tavria State agrotechnological University; No. 0118U000345 "Development of innovative aspects of public-private partnership" of Financial and Administrative Department of Lviv Polytechnic National University; and scientific topic of Department of marketing, Innovation and Regional Development at Yuriy Fedkovych Chernivtsi National University "Marketing approach in the context of creative economy formation: theoretical aspects and applied solutions".

References

1. Roman, M., Roman, M.: Milk market integration between Poland and the EU countries. Agriculture **10**(11), 561 (2020). https://doi.org/10.3390/agriculture10110561
2. Priss, O., et al.: Effect of abiotic factors on the respiration intensity of fruit vegetables during storage. Eastern-Eur. J. Enterp. Technol. **6**(11), 27–34 (2017). https://doi.org/10.15587/1729-4061.2017.117617
3. Lindström, H.: The Swedish consumer market for organic and conventional milk: a demand system analysis. Agribusiness **38**(3), 505–532 (2022). https://doi.org/10.1002/agr.21739
4. Nderitu, P.C., Ndiritu, S.W.: Effects of mergers on processed milk market in Kenya. J. Agribus. Dev. Emerg. Econ. **8**(3), 480–500 (2018). https://doi.org/10.1108/jadee-04-2016-0021
5. Daodu, L., Bhaumik, A., Morakinyo, A.: Business analytics and market adaptation in the e-commerce industry in Nigeria. Future Bus. Adm. **1**(1), 23–40 (2022). https://doi.org/10.33422/fba.v1i1.309

6. Abrahams, S.W.: Milk and social media. J. Human Lact. **28**(3), 400–406 (2012). https://doi. org/10.1177/0890334412447080
7. Legeza, D.G., et al.: A model of consumer buying behavior in relation to eco-intelligent products in catering. Innov. Mark. **15**(1), 54–65 (2019). https://doi.org/10.21511/im.15(1).2019.05
8. Kotyza, P., et al.: Dairy processing industry in the European Union: country-group clustering in pre and post-milk quota abolishment periods. J. Central Eur. Agric. **23**(3), 679–691 (2022). https://doi.org/10.5513/jcea01/23.3.3503
9. Merlino, M.V., et al.: Differences between Italian specialty milk in large-scale retailing distribution. Economia Agro-alimentare (2), 1–28 (2022). https://doi.org/10.3280/ecag2022o a13173
10. Gassler, B., et al.: Keep on grazing: factors driving the pasture-raised milk market in Germany. Br. Food J. **120**(2), 452–467 (2018). https://doi.org/10.1108/bfj-03-2017-0128
11. Knuck, J., Hess, S.: Who buys regional fresh milk brands? An analysis of German household data. Agribusiness [Preprint] (2022). https://doi.org/10.1002/agr.21776
12. Olipra, J.: Cycles in the global milk market. J. Agribus. Rural Dev. **52**(2) (2019). https://doi. org/10.17306/j.jard.2019.01100
13. Roman, M.: Spatial integration of the milk market in Poland. Sustainability **12**(4), 1471 (2020). https://doi.org/10.3390/su12041471
14. Syrůček, J., Bartoň, L., Burdych, J.: Break-even point analysis for milk production—selected EU countries. Agric. Econ. (Zemědělská ekonomika) **68**(6), 199–206 (2022). https://doi.org/ 10.17221/40/2022-agricecon
15. Bor, Ö., Smihan, M., Bayaner, A.: Asymmetry in farm-retail price transmission in the Turkish fluid milk market. New Medit. **13**(2), 2–8 (2014)
16. Chen, X., Liu, Y.: Private labels strategy, retail profitability and bargaining power in the fluid milk market. J. Agric. Food Ind. Org. [Preprint] (2022). https://doi.org/10.1515/jafio-2022-0002
17. Woś, K., et al.: Preliminary analysis of voluntary information on organic milk labels in four European Union countries. Sustainability **14**(24), 16901 (2022). https://doi.org/10.3390/su1 42416901
18. Capelli, S., Thomas, F.: To look tasty, let's show the ingredients! Effects of ingredient images on implicit tasty–healthy associations for packaged products. J. Retail. Consum. Serv. **61**, 102061 (2021). https://doi.org/10.1016/j.jretconser.2020.102061
19. Zhao, J., et al.: Uncovering milk formula advertisements on the Chinese Social Media Platform Weibo (preprint) (2021). https://doi.org/10.2196/preprints.34376
20. Schiano, A.N., et al.: Parents' implicit perceptions of dairy milk and plant-based milk alternatives. J. Dairy Sci. **105**(6), 4946–4960 (2022). https://doi.org/10.3168/jds.2021-21626
21. Fishbein, M.: A behavior theory approach to the relations between beliefs about an object and the attitude toward the object. Lect. Notes Econ. Math. Syst. 87–88 (1976). https://doi.org/10. 1007/978-3-642-51565-1_25
22. Avramidis, E., et al.: Using sociometric techniques to assess the social impacts of inclusion: some methodological considerations. Educ. Res. Rev. **20**, 68–80 (2017). https://doi.org/10. 1016/j.edurev.2016.11.004
23. Krivošíková, A., et al.: Consumer preferences on milk market: evidence from Slovak Republic. Potravinarstvo Slovak J. Food Sci. **13**(1), 961–970 (2019). https://doi.org/10.5219/1221
24. List of Supplying Markets for the Product Imported by Ukraine in 2021 (no date) Trade Statistics for International Business Development. ITC. https://www.trademap.org. Accessed 4 Feb 2023
25. Sokil, O., et al.: The context of 'globalization versus localization' after the world pandemic and quarantine. In: Digital Economy, Business Analytics, and Big Data Analytics Applications, pp. 77–85 (2022). https://doi.org/10.1007/978-3-031-05258-3_8

Modelling and Information Support for Assessing the Potential for Increasing the Financial Stability of Enterprises

Olexandr Yu. Yemelyanov⬤, Ihor M. Petrushka⬤, Tetyana O. Petrushka⬤,
Ilona O. Tuts, and Viacheslav Dzhedzhula⬤

Abstract Currently, there is a need to improve the methodological bases of modeling the process of assessing the potential of increasing the financial stability of enterprises, as well as the need to improve the information support of the process of such an assessment. Solving the specified tasks with subsequent testing of the results obtained by a sample of enterprises is the primary goal of this study. It was established that the potential for increasing the financial stability of the enterprise contains three components: the potential for increasing the amount of available income, which the enterprise can use to repay the loans and to pay interest, the potential for reducing the amount of loan capital, and the potential for increasing the degree of flexibility of the enterprise. The sequence of evaluating the potential for increasing the financial stability of the enterprise is proposed. An array of information is defined, based on which the potential for increasing the financial stability of the enterprise is estimated. Interrelationships between blocks of information have been established, based on which the current level of the financial stability of the enterprise and the potential for its improvement are assessed. A sample of one hundred enterprises in the western region of Ukraine was formed to test the developed principles of modeling and information provision to assess the potential of increasing the financial stability of enterprises. It turned out that an insufficient level of financial stability characterized most of the studied enterprises. In particular, the share of enterprises engaged

O. Yu. Yemelyanov · I. M. Petrushka · T. O. Petrushka
Lviv Polytechnic National University, Lviv, Ukraine
e-mail: Oleksandr.Y.Yemelianov@lpnu.ua

I. M. Petrushka
e-mail: Ihor.M.Petrushka@lpnu.ua

T. O. Petrushka
e-mail: tetiana.o.petrushka@lpnu.ua

I. O. Tuts
Ivano-Frankivsk National Technical University of Oil and Gas, Ivano-Frankivsk, Ukraine

V. Dzhedzhula (✉)
Vinnytsia National Technical University, Vinnytsia, Ukraine
e-mail: djedjula@vntu.edu.ua

© The Author(s), under exclusive license to Springer Nature Switzerland AG 2024
A. Semenov et al. (eds.), *Data-Centric Business and Applications*, Lecture
Notes on Data Engineering and Communications Technologies 195,
https://doi.org/10.1007/978-3-031-54012-7_9

in wholesale, in which the level of financial stability was low and very low, was 28% and 24%, respectively. The share of enterprises in the food industry, in which the level of financial stability was low and very low, was 22% and 18%, respectively. At the same time, the following general trend can be observed for enterprises of both types of economic activity: with an increase in the profitability of total capital and a decrease in the share of loan capital, the level of financial stability of companies increases. The value of the potential for increasing the financial stability of the studied enterprises was also calculated. The main ways of realizing this potential have been determined. In particular, based on the results of the empirical analysis, the importance of information provision in the process of assessing the potential for increasing the financial stability of the enterprise was established.

Keywords Financial stability · Potential · Modelling · Information support · Loan capital · Enterprise · Flexibility

1 Introduction

Currently, in the economy of many countries, the level of uncertainty regarding the expected results from conducting business activities has increased. This increase was primarily due to the rise of political tension in relations between different countries and regions in the world. In some cases, this tension resulted in a direct military confrontation. In particular, the sharp intensification of military operations in Ukraine, which took place at the end of February 2022, significantly increased instability in a large number of global markets for both raw materials and final products.

In general, the subjects of business activity in many branches of the economy in current economic conditions are forced to pay increased attention to the risk factor [1], in particular, to conduct its assessment [2] and develop measures to manage such risk [3]. At the same time, the main reasons for increasing the riskiness of economic activity include a decrease in the level of predictability of prices for some types of production resources, in particular energy [4], significant volatility of conditions in many commodity markets [5], acceleration of the processes of technological changes in the economy [6] and the level of innovative activity in its spheres, etc. [7].

An important area of risk management of business entities is the development of measures aimed at increasing their economic sustainability. This stability reflects the ability of enterprises to resist the harmful effects of the external environment effectively. In turn, one of the main types of economic stability of enterprises is their financial stability [8]. The low level of financial stability of enterprises is often a decisive sign of their being in a state of financial crisis, and it may indicate a significant tendency for such enterprises to go bankrupt [9]. Therefore, improving the financial stability of companies can be a necessary condition for preventing their liquidation in the future. At the same time, such improvement, in many cases, is a very hard task since the financial stability of enterprises is influenced by a significant number of factors. At the same time, the value of some of these factors, in particular future prices for production and financial resources, is quite difficult to forecast and estimate. Therefore, the management of the financial stability of enterprises requires, on the one hand, the development of scientifically based models that can be effectively used during such management. On the other hand, making managerial decisions to ensure the appropriate level of financial stability of enterprises requires the availability of a wide array of information based on which these decisions can be made. These two conditions to ensure the process of managing the financial stability of enterprises are interrelated since modelling makes it possible to simultaneously structure an array of the necessary information.

2 Literature Review and Setting Research Objectives

A significant number of scientific works are devoted to the management of economic stability and, in particular, the financial stability of enterprises. At the same time, scientists investigated the essence and signs of economic stability [10], established the factors and regularities of its formation [11], and also proposed indicators for measuring the level of this stability [12]. Particular attention is paid to consideration of the adaptability of enterprises to changes in their external environment as an important condition for ensuring the appropriate level of economic stability of companies [13].

Investigating the peculiarities of assessing and managing the financial stability of enterprises, some scientists associate the impact of its low level with the occurrence of a financial crisis at enterprises. In particular, in work [14], the financial crisis is considered in the context of the deterioration of the solvency of enterprises. In [15], its authors study the profitability crisis as an important type of financial crisis at companies. In turn, the majority of scientists, in particular the authors [16], consider the decrease in revenues and profits of enterprises as one of the main reasons for the deterioration of their financial stability. Another essential factor in this deterioration is that companies have significant loans for financing their assets [17]. In general, it is possible to state that the most noticeable decrease in the level of financial stability of enterprises occurs when the reduction in income and profits of enterprises is combined with large amounts of loan funds in the capital of companies. At the same time, it is necessary to consider that the possibilities of many enterprises to prevent a decrease in their profitability are quite limited. This is because such a decrease is often caused by external factors [18]. At the same time, the process of forming the loan capital of enterprises is under their direct influence [19]. Therefore, the regulation of the amount and structure of their loan capital should be recognized as the most important direction for managing the financial stability of companies [20]. Such regulation, in turn, requires an assessment of the effectiveness of lending to enterprises based on its various amounts and other parameters.

It should be noted that credit efficiency is a complex and multifaceted concept. As noted in [21], such efficiency should be equated with the economic, financial, social, environmental, and other consequences caused by crediting companies' activities. However, in the literature, the economic results of loan financing of firms are interpreted differently. Most often, scientists single out three types of these results: the amount of economic growth, the increase in profits and profitability, and the increase in the riskiness of economic activity. The listed results of enterprise lending affect the level of their financial stability.

Regarding the influence of loan financing of companies on the rates of their economic growth, the existence of such an influence has been established in some scientific works, in particular in [22]. At the same time, the criteria for the economic growth of enterprises proposed by different scientists differ slightly. Thus, in [23], such a criterion is the increase in the volume of products manufactured by enterprises. At the same time, in work [24], the volume of economic results of loan financing of companies is understood as the increase in their net assets and income. With regard to the second type of economic results of lending to firms, namely, changes in their profit and profitability, the influence of the amount of loan financing on these results was performed, in particular, in [25]. It is necessary to note the ambiguity of the conclusions reached by the authors of various studies devoted to this issue. First of all, it concerns the analysis of the effect of financial leverage, since this effect can be considered as a potential mechanism of influence of debt financing of firms on their profit and profitability. For example, in [26], the positive nature of the specified effect was established. At the same time, the authors [27] revealed some negative consequences of financial leverage for ensuring the proper financial condition of firms. After all, in [28], no influence of this effect was found at all. However, in some publications devoted to financial leverage, in particular in [29], the authors suggest applying an optimization approach. At the same time, it should be noted that there is a close relationship between financial leverage and the level of financial stability of companies.

One of the most important reasons for the differences in estimates of the effect of financial leverage is the lack of a generally accepted tool for determining the impact of the amount and structure of the loan capital of companies on the riskiness of their activities. Such a definition is complicated because there is not only a direct but also an inverse relationship between the parameters of loan financing and the level of riskiness of entrepreneurial activity. Thus, in [30], the negative influence of the uncertainty of enterprise managers in the future results of their activities on the amount of crediting of these enterprises was established. Similar results were presented by other researchers, in particular the authors of [31]. At the same time, as shown in [32], it is necessary to carefully substantiate those indicators, thanks to which it is expected to receive information about the riskiness of the activity of a particular company.

It is necessary to note the existence of objective obstacles to loan financing of companies' activities [33]. These obstacles, in their turn, play a certain preventive role concerning the possible deterioration of the financial stability of companies due to a significant increase in the amount of their loan capital [34]. In this case, we are talking about the limited availability of loan financing for certain groups of enterprises [35, 36]. Thus, the study of the financial activities of small European firms conducted in [37] showed that it was rather difficult for them to gain access to the services of credit institutions. Similar results were obtained for firms in countries with transition economies [38]. At the same time, as established in [39], it is characteristic of companies with smaller assets to finance mainly from their funds, while larger companies use much more external sources of funds. Under such conditions, preferential state lending can be a possible method of external financial support for enterprises, especially small ones [40].

Thus, the issue of justifying the reasoned amount of loan financing of enterprises has not been fully resolved in the scientific literature. Its solution would make it possible to overcome many obstacles to the technological development of enterprises [41] and, accordingly, ensure an increase in their competitiveness. At the same time, the concept of the loan potential of enterprises should be considered [42]. However, the material presented in [42] mainly concerns the financing of energy-saving projects. In addition, the loan potential of companies is only one, although a very important component of the potential for increasing the financial stability of enterprises. The evaluation of the latter can be considered the key to solving the problem of managing the financial stability of enterprises.

Although in modern literature, in particular in work [43], some models for managing the financial stability of enterprises are presented, modelling the potential for increasing this stability still needs to be properly performed. Accordingly, there is a need to form an array of information based on which the specified modelling should be performed. This array should constitute a separate block of information about the internal and external environment of enterprises [44].

At present, there is a need to improve the methodological bases of modelling the process of assessing the potential of increasing the financial stability of enterprises, as well as the need to improve the information support of the process of such an assessment. Solving the specified tasks with subsequent testing of the obtained results by a sample of enterprises is the primary goal of this study.

3   Modelling of the Processes of Formation and Assessment of the Potential for Increasing the Financial Stability of Enterprises

The potential for increasing the financial stability of the enterprise characterizes its ability to change the internal environment thanks to the implementation of relevant organizational, economic, and technical–technological measures, as a result of which an increase in the level of financial stability of the enterprise will be ensured. The general model of the formation of the potential for increasing the financial stability of the enterprise is presented in Fig. 1. As can be seen from this figure, the main direction of the growth of the company's financial stability is an increase in the amount of available income, which the company can use to repay the loans and pay the interest; a decrease in the size of the company's loan capital; an increase

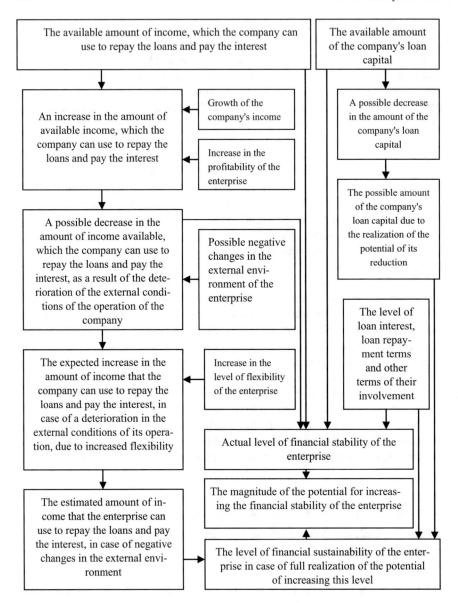

Fig. 1 Model of formation of the potential for increasing the financial stability of the enterprise. *Source* By authors

in the degree of flexibility of the firm. Accordingly, the potential for increasing the financial stability of the enterprise contains three components: the potential for increasing the amount of available income, which the enterprise can use to repay the loans and to pay interest, the potential for reducing the amount of loan capital, and the potential for increasing the degree of flexibility of the enterprise. Each of the listed components of the potential for increasing the financial stability of the enterprise, in turn, is determined by several types of the economic potential of the business entity. In particular, the growth potential of the available amount of income, which the enterprise can use to repay the loans and pay the interest, is determined by the potential of increasing the amount of income and the growth of profitability of the enterprise.

Assessment of the potential for increasing the financial stability of enterprises requires a preliminary determination of its criteria and evaluation indicators. At the same time, it is expedient to choose as a generalized criterion of the financial stability of the enterprise the condition when the amount of income that the enterprise can use to repay the loans and pay the interest exceeds the amount of repayment and servicing of these loans in any interval of the forecast period.

The main problem that arises when assessing the financial stability of an enterprise is the difficulty of forecasting the flow of the above-mentioned revenues. Indeed, if the forecast values of this flow are overestimated, the enterprise will be considered financially stable, although, in reality, it will not be so. Therefore, when carrying out such forecasting, it is advisable to focus on the possible deterioration of the company's operating conditions, in particular, the deterioration of the market conditions for its products and the increase in prices for resources that the company uses in its economic activities.

Modelling the process of assessing the financial stability of an enterprise is facilitated if we proceed to the averaged values of individual parameters that determine the level of this stability. These parameters should include the amount of income the company can use to repay the loans and pay the interest, as well as the time for which the loans were taken and the loan interest rate. At the same time, the first parameter should be averaged over the period during which the loans will be repaid, and the second and third parameters should be averaged over the entire set of components of the enterprise's loan capital. Under such conditions, the enterprise will be considered financially stable if the following inequality is fulfilled (see Eq. 1):

$$\sum_{t=1}^{T} \frac{F_a}{(1+r)^t} = \frac{F_a}{r} \cdot \left(1 - \frac{1}{(1+r)^T}\right) \geq C, \qquad (1)$$

where T is the average value of the repayment term of the loans, years (or other periods); F_a is the average value of the flow of income, which the enterprise can use to repay the loans and pay the interest while ensuring the normal conduct of its economic activity, monetary units; r is the average value of the loan interest rate for the use of loans, in fractions of a unit; C is the total amount of loans taken by the enterprise, monetary units.

Thus, in the case of fulfillment of inequality (1), the enterprise will have time to repay the loans promptly while paying the necessary amounts of interest for their use.

The use of criterion (1) makes it possible to make a quantitative assessment of the level of financial stability of the enterprise. For this purpose, let's represent expression (1) as equality and determine the indicator F_a (see Eq. 2):

$$F_{ma} = \frac{C \cdot r \cdot (1+r)^T}{(1+r)^T - 1},$$ (2)

where F_{ma} is the minimum possible average value of the flow of income, which the enterprise can use to repay the loans and pay the interest so that inequality (1) is fulfilled, monetary units.

Under such conditions, a quantitative assessment of the level of financial stability of the enterprise can be obtained by comparing the values of the F_a and F_{ma} indicators (see Eq. 3):

$$S_{fe} = \frac{F_a}{F_{ma}} = \frac{F_a \cdot \left((1+r)^T - 1\right)}{C \cdot r \cdot (1+r)^T},$$ (3)

where S_{fe} is the relative level of financial stability of the enterprise.

As follows from expression (3), the enterprise can be considered financially stable if the value of the indicator (3) is at least one. Otherwise, the enterprise is not financially stable. At the same time, the higher the value of the indicator (3), the higher the level of financial stability of the enterprise.

Based on Fig. 1 of the model of the formation of the potential for increasing the financial stability of the enterprise and taking into account the proposed indicator for evaluating this stability, it is possible to propose the following sequence of evaluating the specified potential:

(1) collection of necessary information and assessment of the current level of financial stability of the enterprise. In particular, in the case of using the averaged values of the relevant parameters, the assessment of the existing level of financial stability of the enterprise can be carried out with the help of formula (3);

(2) assessment of the growth potential of the available income, which the enterprise can use to repay the loans and pay the interest. For this purpose, it is necessary to consider the possibilities of ensuring the growth of the total amount of the enterprise's income, in particular, due to the increase in the degree of its business activity and the increase in the level of profitability of the enterprise's activity, first of all, due to the decrease in the relative level of its operating costs;

(3) assessment of the potential for reducing the company's loan capital. At the same time, the possible ways of such a reduction are the reduction of the amount of borrowed funds of the enterprise due to the simultaneous reduction of certain types of non-current assets of the enterprise, in particular, its fixed assets (as a result of the sale of excess elements of these assets), production stocks (as a

result of an increase in the number of supplies of materials), work in progress (as a result of reduction of the duration of the production cycle), accounts receivable for goods (as a result of providing price discounts to buyers on the condition of speeding up payment of purchased goods);

(4) assessment of the potential for increasing the flexibility of the enterprise. Such an increase requires, first of all, the improvement of the management competencies of the company's employees in terms of responding and adapting to changes in the external environment. In particular, this concerns the ability to regulate the level of prices for the enterprise's products, change the assortment and volumes of its production and sales, introduce new progressive technological processes, etc.;

(5) forecasting the level of financial stability of the enterprise in case of full realization of the potential of increasing this level. For this purpose, it is possible to modify formula (3), presenting it in the following form (see Eq. 4):

$$S_{pfe} = \frac{(F_0 + \Delta F_1 - \Delta F_2 + \Delta F_3) \cdot \left((1+r)^T - 1\right)}{(C_0 - \Delta C) \cdot r \cdot (1+r)^T}, \tag{4}$$

where S_{pfe} is the relative forecast level of financial stability of the enterprise; F_0 is the available value of the flow of income, which the enterprise can use to repay the loans and pay the interest, monetary units; ΔF_1 is the expected increase in the available amount of the income stream, which the enterprise can use to repay the loans and pay the interest, monetary units; ΔF_2 is a possible decrease in the available amount of the flow of income, which the enterprise can use to repay the loans and pay the interest as a result of the deterioration of the external conditions of the functioning of the enterprise, monetary units; ΔF_3 is the expected increase in the amount of the income stream, which the enterprise can use to repay the loans and pay the interest due to the realization of the potential of increasing the flexibility of the enterprise in case of deterioration of the external conditions of its functioning, monetary units; C_0 is the available amount of the company's loan capital, monetary units; ΔC is a possible decrease in the amount of the loan capital of the enterprise, monetary units;

(6) obtaining a quantitative assessment of the potential for increasing the level of financial stability of the enterprise. Such an estimate can be obtained by finding the difference between the projected and current levels of the financial stability of the enterprise (see Eq. 5):

$$\Delta S_{fe} = S_{pfe} - S_{fe}, \tag{5}$$

where ΔS_{fe} is the value of the potential for increasing the level of financial stability of the enterprise.

It is also possible to estimate the relative value of the potential for increasing the level of financial stability of the enterprise by comparing the result of calculations according to formula (5) with the current value of the level of this stability, which is calculated using expression (3).

It is important to note that the above-proposed approach to modelling the potential for increasing the level of the financial stability of enterprises involves determining the maximum possible value of this potential. At the same time, the criterion of maximizing the level of financial stability should be applied mainly to enterprises where the current value of this level is not high. Such enterprises need to achieve a level of financial stability equal to at least one. Regarding enterprises with a high current level of financial stability, this level may not be a criterion indicator but a definite limitation. Then, it is possible to build the following model of making a set of economic decisions that will have an impact on the financial stability of the enterprise (see Eqs. 6 and 7):

$$W = f(\alpha_1, ..., \alpha_i, ..., \alpha_k) \rightarrow \max; \tag{6}$$

$$S_{nfe} = u(\alpha_1, ..., \alpha_i, ..., a_k) \geq S_{mfe}, \tag{7}$$

where W is the criterion chosen by the managers of the enterprise for making economic decisions (maximum net profit, the maximum market value of the enterprise, etc.) as a function of their parameters, monetary units; α_i is the ith parameter of economic decisions that will have an impact on the financial stability of the enterprise, in the appropriate units of measurement; k is the number of parameters of economic decisions that will influence the financial stability of the enterprise; S_{nfe} is the forecast level of the financial stability of the enterprise as a function of the parameters of economic decision-making; S_{mfe} is the minimum acceptable forecast level of financial stability from the point of view of enterprise managers.

Thus, assessing the potential for increasing the financial stability of enterprises makes it possible to establish the total amount of reserves for increasing the level of this stability. At the same time, the issue of the feasibility of full implementation of these reserves needs separate consideration, taking into account the specifics of each enterprise.

4 Information Provision of the Process of Assessing the Potential for Increasing the Financial Stability of Enterprises

The complexity of assessing the potential for increasing the financial stability of enterprises is caused, in particular, by the fact that, as already noted above, many factors influence the level of this stability. Such factors, in particular, include the following types:

(1) the conditions of borrowings by the enterprise, in particular, the amount, structure, terms, and level of loan interest rates for various types of loans taken by the enterprise;

(2) factors that determine the amount of income that the enterprise can use to repay the loans and pay the interest. In particular, these factors include the volume of sales of the enterprise's products, the degree of its business activity, the relative level of the enterprise's operating expenses, etc.

(3) factors that determine the strength of possible negative effects of the deterioration of the conditions of the external environment of the enterprise on the financial and economic results of its economic activity. In particular, such factors can include a possible relative decrease in prices for the company's products, a reduction in demand for these products, an increase in prices for material, energy, and other resources used by the company, etc.;

(4) factors characterizing the enterprise's ability to resist the negative effects of the external environment. In particular, these factors should include the level of adaptation capabilities and flexibility of the enterprise in each direction of its adaptation to the deterioration of external operating conditions.

It should be noted that the factors of the formation of the financial stability of the enterprise can be grouped according to the following characteristics: the ability of the enterprise to influence them (managed and unmanaged), the place of occurrence (internal and external), the nature of the influence (direct and indirect), etc.

The accuracy of assessing the potential for increasing the financial stability of enterprises will largely depend on the extent to which the main factors affecting this stability are taken into account. It is necessary to properly organize information support for assessing the potential for increasing the financial stability of enterprises. Among other things, the hierarchical nature of the system of factors forming the potential for the financial stability of business entities should be considered. Accordingly, the array of information needed to estimate the value of this potential should also have a hierarchical structure. In particular, it is worth highlighting four levels of such information: input (primary), obtained directly from relevant sources of information; secondary, obtained due to the processing of primary information; generalizing, obtained as a result of processing secondary information, and containing information directly used in the calculation of final data on the level of financial stability of the enterprise and on the potential for increasing this level; the final one, containing the specified data.

At the same time, it is worth highlighting two sets of information (Tables 1 and 2), namely, the information based on which the current level of the financial stability of the enterprise is assessed and the information based on which the potential for increasing the financial stability of the business entity is assessed. However, it should be noted that assessing the potential for increasing the financial stability of enterprises also requires separate information, which is presented in the array of information based on which the current level of such stability is assessed. This conclusion is made from expressions (3)–(5) and the scheme presented in Fig. 2. At the same time,

Table 1 Information, based on which the existing level of the financial stability of the enterprise is evaluated

Types of information by its place in the hierarchy	Names of blocks of information	Markings of information blocks
1. Input (primary) information	1.1. A block of input information on the internal environment of the enterprise, necessary for assessing the current level of its financial stability	B 1.1.1
	1.1.1. Including the information on the current volume of loan capital of the enterprise	B 1.1.1.1
	1.2. A block of input information on the external environment of the enterprise, necessary for assessing the current level of its financial stability	B 1.1.2
2. Secondary (intermediate) information	2.1. A block of information on expected flows of income and expenses of the enterprise	B 1.2.1
	2.2. A block of information on the structure of the company's loan capital based on indicators characterizing the conditions for obtaining loans	B 1.2.2
3. General information	Data on the average amount of the flow of income, which the company can use to repay the loans and pay the interest	B 1.3.1
	3.2. Data on the average values of the repayment period of the loans and the loan interest for the use of the loans	B 1.3.2
4. Final information	4. Data on the current level of the financial stability of the enterprise	B 1.4

Source By authors

it should be emphasized that the markings of the blocks of the information shown in Fig. 2 correspond to the markings given in Tables 1 and 2.

The blocks of input (primary) information, presented in Tables 1 and 2, must contain a sufficiently complete amount of data on the value of all indicators that indirectly determine the level of financial stability of the company. First of all, it concerns the expected values of indicators that affect the flow of income and expenses of the enterprise. It is the completeness of the arrays of input information that largely depends on the quality of information provision to assess the potential for increasing the financial stability of the business entity.

Table 2 Information, based on which the potential for increasing the financial stability of the enterprise is estimated

Types of information by its place in the hierarchy	Names of blocks of information	Markings of information blocks
1. Input (primary) information	1.1. A block of input information on the internal environment of the enterprise, necessary for assessing the maximum possible level of its financial stability	B 2.1.1
	1.2. A block of input information on the external environment of the enterprise, necessary for assessing the maximum possible level of its financial stability	B 2.1.2
2. Secondary (intermediate) information	2.1. A block of information on the optimal parameters of business decisions to increase the financial stability of the enterprise	B 2.2.1
	2.2. A block of information on expected changes in the company's profit flow before interest and taxes are paid due to the implementation of each decision	B 2.2.2
	2.3. A block of information on expected changes in the volume and structure of the company's loan capital as a result of the implementation of each decision	B 2.2.3
3. General information	3.1. Data on the average value of the flow of income, which the company can use to repay the loans and pay the interest after the implementation of economic decisions	B 2.3.1
	3.2. Data on the average values of the repayment period of the loans and the loan interest for the use of the loans after the implementation of economic decisions	B 2.3.2
	3.3. Data on the total amount of loan capital of the enterprise after the implementation of economic decisions	B 2.3.3
4. Final information	4.1. Data on the maximum possible level of the financial stability of the enterprise	B 2.4.1
	4.2. Data on the potential for increasing the financial stability of the enterprise	B 2.4.2

Source By authors

5 Practical Applications of the Developed Principles of Modelling and Information Provision of Assessing the Potential for Increasing the Financial Stability of Enterprises

A sample of one hundred enterprises in the western region of Ukraine was formed to test the developed principles of modelling and information provision to assess the potential of increasing the financial stability of enterprises. Half of these enterprises operate in the wholesale trading of industrial goods, and the other half—in the food industry. Applying formula (3), the current financial stability of the studied enterprises was assessed. At the next stage of the research, these enterprises were graded

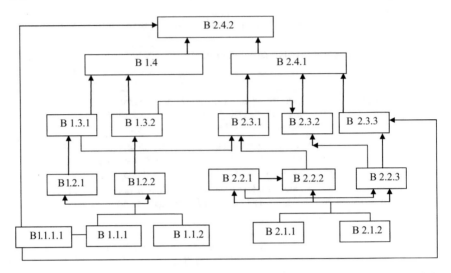

Fig. 2 Interrelationships between blocks of information, based on which the current level of the financial stability of the enterprise and the potential for its improvement are assessed. *Source* By authors

according to the level of their financial stability. The results of such gradation are presented in Table 3.

According to the data presented in Table 1, the share of the number of enterprises from each of their groups in the total number of investigated enterprises was calculated. As seen from the data presented in Figs. 3 and 4, an insufficient level of financial stability characterized most of the studied enterprises of both types of economic activity. In particular, the share of enterprises engaged in wholesale, in which the level of financial stability was low and very low, was 28% and 24%, respectively. The share of enterprises in the food industry, in which the level of financial stability was low and very low, was 22% and 18%, respectively.

To assess the influence of individual factors on the level of financial stability, we divided the studied enterprises by the share of loan capital in the total capital of companies and by the level of profitability of their total capital. According to the first indicator, the following gradation of its level was applied: a low share of loan capital (up to 33%), a medium share of loan capital (from 33 to 66%), and a high share of loan capital (over 66%). Regarding the second indicator, the following gradation was applied: low return on total capital (up to 5% per year), medium return on total capital (from 5 to 10% per year), and high return on total capital (over 10% per year). The results of the distribution of the studied enterprises by these two indicators according to the established gradation are presented in Tables 4 and 5.

As follows from the data presented in Tables 4 and 5, a low or medium level of return on total capital and a medium or high share of loan capital characterized both groups of the studied enterprises. A slightly larger number of enterprises with higher

Table 3 Distribution of the investigated enterprises according to the level of their financial stability as of 01.01.2022

Numbers of enterprise groups	Groups of enterprises, according to the level of their financial stability	Value ranges of the level of the financial stability of enterprises	Number of enterprises	
			Wholesale enterprises	Food industry enterprises
1	Enterprises with a very low level of financial stability	Up to 0.333	14	11
2	Enterprises with a low level of financial stability	From 0.333 to 0.667	12	9
3	Enterprises with an insufficient level of financial stability	From 0.667 to 1.000	10	12
4	Enterprises with a sufficient level of financial stability	From 1.000 to 1.333	11	13
5	Enterprises with a high level of financial stability	More than 1.333	3	5

Source By authors

Fig. 3 Shares of the studied trading enterprises with different levels of financial stability as of January 1, 2022, % (the numbers in the figure correspond to the numbers of enterprise groups in Table 3). *Source* By authors

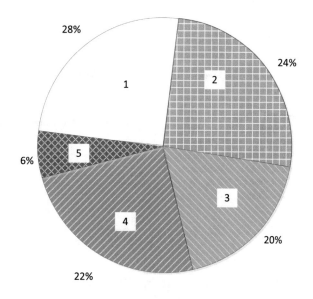

Fig. 4 Shares of the studied enterprises in the food industry with different levels of financial stability as of January 1, 2022, % (the numbers in the figure correspond to the numbers of enterprise groups in Table 3). *Source* By authors

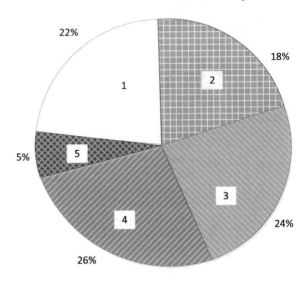

Table 4 The results of the distribution of the studied trading enterprises by the share of their loan capital and profitability of the total capital as of 01.01.2022

Profitability of total capital	The number of enterprises with a share of loan capital in the structure of total capital		
	Low	Medium	High
Low	6	6	8
Medium	4	5	7
High	5	4	5

Source By authors

Table 5 The results of the distribution of the studied enterprises in the food industry by the share of their loan capital and profitability of the total capital as of 01.01.2022

Profitability of total capital	The number of enterprises with a share of loan capital in the structure of total capital		
	Low	Medium	High
Low	4	5	6
Medium	6	7	5
High	5	6	6

Source By authors

profitability and a larger share of loan capital are characteristic of trading enterprises compared to enterprises in the food industry.

As a result of the distribution of enterprises within each type of economic activity, nine groups of companies were formed. We calculated the average current level of the financial stability of enterprises for each of these groups. The results of such calculations are presented in Tables 6 and 7.

Table 6 The average current level of the financial stability of the studied trading enterprises depending on the profitability of their total capital and the share of loan capital as of 01.01.2022

Profitability of total capital	The share of loan capital in the structure of total capital		
	Low	Medium	High
Low	0.371	0.293	0.195
Medium	0.698	0.543	0.403
High	1.390	1.319	1.264

Source By authors

Table 7 The average current level of the financial stability of the studied enterprises in the food industry, depending on the profitability of their total capital and the share of loan capital as of 01.01.2022

Profitability of total capital	The share of loan capital in the structure of total capital		
	Low	Medium	High
Low	0.405	0.276	0.203
Medium	0.779	0.563	0.435
High	1.432	1.390	1.341

Source By authors

As follows from the data presented in Tables 6 and 7, the following general trend can be observed for enterprises of both types of economic activity: with an increase in the profitability of total capital and a decrease in the share of loan capital, the level of financial stability of companies increases. To partially assess the statistical significance of this pattern, a variance analysis was performed. Its results are presented in Table 8.

Comparison of the data presented in Table 8, with critical values of the F-criterion, allows us to draw two main conclusions. First, the relationship between the share of loan capital of the studied enterprises and the current level of their financial stability is statistically significant, with a significance level of $\alpha = 0.05$ for enterprises with low and medium levels of return on total capital. Secondly, the specified dependence is not statistically significant for enterprises with a high return on their total capital. At the same time, these conclusions are valid for enterprises of both types of economic activity.

Table 8 Actual values of the F-criterion based on the results of the variance analysis of the dependence between the share of loan capital of the studied enterprises and the current level of their financial stability

The level of profitability of the total capital of enterprises	Actual values of the F-criterion	
	Trading enterprises	Food industry enterprises
Low	5.62	6.29
Medium	6.01	5.57
High	1.46	1.88

Source By authors

At the next stage of empirical research, several measures were developed for enterprises with a very low, low, and insufficient level of financial sustainability aimed at the increase of this level. Then, applying formula (4), the relative forecast level of its financial stability was calculated for each enterprise. The results of such calculations, averaged by groups of enterprises, are presented in Table 9.

By comparing the values of the forecast level of financial stability averaged by the groups of the studied enterprises with the values of the current level of this stability, it is possible to assess the possibilities of improving the specified level. In particular, using formula (5), it is possible to calculate the potential for increasing the financial stability of the enterprises under study. The results of such calculations for trading enterprises are presented in Table 10, and for food industry enterprises—in Table 11.

As follows from the data presented in Tables 10 and 11, the following patterns are inherent for enterprises of both types of economic activity:

(1) the absolute value of the potential for increasing the financial stability of enterprises increases with the deterioration of the current level of this stability;

Table 9 Values of the forecast level of financial stability, averaged by groups of the studied enterprises

Current level of financial stability of enterprises	Averaged values of the forecast level of financial stability	
	Trading enterprises	Food industry enterprises
Very low	0.256	0.231
Low	0.503	0.454
Insufficient	0.765	0.802

Source By authors

Table 10 Results of calculating the value of the potential for increasing the financial stability of the studied trading enterprises

The current level of the financial stability of enterprises	The level of financial stability averaged by groups of enterprises		The absolute value of the potential for increasing the financial stability, averaged by groups of enterprises			The relative value of the potential for increasing the financial stability, averaged by groups of enterprises
			General	Conditioned by		
	Current	Forecast		A possible increase in the profit of enterprises	A possible decrease in the amounts of loan capital of enterprises	
Very low	0.256	0.685	0.429	0.088	0.272	1.676
Low	0.503	0.848	0.345	0.107	0.147	0.686
Insufficient	0.765	1.024	0.259	0.131	0.078	0.339

Source By authors

Table 11 Results of calculating the value of the potential for increasing the financial stability of the studied enterprises in the food industry

The current level of the financial stability of enterprises	The level of financial stability averaged by groups of enterprises		The absolute value of the potential for increasing the financial stability, averaged by groups of enterprises			The relative value of the potential for increasing financial stability, averaged by groups of enterprises
	Current	Forecast	General	Conditioned by		
				A possible increase in the profit of enterprises	A possible decrease in the amounts of loan capital of enterprises	
Very low	0.231	0.685	0.454	0.073	0.261	1.965
Low	0.454	0.848	0.394	0.099	0.203	0.868
Insufficient	0.802	1.024	0.222	0.119	0.067	0.277

Source By authors

(2) the largest source of potential for increasing the financial stability of the enterprises under study is the reduction of the amount of their loan capital. At the same time, the importance of this direction of increasing the financial stability of enterprises increases with the deterioration of its current level;

(3) the role of the increase in the profit of enterprises as the way of forming the potential for increasing their financial stability increases with the improvement of the current level of this stability.

However, the conducted research showed that the main method of reducing the amount of loan capital of the enterprises in the trading sector is a simultaneous reduction in the value of the stock of goods and receivables by the same amount. At the same time, the main method of reducing the amount of loan capital of the enterprises in the food industry is a simultaneous reduction in the cost of material stocks and other production stocks by the same amount.

As already noted above, the correctness and completeness of the assessment of the potential for increasing the financial stability of enterprises largely depend on the level of information provision in the process of such assessment. Grading this level into low, medium, and high, the distribution of the studied enterprises was carried out according to these levels. The results of such distribution are presented in Table 12, and the indicators of its structure are shown in Table 13.

As follows from the data in Table 13, 62% of the studied trading enterprises and 54% of the food industry enterprises are characterized by a low level of information provision in the process of assessing the potential for increasing their financial stability. At the same time, only 8% and 12% of enterprises have a high level of this provision, respectively. In this regard, it is advisable to check the influence of the level of information provision in the process of assessing the potential for increasing the financial stability of enterprises at the current level of this stability (Table 14).

Table 12 Results of the distribution of the studied enterprises by the level of information provision of the process of assessing the potential for increasing their financial stability

The level of information provision of the process of assessing the potential for increasing the financial stability of enterprises	General characteristics of the level of information provision of the process of assessing the potential for increasing the financial stability of enterprises	The number of enterprises that belong to the corresponding level of information provision	
		Trading enterprises	Food industry enterprises
Low	There is no information provision on the process of assessing the potential for increasing financial stability in enterprises	31	27
Medium	Information provision of the process of assessing the potential for increasing financial stability at enterprises is partially available	15	17
High	Information provision on the process of assessing the potential for increasing financial stability at enterprises is available	4	6

Source By authors

Table 13 Indicators of the structure of the distribution of the studied enterprises by the level of information provision of the process of assessing the potential for increasing their financial stability

The level of information provision in the process of assessing the potential for increasing their financial stability	The share of enterprises that belong to the appropriate level of information provision of the total number of the studied enterprises, %	
	Trading enterprises	Food industry enterprises
Low	62	54
Medium	30	34
High	8	12

Source By authors

As follows from the data presented in Table 14, the dependence of the current level of financial stability of the studied enterprises on the level of information provision to assess the potential for increasing such stability is statistically significant. This is explained by the fact that the actual value of the F-criterion exceeds its critical value with a significance level of $\alpha = 0.05$.

Table 14 Results of verification of the influence of the level of information provision of assessing the potential for increasing the financial stability of enterprises at the current level of this stability

Indicator names	Indicator values	
	Trading enterprises	Food industry enterprises
1. Averaged values of the current financial stability of enterprises by their groups, depending on the level of information provision of assessing the potential for increasing this stability:		
1.1. For enterprises with a high level of information provision	1.197	1.249
1.2. For enterprises with a medium level of information provision	0.853	0.911
1.3. For enterprises with a low level of information provision	0.476	0.560
The actual value of the F-criterion	6.34	5.92

Source By authors

6 Conclusions

Establishing reserves for increasing the financial stability of enterprises requires the ability to assess the potential for its increase. In turn, such assessment should be based on the construction of appropriate models and the availability of appropriate information provision. At the same time, the models must take into account the conditions under which the company will have time to repay the loans in a timely manner while paying the necessary amounts of interest. Then the level of the financial stability of the enterprise can be estimated by the degree of approximation of a certain enterprise to the specified conditions. In turn, the possibility of growth of this measure can be an estimated value of the potential for increasing the financial stability of business entities.

The potential for increasing the financial stability of the enterprise characterizes its ability to change the internal environment thanks to the implementation of relevant organizational, economic, and technical–technological measures, as a result of which an increase in the level of financial stability of the enterprise will be ensured. The main directions of the growth of the company's financial stability are an increase in the amount of available income, which the company can use to repay the loans and pay the interest; a decrease in the size of the company's loan capital; an increase in the degree of flexibility of the firm. Accordingly, the potential for increasing the financial stability of the enterprise contains three components: the potential for increasing the amount of available income, which the enterprise can use to repay the loans and to pay interest, the potential for reducing the amount of loan capital, and the potential for increasing the degree of flexibility of the enterprise. Each of the listed components of the potential for increasing the financial stability of the enterprise, in turn, is determined by several types of the economic potential of the business entity. In particular, the growth potential of the available amount of income, which the enterprise can use to repay the loans and pay the interest, is determined by the potential of increasing the amount of income and the growth of profitability of the enterprise.

The accuracy of assessing the potential for increasing the financial stability of enterprises will largely depend on the extent to which the main factors affecting this stability are taken into account. It is necessary to properly organize information support for assessing the potential for increasing the financial stability of enterprises. Among other things, the hierarchical nature of the system of factors forming the potential for the financial stability of business entities should be considered. Accordingly, the array of information needed to estimate the value of this potential should also have a hierarchical structure. In particular, it is worth highlighting four levels of such information: input (primary), obtained directly from relevant sources of information; secondary, obtained due to the processing of primary information; generalizing, obtained as a result of processing secondary information, and containing information directly used in the calculation of final data on the level of financial stability of the enterprise and on the potential for increasing this level; the final one, containing the specified data.

The conducted empirical analysis of a sample of enterprises also made it possible to establish the following patterns:

(1) the absolute value of the potential for increasing the financial stability of enterprises increases with the deterioration of the current level of this stability;
(2) the largest source of potential for increasing the financial stability of the enterprises under study is the reduction of the amount of their loan capital. At the same time, the importance of this direction of increasing the financial stability of enterprises increases with the deterioration of its current level;
(3) the role of the increase in the profit of enterprises as the way of forming the potential for increasing their financial stability increases with the improvement of the current level of this stability.

However, the conducted research showed that the main method of reducing the amount of loan capital of the enterprises in the trading sector is a simultaneous reduction in the value of the stock of goods and receivables by the same amount. At the same time, the main method of reducing the amount of loan capital of the enterprises in the food industry is a simultaneous reduction in the cost of material stocks and other production stocks by the same amount.

The analysis also showed that 62% of the studied trading enterprises and 54% of the food industry enterprises are characterized by a low level of information provision in the process of assessing the potential for increasing their financial stability. At the same time, only 8% and 12% of enterprises have a high level of this provision, respectively. At the same time, the dependence of the current level of financial stability of the studied enterprises on the level of information provision to assess the potential for increasing such stability is statistically significant.

References

1. Fraser, J.R.S., Quail, R., Simkins, B.J.: Questions that are asked about enterprise risk management by risk practitioners. Bus. Horiz. **65**(3), 251–260 (2022). https://doi.org/10.1016/j.bushor.2021.02.046

2. Fu, J.Q.: Enterprise risk assessment of based on data mining. In: 14th International Conference on Measuring Technology and Mechatronics Automation (ICMTMA), Changsha, China, pp. 1052–1057 (2022). https://doi.org/10.1109/ICMTMA54903.2022.00212

3. Arshi, N.: Role of artificial intelligence in business risk management. Am. J. Bus. Manag. Econ. Bank. **1**, 55–66 (2022). https://www.americanjournal.org/index.php/ajbmeb/article/view/37

4. Arif, A., Yu, H.M., Cong, M., Wei, L.H., Islam, M., Niedbala, G.: Natural resources commodity prices volatility and economic performance: evaluating the role of green finance. Resour. Policy **76**, 102–143 (2022). https://doi.org/10.1016/j.resourpol.2022.102557

5. Prokopczuk, M., Stancu, A., Symeonidis, L.: The economic drivers of commodity market volatility. J. Int. Money Financ. **98**, 102063 (2019). https://doi.org/10.1016/j.jimonfin.2019.102063

6. Opazo-Basáez, M., Vendrell-Herrero, F., Bustinza, O.F.: Digital service innovation: a paradigm shift in technological innovation. J. Serv. Manag. **33**(1), 97–133 (2022). https://doi.org/10.1108/JOSM-11-2020-0427

7. Wen, H., Zhong, Q., Lee, C.C.: Digitalization, competition strategy and corporate innovation: evidence from Chinese manufacturing listed companies. Int. Rev. Financ. Anal. **82**, 102166 (2022). https://doi.org/10.1016/j.irfa.2022.102166

8. Drobyazko, S., Barwinska-Malajowicz, A., Slusarczyk, B., Chubukova, O., Bielialov, T.: Risk management in the system of financial stability of the service enterprise. J. Risk Financ. Manag. **13**(12), 300 (2020). https://doi.org/10.3390/jrfm13120300

9. Pilch, B.: An analysis of the effectiveness of bankruptcy prediction models—an industry approach. Folia Oeconomica Stetinensia **21**(2), 76–96 (2021). https://doi.org/10.2478/foli-2021-0017

10. Husainova, E.A., Urazbahtina, L.R., Serkina, N.A., Dolonina, E.A., Derbeneva, A.A., Filina, O.V.: Features of management and factors of economic stability of an industrial enterprise in the region. E3S Web Conf. **124**, 05025 (2019). https://doi.org/10.1051/e3sconf/201912405025

11. Andersson, F.N.G.: The quest for economic stability: a study on Swedish stabilisation policies 1873–2019. Scand. Econ. Hist. Rev. (ahead-of-print) 1–29 (2021). https://doi.org/10.1080/03585522.2021.1984300

12. Gospodarchuk, G.G., Zeleneva, E.S.: Stability of financial development: problems of measurement, assessment and regulation. PLoS ONE **17**(11), e0277610 (2022). https://doi.org/10.1371/journal.pone.0277610

13. Yemelyanov, O., Symak, A., Petrushka, T., Lesyk, R., Lesyk, L.: Evaluation of the adaptability of the Ukrainian economy to changes in prices for energy carriers and to energy market risks. Energies **11**, 3529 (2018). https://doi.org/10.3390/en11123529

14. Akbulaev, N., Guliyeva, N., Aslanova, G.: Economic analysis of tourism enterprise solvency and the possibility of bankruptcy: the case of the Thomas Cook Group. African J. Hosp. Tour. Leis. **9**(2), 1–12 (2020). https://www.ajhtl.com/uploads/7/1/6/3/7163688/article_26_vol_9_2_2020_azerbaijan.pdf

15. Izmailova, K., Zapiechna, Y.: Study of unprofitability of Ukraine's large construction enterprises by the DuPont method. Three Seas Econ. J. **1**(4), 84–89 (2020). https://doi.org/10.30525/2661-5150/2020-4-12

16. Yemelyanov, O., Petrushka, T., Symak, A., Lesyk, L., Musiiovska, O.: Modelling the impact of energy-saving technological changes on the market capitalization of companies. In: Systems, Decision and Control in Energy III, pp. 106–189. Springer (2022). https://doi.org/10.1007/978-3-030-87675-3_5

17. Korepanov, G., Yatskevych, I., Popova, O., Shevtsiv, L., Marych, M., Purtskhvanidze, O.: Managing the financial stability potential of crisis enterprises. Int. J. Adv. Res. Eng. Technol. **11**(4), 359–371 (2020). https://ssrn.com/abstract=3599794

18. Cheong, C., Hoang, H.V.: Macroeconomic factors or firm-specific factors? An examination of the impact on corporate profitability before, during and after the global financial crisis. Cogent Econ. Finance **9**, 1–24 (2021). https://doi.org/10.1080/23322039.2021.1959703

19. Gajdosikova, D., Valaskova, K., Kliestik, T., Kovacova, M.: Research on corporate indebtedness determinants: a case study of Visegrad Group countries. Mathematics **11**(2), 299 (2023). https://doi.org/10.3390/math11020299

20. Dinh, H.T., Pham, C.D.: The effect of capital structure on the financial performance of Vietnamese listing pharmaceutical enterprises. J. Asian Finance Econ. Bus. **7**(9), 329–340 (2020). https://doi.org/10.13106/jafeb.2020.vol7.no9.329

21. Yemelyanov, O., Petrushka, T., Symak, A., Trevoho, O., Turylo, A., Kurylo, O., Danchak, L., Symak, D., Lesyk, L.: Microcredits for sustainable development of small Ukrainian enterprises: efficiency, accessibility, and government contribution. Sustainability **12**, 6184 (2020). https://doi.org/10.3390/su12156184

22. Rostamkalaei, A., Freel, M.: The cost of growth: small firms and the pricing of bank loans. Small Bus. Econ. **46**, 255–272 (2016). https://doi.org/10.1007/s11187-015-9681-x

23. Yang, W.: Empirical study on effect of credit constraints on productivity of firms in growth enterprise market of China. J. Financ. Econ. **6**, 173–177 (2018). https://doi.org/10.12691/jfe-6-5-2

24. Lin, Y., Li, L.: Empirical analysis of microcredit in Western China: based on empirical analysis. J. Chongqing Technol. Bus. Univ. **5**, 1 (2018). https://en.cnki.com.cn/Article_en/CJFDTotal-CQYZ201805001.htm

25. Akinleye, G.T., Olarewaju, O.O.: Credit management and profitability growth in Nigerian manufacturing firms. Acta Univ. Danub. Oecon. **15**, 445–456 (2019). http://journals.univ-danubius.ro/index.php/oeconomica/article/view/5281/5232

26. Gill, A.S., Mand, H.S., Sharma, S.P., Mathur, N.: Factors that influence financial leverage of small business firms in India. Int. J. Econ. Financ. **4**, 33 (2012). https://doi.org/10.5539/ijef.v4n3p33

27. Javed, Z.H., Rao, H.H., Akram, B., Nazir, M.F.: Effect of financial leverage on performance of the firms: empirical evidence from Pakistan. SPOUDAI J. Econ. Bus. **65**, 87–95 (2015). https://EconPapers.repec.org/RePEc:spd:journl:v:65:y:2015:i:1-2:p:87-95

28. Hoque, M.A.: Impact of financial leverage on financial performance: evidence from textile sector of Bangladesh. IIUC Bus. Rev. **6**, 75–84 (2017). http://dspace.iiuc.ac.bd:8080/xmlui/handle/88203/687

29. Adenugba, A.A., Ige, A.A., Kesinro, O.R.: Financial leverage and firms' value: a study of selected firms in Nigeria. Eur. J. Res. Reflect. Manag. Sci. **4**, 14–32 (2016). https://www.idpublications.org/wp-content/uploads/2016/01/Full-Paper-FINANCIAL-LEVERAGE-AND-FIRMS'-VALUE-A-STUDY-OF-SELECTED-FIRMS-IN-NIGERIA.pdf

30. Agarwal, S., Chomsisengphet, S., Driscoll, J.C.: Loan Commitments and Private Firms. FEDS Working Paper No. 2004-27 (2004). https://doi.org/10.2139/ssrn.593862. https://ssrn.com/abstract=593862

31. Choi, S., Furceri, D., Huang, Y., Loungani, P.: Aggregate uncertainty and sectoral productivity growth: the role of credit constraints. J. Int. Money Financ. **88**, 314–330 (2018). https://doi.org/10.1016/j.jimonfin.2017.07.016

32. Lesinskyi, V., Yemelyanov, O., Zarytska, O., Symak, A., Koleshchuk, O.: Substantiation of projects that account for risk in the resource-saving technological changes at enterprises. East. Eur. J. Enterp. Technol. **6**, 6–16 (2018). https://doi.org/10.15587/1729-4061.2018.149942

33. Kryvinska, N., Strauss, C.: Conceptual model of business services availability vs. interoperability on collaborative IoT-enabled eBusiness platforms. In: Internet of Things and Intercooperative Computational Technologies for Collective Intelligence, pp. 167–187 (2013). https://doi.org/10.1007/978-3-642-34952-2_7

34. Fauska, P., Kryvinska, N., Strauss, C.: The role of e-commerce in B2B markets of goods and services. IJSEM **5**, 41 (2013). https://doi.org/10.1504/IJSEM.2013.051872
35. Fauska, P., Kryvinska, N., Strauss, C.: Agile management of complex goods & services bundles for B2B E-commerce by global narrow-specialized companies. Glob. J. Flex. Syst. Manag. **15**, 5–23 (2014). https://doi.org/10.1007/s40171-013-0054-5
36. Fauska, P., Kryvinska, N., Strauss, C.: E-commerce and B2B services enterprises. In: 2013 27th International Conference on Advanced Information Networking and Applications Workshops (2013). https://doi.org/10.1109/WAINA.2013.98
37. Angori, G., Aristei, D.: A Panel Data Analysis of Firms' Access to Credit in the Euro Area: Endogenous Selection, Individual Heterogeneity and Time Persistence 2018. https://doi.org/10.2139/ssrn.3254358. https://ssrn.com/abstract=3254358
38. Rostamkalaei, A., Freel, M.S.: The cost of growth: small firms and the pricing of bank loans. Small Bus. Econ. **46**(2), 255–272 (2016). https://doi.org/10.1007/s11187-015-9681-x
39. Bhalli, M., Hashmi, S., Majeed, A.: Impact of credit constraints on firms growth: a case study of manufacturing sector of Pakistan. J. Quant. Methods **1**(1), 4–40 (2017). https://doi.org/10.29145/2017/jqm/010102
40. Yemelyanov, O., Petrushka, T., Lesyk, L., Symak, A., Vovk, O.: Modelling and information support for the development of government programs to increase the accessibility of small business lending. In: IEEE 15th International Conference on Computer Sciences and Information Technologies (CSIT), pp. 229–232 (2020). https://doi.org/10.1109/CSIT49958.2020.9322040
41. Lesinskyi, V., Yemelyanov, O., Zarytska, O., Petrushka, T., Myroshchenko, N.: Designing a toolset for assessing the organizational and technological inertia of energy consumption processes at enterprises. East. Eur. J. Enterp. Technol. **6**(13), 29–40 (2022). https://doi.org/10.15587/1729-4061.2022.267231
42. Lesinskyi, V., Yemelyanov, O., Zarytska, O., Symak, A., Petrushka, T.: Devising a toolset for assessing the potential of loan financing of projects aimed at implementing energy-saving technologies. East. Eur. J. Enterp. Technol. **4**(13), 15–33 (2021). https://doi.org/10.15587/1729-4061.2021.238795
43. Sylkin, O., Kryshtanovych, M., Zachepa, A., Bilous, S., Krasko, A.: Modeling the process of applying anti-crisis management in the system of ensuring financial security of the enterprise. Bus.: Theory Pract. **20**, 446–455 (2019). https://doi.org/10.3846/btp.2019.41
44. Ren, S.: Optimization of enterprise financial management and decision-making systems based on big data. J. Math. **2022**, 1708506 (2022). https://doi.org/10.1155/2022/1708506

Transformation of a Regional IT Cluster into a Cross-Border IT Cluster as a Direction of It Business Development Under the Conditions of Negative Influence of External Factors

Mariya Kirzhetskaⓘ, **Ihor Novakivskyi**ⓘ, **Oksana Musiiovska**ⓘ,
Yuriy Kirzhetskyyⓘ, and **Iryna Yepifanova**ⓘ

Abstract The problem of the transformation of a regional IT cluster (Lviv IT cluster) into a cross-border IT cluster under the negative influence of external factors was considered, and the stability of the functioning of the combined structures was investigated to determine the level of synergistic effect. The apparatus of nonlinear differential equations is used for this purpose, making it possible to analyze at an abstract level the complex economic market relations of enterprises that arise during a merger.

Keywords Cross-border cluster · Regional IT cluster · IT sector · Singular effect · Labor potential

1 Introduction

The importance of information business, which lies at the intersection of information, business, and technology, is constantly growing—both at the national level and, in comparison with the competitiveness of countries, at the international level. This

M. Kirzhetska · I. Novakivskyi · O. Musiiovska
Lviv Polytechnic National University, Lviv, Ukraine
e-mail: mariya.s.kirzhetska@lpnu.ua

I. Novakivskyi
e-mail: inovak@ukr.net

O. Musiiovska
e-mail: oksana.b.musiiovska@lpnu.ua

Y. Kirzhetskyy
Lviv State University of Internal Affairs, Lviv, Ukraine

I. Yepifanova (✉)
Vinnytsia National Technical University, Vinnytsia, Ukraine
e-mail: yepifanova@vntu.edu.ua

© The Author(s), under exclusive license to Springer Nature Switzerland AG 2024
A. Semenov et al. (eds.), *Data-Centric Business and Applications*, Lecture
Notes on Data Engineering and Communications Technologies 195,
https://doi.org/10.1007/978-3-031-54012-7_10

business became a strategic direction for increasing competitiveness, productivity, and employment in the conditions of war in Ukraine and all directions will be considered in our work. The growth of sales volume, the focus not only on internal, but also on external consumers, preservation, and in some cases, creation of new jobs, and resistance to the economic crisis caused by the war make the information business industry an attractive area of investment both domestically: private and public, and on the foreign market. On the other hand, the field of knowledge regarding the economic role of information business and information entrepreneurship and their impact on the economy of a country bearing economic losses from armed aggression is quite limited.

In Ukraine, information entrepreneurship has reached a significant scale and occupies a special place in the multifunctional economy. The information business sector of Ukraine, by its essence, is a pool of industries that provide services in the field of informatization and telecommunication technologies; produce software and hardware tools, information products, and databases that logically form a single whole and interact for the active formation and development of the information space and information technologies based on the knowledge and values of the information society [1].

In pre-war 2021 the information business sector of Ukraine generated gross added value in the amount of UAH 255.635 million, which is 5.5% of Ukraine's GVA. In 2021, the industry was in the TOP 3, according to the index of gross added value by types of economic activity (Table 1).

When analyzing the generalized statistical data we see that the current state of development of information entrepreneurship is characterized by positive dynamics of profitability indicators: net profit and profitability of operating activity, despite the negative impact of the war on all sectors of economic activity in Ukraine.

During January–June 2022 profitability of the operating and all activities of large and medium-sized enterprises is positive (Table 2). The information business sector of Ukraine generated 17.5% profitability of operational activity of enterprises, in our opinion is quite a significant indicator of the functioning sector of economic activity in Ukraine during wartime.

A major part of the added value in the information business sector of Ukraine is generated by the IT industry: 52%.

Table 1 Index of gross added value by types of economic activity in 2021 in Ukraine (TOP 5 sectors of the economy)

Sectors of the economy	Volume indices of gross, %
Construction	189.1
Financial and insurance activities	159.2
Information and communication	145.8
Accommodation and food service activities	143.9
Real estate activities	120.5

Source State Statistics Service of Ukraine

Table 2 Profitability indicators by types of economic activity

Sectors of the economy	Profitability of operational activity of enterprises	Profitability of all activity of enterprises
Total	**1.8**	**−3.3**
Agriculture, forestry and fishing	−2.6	−5.8
Industry	1.5	−2.6
Construction	−4.1	−5.1
Wholesale and retail trade; repair of motor vehicles and motorcycles	10.8	−2.7
Transportation and warehousing, postal and courier activities	3.6	−3.3
Accommodation and food service activities	−17.2	−21.2
Information and communication	17.5	5.8
Financial and insurance activities	−2.2	0.8
Real estate activities	5.4	−16.7
Professional, scientific and technical activities	−13.6	−15.6
Administrative and support service activities	6.5	3.7
Education	44.1	22.3
Human health and social work activities	0.9	4.4
Arts, sport, entertainment and recreation	−48.5	−50.5

Source State Statistics Service of Ukraine

That is why the IT industry and the peculiarities of its functioning and development in Ukraine under the influence of an external factor: the war will be the object of this research.

Figure 1 shows the index of gross added value in the information business sector in 2022 in Ukraine.

If we look at the ratio of people employed in the IT sector in Ukraine and the countries bordering (Table 3) we can see that the share of specialists involved in this industry is the lowest in Ukraine [2].

In the short term, the war factor did not have a decisive impact on the total number of specialists in the IT sector. However, there is a slight decrease in non-technical specialists in the sector (a 3% decrease) in the first half of 2022. The total number of specialists in the top 50 IT companies of Ukraine reached 100,000 for the first time

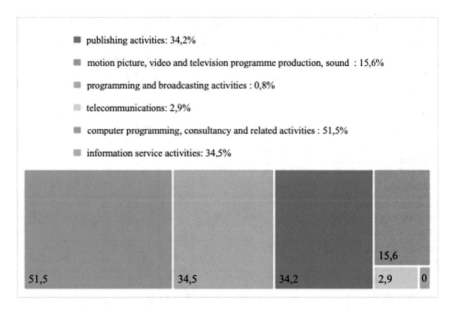

- publishing activities: 34,2%
- motion picture, video and television programme production, sound : 15,6%
- programming and broadcasting activities : 0,8%
- telecommunications: 2,9%
- computer programming, consultancy and related activities : 51,5%
- information service activities: 34,5%

Fig. 1 Index of gross added value in the information business sector in 2022 in Ukraine (*Source* State Statistics Service of Ukraine)

Table 3 The employment rate in the IT sector in Ukraine and the CEE countries of the region

Country	Population, millions	Number of software developers, thousand	The coefficient of employed in IT, %
Poland	38,400	0.2798	0.73
Ukraine	38,600	0.1847	0.48
Romania	19,600	0.1161	0.59
Czech	10,600	0.0953	0.9
Slovakia	9800	0.0801	0.82

in January. But in the first half of the year, it decreased by 3%, and currently 97,600 specialists work in the 50 largest companies of Ukraine (Fig. 2).

The landscape of the functioning of the IT sector in Ukraine is characterized by the unification of IT companies in 22 economic clusters, which were created according to the regional component. The largest among them are Kyiv (39%); Kharkiv (28% in 2021–10% in 2022); Lviv (24% in 2021 and 16.5% in 2022); Dnipro (8.4% in 2021 and 22% in 2022) and Odesa (8% in 2021 and 5% in 2022).

The current trends in the functioning of the IT sector of Ukraine, under the influence of the security component, are also influenced by the relocation of IT specialists and IT companies to regions that are territorially distant from hostilities and outside the country. We can see that the employees of the Kharkiv IT cluster were mostly relocated to the Central and Western regions of Ukraine, and the share of employees

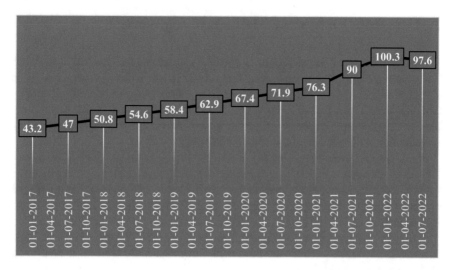

Fig. 2 The total number of specialists in the top 50 IT companies of Ukraine, in thousands [3]

from the Western regions of Ukraine, especially, Lviv and Volyn regions, to Poland. Due to the internal migration of IT specialists from the Eastern and Southern regions, the number of IT sector specialists in the Western regions increased by 24% at the beginning of summer, and the number of IT companies that opened their offices in Lviv region increased by 10%. At least 30 companies opened new offices in Western Ukraine in the first half of 2022. There is also a trend to open offices in foreign locations, among which Poland has become the most popular. Thus, 15 of the largest companies opened their offices in Krakow, Poznan, Bydgoszcz, Gdansk, Lodz, and Warsaw.

However, the nature of IT business development in Ukraine, the efficiency of operation, and the level of competitiveness, both before the military aggression and during 2022, are somewhat lower than in the EU countries. This is the result of the unsystematic, deformed, and sometimes contradictory development of information entrepreneurship, in particular IT business in Ukraine. The main reason for such situation is the lack of modern fundamental scientific research in this direction and the uncertainty of strategic priorities for the development of IT business in Ukraine under the conditions of the negative influence of war factors, which objectively actualizes the problem of strategic development of information entrepreneurship, in particular IT business, on a scientific basis within the limits of modern concepts, programs, and effective economic policy.

The purpose of the study is the scientific substantiation of the strategic priorities of the development and functioning of IT business in Ukraine under the influence of determining factors, in particular the influence of the war factor on the transformation of a regional IT cluster into a cross-border IT cluster.

2   Literature Review and Methodical Approaches

The analysis of scientific research on the problem found that the work of M. E. Porter (for example, Porter [4]) is fundamental in this direction, which defines a cluster as a geographically close group of interconnected companies and related institutions (suppliers, etc.) in a certain field, which compete and cooperate at the same time. However, based on this research, there are several new concepts that allow a deeper investigation of the features of such associations.

Thus, Gupta and Subramanian [5], Rosenfeld [6], Chapain and Sagot-Duvauroux [7] and Yepifanova et al. [8] investigated information sharing in the development of labor potential between cluster participants. In the works of Chan and Zheng [9], the positive and significant impact of migrants on cross-border mergers and acquisitions is confirmed.

The territorial component and its impact: both positive and negative on the development of cluster associations are thoroughly investigated by Gordon and McCann [10], Yemelyanov et al. [11].

The influence of cultural factors and target legal environments on the development of the cross-border cluster was studied by Brada and Iwasaki [12].

The concept of the cross-border cluster is quite clearly defined in economic literature. Thus, scientists investigating this issue note that a cross-border cluster is one type of economic cluster. And Ukrainian scientists S. Kovalenko, N. Dobreva define this form of cooperation as a form of network interaction of specialized enterprises of the Euroregion, integrated on both sides of the border [13, 14]. Mikula and Pasternak [15], Mrykhina et al. [15, 16] determined that the formation of cross-border clusters involves obtaining synergistic and network effects, diffusion of knowledge and skills [16, 17]. Kryvinska et al. in [18–20] describe a foundation of the definitions and concepts regarding Servitization in the modern IT Business.

Amiti and Pissarides [21], Brenton and Manzocchi [22] in their scientific works highlight factors of stimulation and disincentive for the region in the formation of cross-border clusters. The authors attribute a larger labor market and the possibility of choosing more qualified employees and reducing operating costs to the incentives. Disincentives include the existence of a border as a potential barrier that increases the cost of resource transit, including labor.

Zimmer [23] investigates the problem of the possibility of cross-border cluster participants obtaining a synergistic effect or, on the contrary, an additional competitive economic force that absorbs resources from regional competitors and centers.

By transforming Sebastian Zimmer's visual model of cooperation, we will get cooperation between two regional clusters through the formation of a cross-border cluster, within which participants indirectly cooperate with each other, but they are competitors outside the cluster (Fig. 3).

Based on the analysis of literary sources and the analysis of statistical information, we will summarize the personal and business factors that, in our opinion, are a prerequisite for the formation of the Ukrainian-Polish cross-border IT cluster and were formed under the influence of the war in Ukraine.

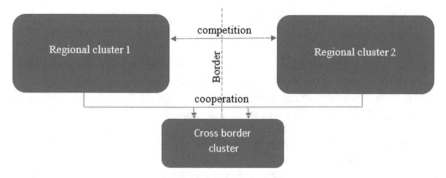

Fig. 3 Model of interaction of regional border clusters and cross-border cluster

Among the personal factors that were singled out based on a survey of specialists in the IT sector of Ukraine [24] and which create prerequisites for the transformation of regional IT clusters into cross-border IT clusters under the influence of military actions, it is appropriate to identify the following:

(1) Security factor: failure to meet the basic human need for security: shelling and destruction of residential and energy infrastructure. A survey of Ukrainian IT specialists conducted by the Lviv IT cluster among cluster participants shows that 50% of IT specialists intend to migrate abroad in the event of freezing of military actions; loss of housing at the place of permanent residence, general mobilization of the population.

(2) An economic factor, among which: job loss or, on the contrary, the introduction of additional incentives for relocation, shortage of vacancies in Ukraine, the introduction of strict regulation of the economy, or a significant increase in the cost of living is a factor in the emigration of IT specialists.

(3) Social factor: reconstruction of Ukraine without the support of Western countries; freezing of the military situation in the war with Russia, emigration sentiments of Ukrainian specialists, despite the war (20% of specialists have a desire to migrate abroad, despite the war factor).

(4) Cultural factor. For Ukrainians, Poland is a country with a very attractive culture. 43% of respondents spontaneously indicated Poland as the most culturally interesting country, and 56% of Ukrainians declared an interest in Polish culture, such as books or films. The feeling of cultural closeness to the Poles is more pronounced in the regions of historical Volyn, Podillia, and Kyiv region (81% of indications) and Halychyna (91%), which were connected to Poland for several centuries. The survey also showed that there is a great demand in Ukraine to get to know Poland and Polish culture: 27% of Ukrainians, including 35% of young people under the age of 24, said that they became more interested after the start of the war [25].

As for the business factors that were formed by the war and had an impact on the formation of the Ukrainian-Polish cross-border IT cluster, we can highlight the following among them:

(1) Mobilization of employees of IT companies: in particular, 7,000 specialists have been recruited into the ranks of the Armed Forces since the beginning of the year.

(2) The largest share of employees who are determined to leave are professionals with 3–6 years of work experience and Middle/Senior qualifications.

(3) A quite young age of the employees and the majority of IT specialists of the company are in the age range of 21–30 years.

(4) The determining factor for work in the IT sector in Ukraine is the level of wages.

(5) The number of IT specialists in the regions of Western Ukraine in 2022 is more than 52,000 people and 80% of them are technical specialists.

(6) Educational factor: in 2023, 100,000 Ukrainian secondary and high school students continued their studies in Polish schools, especially in the regions bordering Ukraine: Krakow, Łódź, and Poznan. Before the war, about 40,000 students from Ukraine studied in Polish universities; about 600 students who are registered as Ukrainian refugees have continued their studies or are in the process of being recruited to study at Polish universities, and 2,800 people from Ukraine who are on the territory of Poland have submitted documents to universities expressing interest in enrolling as students [26].

(7) The level of knowledge of the Polish language is still at a low level, but a survey conducted among Ukrainians shows that 37% of respondents are interested in learning the Polish language [27].

(8) A survey of CEOs of IT companies in Ukraine regarding plans to open new offices shows that there is a high interest in opening new offices abroad. And 75% of respondents out of the 15% of IT companies that plan to open an office abroad are considering Poland.

(9) As a result of the war, 20% of contracts were lost, especially due to the high risk of force majeure events caused by the territorial affiliation of companies to Ukrainian jurisdiction.

Taking into consideration the analysis of the market situation, which developed under the influence of the factor of military aggression against Ukraine, we can conclude that there is relevance in the study of the development of domestic IT structures. The importance of solving this problem is determined by the following factors:

– the majority of Ukrainian enterprises in the field of material production were forced to reduce and localize their activities, which significantly limited production transnational connections, and at the same time, IT structures mainly preserved their potential even in the conditions of the migration of their offices;

– IT enterprises are less dependent on the use of material resources, and even the introduction of restrictions on the use of energy resources affects their activities to a lesser extent;

– the survival of Ukraine's economy primarily depends on maintaining its manageability, on the ability to adapt and be flexible, which largely depends on the development of national IT structures;

– the further integration of Ukraine into the EU mainly lies in the sphere of intangible activity, and the survival and further development of the country, in general, will depend on the level of integration of the national information infrastructure into the European transnational one.

IT structures in general have an impact on the entire economy of Ukraine, and therefore the formation of a wide range of urgent tasks is possible, among which is the transformation of the Lviv IT cluster into a cross-border Polish-Ukrainian IT cluster. The expected result of the functioning of such a cross-border structure is the preservation of the economic growth potential of IT companies in Ukraine, in particular in Western Ukraine. That is the participant: the Lviv IT cluster, which successfully developed as a regional cluster before the start of military aggression after its start experienced a decrease in efficiency due to the negative impact of external factors. That is why, increasing the efficiency of cluster participants is possible due to obtaining a synergistic effect from the growth of labor potential using a positive security component of close, in accordance with cultural features, neighboring cross-border territories, in our example—Poland.

As for the constructive factors of the development of cross-border IT clusters in Poland, Ewa Kraska's research shows that there is also an increased interest in cluster structures in this country in the following directions [27]:

(1) increasing the innovativeness of enterprises;
(2) development of information technologies;
(3) Polish participation in the activities of various international organizations,
(4) aspirations of enterprises to increase their competitiveness and innovation.

3   Research Results

Let's analyze the existing problem of the stability of the development of the Lviv IT cluster through its transformation into a cross-border cluster through the merger with ICT Polska Centralna Klaster with similar characteristics of labor potential. To solve this scientific and practical problem, we will use the apparatus of models of nonlinear differential equations. The authors chose the model of nonlinear differential equations because they conduct research on systems: independently functioning clusters and the labor potential of these clusters changes over time. The use of such models is a partial case when using production models and it will enable to reveal of the dynamics of the internal mechanism of the phenomenon of synergy in the case of mergers and acquisitions of companies, firms, enterprises, and, therefore, it will be possible to highlight the synergistic effect of the merger in case it exists.

The model of extended development of a single isolated system can be presented in the following form (Fig. 4).

In this case, for a system of independently functioning clusters, their independent development can be described by the following system of differential equations:

Fig. 4 Model of functioning
of a single isolated system Y_i

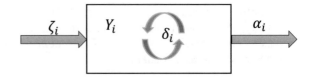

$$\begin{cases} \frac{dY_1}{dt} = \zeta_1 \cdot Y_1(t) - \alpha_1 \cdot Y_1(t) - \delta_1 \cdot Y_1(t)^2; \\ \frac{dY_2}{dt} = \zeta_2 \cdot Y_2(t) - \alpha_2 \cdot Y_2(t) - \delta_2 \cdot Y_2(t)^2, \end{cases}$$

where $Y_1(t)$, $Y_2(t)$ is respectively, the labor potential of the clusters; ζ_1, ζ_2 is the coefficients of accumulation of the power of cluster's labor potentials; α_1, α_2 is the coefficients of realizing the capacity of clusters labor potentials; δ_1, δ_2 is the coefficients of the averaged non-productive internal losses of the capacity of clusters labor potentials; t is the time parameter.

It is not difficult to somewhat simplify this system to the form:

$$\begin{cases} \frac{dY_1}{dt} = (\zeta_1 - \alpha_1) \cdot Y_1(t) - \delta_1 \cdot Y_1(t)^2; \\ \frac{dY_2}{dt} = (\zeta_2 - \alpha_2) \cdot Y_2(t) - \delta_2 \cdot Y_2(t)^2. \end{cases}$$

In the case of stable growth, the solution for each independent differential equation is described by the following equation:

$$\begin{cases} \frac{dY_1}{dt} = 0.04 \cdot Y_1(t) - 0.009 \cdot Y_1(t)^2; \\ \frac{dY_2}{dt} = 0.12 \cdot Y_2(t) - 0.008 \cdot Y_2(t)^2. \end{cases}$$

In our case to describe the development of the Lviv IT cluster under certain accepted assumptions it was calculated that:

$$\begin{cases} Y_1(0) = 4.1; \\ Y_2(0) = 12.2. \end{cases}$$

For this case, the graph of the expected growth of the labor potential of the Lviv IT cluster is shown in Fig. 5.

For this case, the graph of the expected growth of the labor potential of the ICT Polska Centralna Klaster is shown in Fig. 6.

The result of the model for these clusters, which are described is in an analytical form:

$$\begin{cases} Y_1(t) = \dfrac{1640}{369 + 31 \cdot e^{-\frac{t}{25}}}; \\ Y_2(t) = \dfrac{915}{61 + 14 \cdot e^{-\frac{3}{25}}}. \end{cases}$$

Differences in the development of the clusters are due to the scale of these clusters.

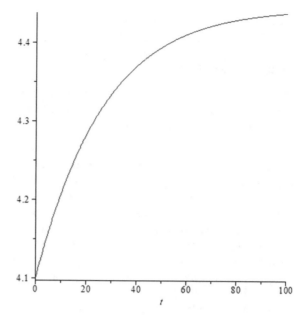

Fig. 5 Graph of growth of labor potential of Lviv IT. *Maple 14 was used to build the graph*

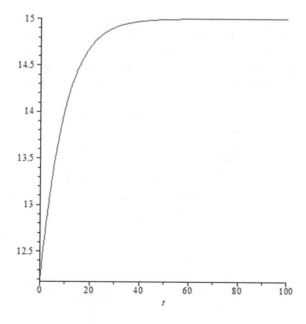

Fig. 6 Graph of growth of labor potential of ICT Polska Centralna Klaster. *Maple 14 was used to build the graph*

The proposed model of association of the Lviv IT cluster and ICT Polska Centralna Klaster into a Ukrainian-Polish cross-border cluster assumes that the unification model must necessarily take into account that the freedom to realize the available labor potential is transferred from individual IT clusters to their association. This assumption should form the basis of the merger, which in turn should ensure the manifestation of the synergistic effect of the combined structure.

The establishment of coordination and partnership relations among the participants of the united cross-border IT cluster will allow:

(1) to expand the circle of persons interested in its use;
(2) jointly use the accumulated experience of scientific and technical developments, optimize the attraction of labor potential,
(3) reduce the terms of order fulfillment, etc.

Mutual cooperation within the framework of the united cross-border cluster expands the horizons of activity, especially for the participants of the Lviv IT cluster.

The unification process is a rather complex process, which will unfold in parallel in various areas of IT cluster activity meaning organizational, legislative, etc.

We will analyze possible options for association in the area of organizing and coordinating the use of the labor potential of IT clusters. When considering the basic principle of synergy, it should be expected that the exchange with the external environment after the unification of the systems should give a tangible result of an increase in labor potential.

For a more detailed study of the functioning of such a combined system, the model is shown in Fig. 7.

In this example, the following notations are used: $Y_1(t)$ is the capacity of the labor potential of Lviv IT Cluster; $Y_2(t)$ is the capacity of the labor potential of ICT Polska Centralna Klaster; $Y_3(t)$ is the capacity of the labor potential of cross-border cluster; $\zeta_1, \zeta_2, \zeta_3$ is the coefficients of growth of labor potential capacities of IT clusters; α_{ij} is the coefficient of transfer of the capacities of the labor potential of the i-th IT cluster

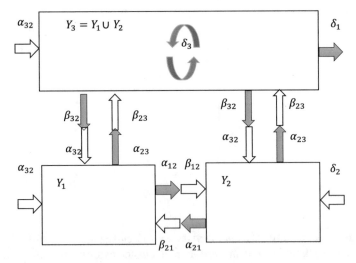

Fig. 7 The model of unification of the capacities of the labor potential of IT clusters under the condition of their stable development

to the j-th cluster (the first index defines the participant-donor and the second—the recipient); β_{ij} is the utilization rate of the labor potential of the i-th IT cluster in the j-th cluster; η is the coefficient of reproduction of the capacities of the labor potential into the external environment.

It is worth mentioning that this model takes into account the possible loss of part of the capacity during the transfer under market conditions, which will be denoted by:

$$\alpha_{ij} - \beta_{ij} = \delta_{ij} \geq 0,$$

δ_{ij} is the coefficient of inconsistency when transferring the capacities of the labor potential of the i-th IT cluster to the j-th cluster.

In Fig. 7, this situation is shown by dark and light arrows. It should be noted that we will not consider internal inconsistency in IT clusters (unlike model (1)) but will only take into account the inconsistency of the united cross-border cluster.

In the most general case, the model of combining IT clusters can be written as follows:

$$
\begin{cases}
\frac{dX_1}{dt} = \zeta_1 \cdot X_1(t) + \beta_{21} \cdot X_2(t) \cdot X_1(t) - \alpha_{12} \cdot X_1(t) \cdot X_2(t) + \beta_{31} \cdot X_3(t) \cdot X_1(t) - \alpha_{13} \cdot X_1(t) \cdot X_3(t) - \alpha_{11} \cdot X_1(t)^2; \\
\frac{dX_2}{dt} = \zeta_2 \cdot X_2(t) + \beta_{12} \cdot X_1(t) \cdot X_2(t) - \alpha_{21} \cdot X_1(t) \cdot X_2(t) + \beta_{32} \cdot z(t) \cdot X_2(t) - \alpha_{23} \cdot X_2(t) \cdot X_3(t) - \alpha_{22} \cdot X_2(t)^2; \\
\frac{dX_3}{dt} = (\zeta_3 - \eta) \cdot X_3(t) + \beta_{13} \cdot X_1(t) \cdot z(t) - \alpha_{31} \cdot X_3(t) \cdot X_1(t) + \beta_{23} \cdot X_2(t) \cdot X_3(t) - \alpha_{32} \cdot X_3(t) \cdot X_2(t) - \alpha_{33} \cdot X_3(t)^2;
\end{cases}
\tag{1}
$$

The left part of the equation represents the amount of growth in the labor potential of IT clusters.

For a better meaningful representation of the capacity transfer of the labor potential and bearing in mind the losses we will introduce the following notations (see Eq. 2):

$$
\begin{cases}
\beta_{12} = \alpha_{12} - \delta_{12}; \\
\beta_{21} = \alpha_{21} - \delta_{21}; \\
\beta_{13} = \alpha_{13} - \delta_{13}; \\
\beta_{31} = \alpha_{31} - \delta_{31}; \\
\beta_{23} = \alpha_{23} - \delta_{23}; \\
\beta_{32} = \alpha_{32} - \delta_{32}
\end{cases}
\tag{2}
$$

In this case, we get the following system of equations (see Eq. 3):

$$
\begin{cases}
\frac{dX_1}{dt} = \zeta_1 \cdot X_1(t) - (\alpha_{12} - \alpha_{21} + \delta_{21}) \cdot X_1(t) \cdot X_2(t) - (\alpha_{13} - \alpha_{31} + \delta_{31}) \cdot X_1(t) \cdot X_3(t) - \delta_{11} \cdot X_1(t)^2; \\
\frac{dX_2}{dt} = \zeta_2 \cdot X_2(t) - (\alpha_{21} - \alpha_{12} + \delta_{12}) \cdot X_2(t) \cdot X_1(t) - (\alpha_{23} - \alpha_{32} + \delta_{32}) \cdot X_2(t) \cdot X_3(t) - \delta_{22} \cdot X_2(t)^2; \\
\frac{dX_3}{dt} = (\zeta_3 - \eta) \cdot X_3(t) - (\alpha_{13} - \alpha_{31} + \delta_{13}) \cdot X_3(t) \cdot X_1(t) - (\alpha_{23} - \alpha_{32} + \delta_{23}) \cdot X_3(t) \cdot X_2(t) - \delta_{33} \cdot X_3(t)^2;
\end{cases}
\tag{3}
$$

To simplify the system of differential equations, we will use the following indicators (see Eq. 4):

$$\begin{cases} \alpha_{12} - \alpha_{21} + \delta_{21} = \lambda_{12}; \\ \alpha_{13} - \alpha_{31} + \delta_{31} = \lambda_{13}; \\ \alpha_{21} - \alpha_{12} + \delta_{12} = \lambda_{21}; \\ \alpha_{23} - \alpha_{32} + \delta_{32} = \lambda_{23}; \\ \alpha_{13} - \alpha_{31} + \delta_{13} = \lambda_{31}; \\ \alpha_{23} - \alpha_{32} + \delta_{23} = \lambda_{32}; \end{cases} \tag{4}$$

As a result (see Eq. 5):

$$\begin{aligned} \frac{dX_1}{dt} &= \zeta_1 \cdot X_1(t) - \lambda_{12} \cdot X_1(t) \cdot X_2(t) - \lambda_{13} \cdot X_1(t) \cdot X_3(t) - \delta_{11} \cdot X_1(t)^2; \\ \frac{dX_2}{dt} &= \zeta_2 \cdot X_2(t) - \lambda_{21} \cdot X_2(t) \cdot X_1(t) - \lambda_{23} \cdot X_2(t) \cdot X_3(t) - \delta_{22} \cdot X_2(t)^2; \\ \frac{dX_3}{dt} &= (\zeta_3 - \eta) \cdot X_3(t) - \lambda_{31} \cdot X_3(t) \cdot X_1(t) - \lambda_{32} \cdot X_3(t) \cdot X_2(t) - \delta_{33} \cdot X_3(t)^2; \end{aligned} \tag{5}$$

The model was calculated for the following model parameters (see Eq. 6):

$$\begin{cases} \frac{dX_1}{dt} = 0.04 \cdot X_1(t) - 0.001 \cdot X_1(t) \cdot X_2(t) - 0.001 \cdot X_1(t) \cdot X_3(t) - 0.001 \cdot X_1(t)^2; \\ \frac{dX_2}{dt} = 0.12 \cdot X_2(t) - 0.001 \cdot X_2(t) \cdot X_1(t) - 0.0009 \cdot X_2(t) \cdot X_3(t) - 0.001 \cdot X_2(t)^2; \\ \frac{dX_3}{dt} = 0.005 \cdot X_3(t) - 0.04 \cdot X_3(t) \cdot X_1(t) - 0.001 \cdot X_3(t) \cdot X_2(t) - 0.0001 \cdot X_3(t)^2; \end{cases} \tag{6}$$

And such initial conditions (see Eq. 7):

$$\begin{cases} X_1(0) = 4.1; \\ X_2(0) = 6.6; \\ X_3(0) = 12.5. \end{cases} \tag{7}$$

The resulting graph of the growth of the power of the labor potential of the united cross-border IT cluster is shown in Fig. 8.

From Fig. 8, we can conclude that the creation of a united cluster can provide a singular effect.

The results of the total independent functioning of IT clusters reach the level of 20 units, while the joint one provides a level of at least 22 units.

However, the behavior of the united cluster is unstable—at first it grows up to 26 units, but then the process stabilizes at the level of 22 units.

It can be assumed that the behavior of the united IT cluster requires constant regulation in such areas as the presence of a collegial center for the cooperation of the participants, the functioning of an open common informational environment, the development of the internal communication environment, compliance with the agreed corporate culture, etc. In our opinion, a high level of the indicated indicators can ensure the stable functioning of the united cross-border IT cluster.

Fig. 8 Reflecting the synergistic effect on the model of unification of IT clusters

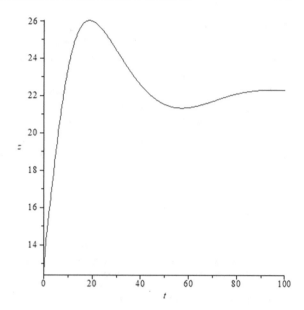

4   Conclusions

Even despite the state of war, Ukraine today retains a great intellectual potential that can form a powerful IT center in Central and Eastern Europe, which will attract international investments and strengthen Ukraine's position on the world stage. Unfortunately, in the conditions of martial law, this potential is far from being used to its full extent.

Restrictions on international cooperation, which arose due to Russia's aggression in Ukraine, significantly limited physical international contacts as well as somewhat narrowed the current opportunities for the deployment of own projects. Along with this, new horizons for cooperation with the developed countries of Europe and America are being opened up for domestic IT structures. This change should be considered and used as much as possible, for example, when forming various cross-border cooperation projects.

Our case, which was based on the assumption of the need to increase the strength of the labor potential of the Lviv IT cluster due to its integration into the EU environment and the creation of a cross-border cluster association with ICT Polska Centralna Klaster, showed that the obtained results confirm the synergistic effect, since after the merger, their joint functioning gives greater effect rather than the effect of their separate functioning.

The used model of nonlinear differential equations enables to highlight of the synergistic effect of combining two clusters. However, it should be noted that specific situations require clarification of the stability of the resulting system of differential equations. After all, in such a case, the development trajectories diverge, and the solutions obtained are unstable and may differ greatly in the event of small perturbations of the system. When considering the basic principle of synergy, it should be expected that the exchange with the external environment after the unification of the systems should give a tangible result of an increase in labor potential.

Along with this, the analysis of the considered models showed that such processes are very complex and require constant adjustment in order to maintain them on a stable development trajectory. It is obvious that in such a case further development of economic and mathematical models is required, which will make it possible to specify organizational decisions regarding the creation of a cross-border cluster association.

References

1. Prodanova, L.V., Sherstyukova, K.Yu.: Osoblyvosti rozvytku informatsiynoho biznesu ta pidpryyemnytstva v Ukrayini. Visnyk KHNAU. Seriya: Ekonomichni nauky **4**(2), 115–126 (2019). https://repo.btu.kharkov.ua/bitstream/123456789/7011/1/statya_prodanova_sherstiuk ova_osoblyvosti_rozvytku_informatsiinoho_biznesu_ta_pidpryiemnytstva_v_Ukraini.pdf
2. Malashniak, M.: Software development in Ukraine: 2019–2020 IT market report [online]. https://www.n-ix.com/software-development-in-ukraine-2019-2020-market-report/ (2020). Accessed 10 Dec 2022
3. Ipolitova, I.: Rynok pratsi pid chas viyny. Kil'kist' vakansiy zmenshylasya vdvichi, kompaniyi shche aktyvnishe shukayut' sen'yoriv i na QA Intern tysyacha vidhukiv [online]. https://dou.ua/lenta/articles/job-market-during-wartime/?from=similar_posts (2020). Accessed 25 Dec 2022
4. Porter, M.E.: The economic performance of regions. Reg. Stud. **37**, 549–578 (2003). https://doi.org/10.1080/0034340032000108688
5. Gupta, V., Subramanian, R.: Seven perspectives on regional clusters and the case of Grand Rapids office furniture city. Int. Bus. Rev. **17**(4), 371–384 (2008). https://doi.org/10.1016/j.ibusrev.2008.03.001
6. Rosenfeld, S.: Bringing business clusters into the mainstream of economic development. Eur. Plan. Stud. **5**, 3–23 (1997). https://doi.org/10.1080/09654319708720381
7. Chapain, C., Sagot-Duvauroux, D.: Cultural and and creative clusters—a systematic literature review and a renewed research agenda. Urban Res. Pract. **13**, 300–329 (2020). https://doi.org/10.1080/17535069.2018.1545141
8. Dzhedzhula, V., Hurochkina, V., Yepifanova, I., Telnov, A.: Fuzzy technologies for modeling social capital in the emergent economy. WSEAS Trans. Bus. Econ. **19**, 915–923 (2022). https://doi.org/10.37394/23207.2022.19.80
9. Chan, J.M.L., Zheng, H.: FDI on the move: cross-border M&A and migrant networks. Rev. World Econ. **158**, 947–985 (2022). https://doi.org/10.1007/s10290-021-00450-1
10. Gordon, I.R., Mccann, P.: Industrial clusters: complexes, agglomeration and/or social networks? Urban Studies **37**(3), 513–532 (2000). https://doi.org/10.1080/0042098002096
11. Yemelyanov, O., Petrushka, K., Tranchenko, O., Ponomariov, O., Parubok, N.: Strategic diagnosis of regional rural areas. J. Adv. Res. Dyn. Control Syst. **12**, 376–389 (2020). https://doi.org/10.5373/JARDCS/V12SP7/20202119
12. Brada, J.C., Iwasaki, I.: Do target-country legal institutions affect cross-border mergers and acquisitions? A quantitative literature survey. Eur. J. Law Econ. (2022). https://doi.org/10.1007/s10657-022-09751-8
13. Kovalenko, S.I.: Transkordonni klasterni systemy—mezoriven mizhnarodnykh intehratsiynykh ob"yednan'. Prychornomorski ekonomichni studiyi **11**, 22–26. http://www.bses.in.ua/journals/2016/11-2016/6.pdf (2016)

14. Dobrieva, N.F.: Osnovni napriamy rozvytku transkordonnykh klasteriv v Ukraini. Efektyvnist derzhavnoho upravlinnia **34**, 246–253 [Online]. http://nbuv.gov.ua/j-pdf/efdu_2013_34_29.pdf (2013)

15. Mikula, N.A., Pasternak, O.I.: Transkordonni klastery. Rehional'na ekonomika **2**, 228–229. https://re.gov.ua/re200902/re200902_za.pdf (2009)

16. Chukhray, N., Shakhovska, N., Mrykhina, O., Lisovska, L., Izonin, I.: Stacking machine learning model for the assessment of R&D product's readiness and method for its cost estimation. Mathematics **10**, 1466 (2022). https://doi.org/10.3390/math10091466

17. Kozyk, V., Lutak, O., Lisovska, L., Mrykhina, O., Novakivskyj, I.: The impact of economic entities' innovative activity on the indicators of sustainable development of Ukraine. In: IOP Conference Series: Earth and Environmental Science, vol. 628 (2021). https://iopscience.iop.org/article/10.1088/1755-1315/628/1/012041/pdf

18. Amiti, M., Pissarides, C.: Trade and industrial location with heterogeneous labor. J. Int. Econ. **67**(2), 392–412 (2005). https://doi.org/10.1016/j.jinteco.2004.09.010

19. Kryvinska, N., Bickel, L.: Scenario-based analysis of IT enterprises servitization as a part of digital transformation of modern economy. Appl. Sci. **10**, 1076 (2020). https://doi.org/10.3390/app10031076

20. Kryvinska, N., Kaczor, S., Strauss, C.: Enterprises' servitization in the first decade—retrospective analysis of back-end and front-end challenges. Appl. Sci. **10**, 2957 (2020). https://doi.org/10.3390/APP10082957

21. Kryvinska, N., Kaczor, S., Strauss, C., Gregus, M.: Servitization strategies and product-service-systems. In: 2014 IEEE World Congress on Services (2014). https://doi.org/10.1109/SERVICES.2014.52

22. Brenton, P., Manzocchi, S.: Enlargement, trade and investment: the impact of barriers to trade in Europe. https://api.semanticscholar.org/CorpusID:152727954 (2002)

23. Zimmer, S.: Cross-border clusters: opportunity or competitive threat? https://www.researchgate.net/publication/282287430_Cross-Border_Clusters_Opportunity_or_Competitive_Threat (2014)

24. IT-Research-Resilience-2022 [online]. https://itcluster.lviv.ua/wp-content/uploads/2022/07/IT_Research_Resilience_2022-public-UA.pdf (2022)

25. Polska kultura oczami Ukrainco [online]. https://mieroszewski.pl/programy/badania-opinii-publicznej/polska-kultura-oczami-ukraincow?fbclid=IwAR289F24UunNnL7GCIOdcmBbHRfYpkrr0JBuGLSrOonuCuLydnnbVlx6O8Y (2022)

26. How many students from Ukraine have been accepted by Polish universities? There are data [online]. https://wmeritum.pl/studentow-polskie-uczelnie/372017 (2022)

27. Kraska, E.: Analysis of the functioning of clusters in Poland. J. Int. Bus. Res. Mark. **7**(1), 29–33 (2021). https://doi.org/10.18775/10.18775/jibrm.1849-8558.2015.71.3004

Formation of Organizational Change Management Strategies Based on Fuzzy Set Methods

Olha Bielienkova⊙, Galyna Ryzhakova⊙, Oleksii Kulikov⊙,
Roman Akselrod⊙, and Yana Loktionova⊙

Abstract Change management is an urgent requirement for the successful functioning of an enterprise, and the search for change factors is an urgent scientific
task. The toolkit of the fuzzy set method is used in the work to identify and justify
change management strategies, considering the degree of development and the financial state of the organization at any moment in time. The model includes factors that
describe the flexibility of the management system, the efficiency of operations, the
enterprise's activity on the market, and its ability to adapt to market conditions. The
factors "rigidity-flexibility of the management system", "market position—the ratio
of the company's market share to the market share of the largest competitor", "production efficiency—profitability of production activities" and "level of adaptability"
were selected for evaluation. Combinations of these factors can affect the success of
the strategy and the ability of the organization to implement changes. Modeling was
carried out based on data from a sample of Ukrainian construction enterprises during
2007–2020. The constructed system of fuzzy rules based on the model of the artificial intelligence system makes it possible to evaluate the processes and factors that
are key to the formation of change management strategies and allows for multi-level
classification of success factors. The developed method reflects the internal connections between the components of the artificial intelligence system. The results of

---

O. Bielienkova (✉) · G. Ryzhakova · R. Akselrod · Y. Loktionova
Kyiv National University of Construction and Architecture, Povitroflotskyi Ave. 31, Kyiv 03037,
Ukraine
e-mail: bielienkova.oiu@knuba.edu.ua

G. Ryzhakova
e-mail: ryzhakova.gm@knuba.edu.ua

R. Akselrod
e-mail: Akselrod.knuba@ukr.net

Y. Loktionova
e-mail: loktionova.yaf@knuba.edu.ua

O. Kulikov
Kyiv National University of Trade and Economics, Kyoto Str. 19, Kyiv 02156, Ukraine
e-mail: k_kpipp@ukr.net

© The Author(s), under exclusive license to Springer Nature Switzerland AG 2024
A. Semenov et al. (eds.), *Data-Centric Business and Applications*, Lecture
Notes on Data Engineering and Communications Technologies 195,
https://doi.org/10.1007/978-3-031-54012-7_11

251

this study can be used to further improve the knowledge and skills necessary for the successful implementation of change management.

Keywords Change management · Business performance · Fuzzy logic · Strategies · McKinsey "7S" model

1 Introduction

Globalization, digitalization, demographic changes, the rapid implementation of innovative technological, and technical business solutions, and other factors, as well as the transformation of the world economic and political systems as a result of the war in Ukraine, have led to the fact that changes have become a constant background for the activities of enterprises. Today, the development and functioning of economic entities is impossible without changes. Moreover, the demands of today are such that organizations that resist change must either perish or change anyway.

The article [1] claims that changes have become the only constant in the activities of enterprises and organizations. This statement is supported by the authors of this work.

Changes can be positive and contribute to further development, or they can have a negative impact on the operational, financial, and investment activities of the enterprise. Therefore, change management and adaptation to changes is a relevant direction of economic science and management, which is designed to adapt enterprises to the fluidity of external and internal operating conditions and ensure the highest efficiency in the conditions of changes and transformations.

However, despite the great attention paid by researchers to this issue, the effectiveness of change management for many organizations is still low, and the percentage of failures is 60–70% for organizational change projects [2, 3].

For modern enterprises, the basis of change management should be a development strategy, which today should adapt to volatility, and uncertainty and enable constant adaptation to new requirements. Although there are opinions that in the new, constantly changing reality, the basis for development should not be strategic (which is rejected as insufficiently effective in changing conditions), the authors adhere to the opinion that strategy should serve as a guide for development, but it should really transform and adapt to new conditions, showing great flexibility. Therefore, a relevant direction of research is the identification of the influence of factors that largely influence the strategy on the adaptive capacity of enterprises in changing conditions.

2 Literature Review and Methodology

Many works are devoted to the study of change management of enterprises of various sectors of the economy since it is impossible to achieve stable development without changes. At the same time, change management is often considered as supportive leadership, personnel, multidisciplinary teamwork, a joint approach to work, strong communicative behavior, successful models, conflict resolution and identification of capable employees capable of acting as drivers of change and participating in further development [4] can be factors of change. A statistically significant influence was found between change management and efficiency, where the factors are the areas of change implementation, the type of changes, impulses to change, indicators for evaluating the consequences of changes, and management [5], management and leadership style [6–10], thinking systems [11], as well as strategies [12] and structural changes [13, 14]. The factors of change can be the leadership of the change manager, effective and constant communication during the change, the involvement of stakeholders and the motivation of employees and change agents, and the lack of leadership skills, low participation of stakeholders, and poor communication of the main changes are the reasons for the failure of change management [3]. At the same time, many works emphasize the role of leadership in the success of transformations [15–19].

Also, in a number of works, digitization is called the basis for effective transformations of business processes and change management [20–22]. The article [1] examines the tools of change management, among which are the definitions of the forces driving change and resisting change, their directions, nature, and strength, as well as how they can be modified for the effective operation of the organization, leadership style, and strategy. The works [23, 24] emphasize the importance of diversification.

In works [5, 25–28] it is stated that to measure the effectiveness of organizations use financial results (profit, profitability, etc.), market indicators (market share, sales volume, etc.), and business profitability for owners (added value, profitability per share, etc.). However depending on the method of measuring the effectiveness and the goal of the organization, the strategy is formed. Therefore, not only the strategy but also the efficiency factors and the market factor were chosen as the factors that influence the success of change management, as such, which serve as a basis for the formation and implementation of the strategy.

Within the framework of the conceptual approach, which is embedded in the McKinsey model, it is proposed to use artificial intelligence tools to determine the ability of enterprises to change, which are used as an effective means of managing enterprises in conditions of uncertainty and environmental variability, and also allow flexibility in identifying the optimal strategy for some combination factors.

The data of 26 construction enterprises collected from open sources (the company's website, generators of data on the financial state and activity of Smida enterprises, YouControl) were used as a sample. A sample of 151 observations was studied. All enterprises according to the classification of types of economic activity

(KVED-2010) belong to section F construction of buildings (division 41), civil engineering (division 42), namely, perform mainly or general construction works or works in the direction of "road construction". Data on the activities of companies from 2007 to 2020 are used.

The methodological approach used to identify the influence of factors that shape strategy and adaptability to changes is not new and is used in various works to identify factors of enterprise competitiveness [25, 29], capital expenditure management [30], management of the financial condition of organizations [31], multi-objective optimization [32], research on the influence of personnel on adaptability to changes [33].

The research sequence consists of the following stages:

1. *Collection and processing of raw data.* It is a mandatory stage when working with statistical data, which is used in all studies, so we will not dwell on it in detail.
2. *Formation of a theoretical scheme of a hierarchical system of fuzzy logical conclusion,* which shows the influence of a number of factors on the formation of organizational change management strategies. It is suggested that factors S_1–S_4 are selected as indicators of the impact on the company's change management strategy, which are considered in detail and substantiated according to the approach laid down in the McKinsey 7S model in work [5], namely:

    S_1—*the flexibility of the enterprise management system* (the assessment was carried out by an expert on a verbal scale according to the "flexible-rigid" rating. Since the assessment was carried out over a four- to seven-year period for each enterprise, at different time intervals for the same organization, the assessment of the flexibility of the management system could differ);

    S_2—*competitiveness* (since there are quite a lot of methods and indicators for assessing competitiveness and they differ in complexity, accuracy, and approaches [29], the relative factor was chosen as the evaluation indicator, which was defined as the ratio of the market share organization to the market share of the largest competitor according to the You Control website).

    S_3—*efficiency of production activity*—a relative indicator that can also be measured in several ways (financial results, profitability, growth or preservation of market share, etc. In this work, efficiency was assessed by the indicator of profitability (unprofitability) of the enterprise's production activity (according to financial reporting data on the websites of the researched enterprises or according to Smida.gov.ua accelerator data).

    S_4—*the level of adaptability of the enterprise* shows the ability to respond to changes in time and adapt to real circumstances. The indicator was defined as the ratio of the profitability of the enterprise's production activity to the average level of profitability of enterprises in the construction sector. A value greater than one means that the company adapts to environmental changes better than most of the companies in the industry, if this indicator is less than one, then the company is unable to respond to changes and transformational processes in the industry in

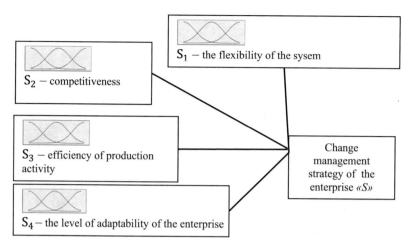

Fig. 1 The model of the influence of factors S_1–S_4 on the choice of change management strategy

a timely and qualitative manner. As a result, we get the "four inputs-one output" model (Fig. 1).

3. *Fuzzification.* At this stage, with the help of membership functions of input variables, the transition from numerical parameters of input variables to fuzzy values of linguistic variables is carried out. That is, membership functions are created for variable terms separately for each group of indicators, which, according to the McKinsey 7S model, describe common values, structure, strategy, management system, style, personnel, and skills).

In this case, we have created the appropriateness functions for indicators of factors S_1–S_4. After formulating membership functions for the terms of all input variables, logical rules of fuzzy inference are drawn up and the weight of each rule is determined.

Membership functions allow for any value from a series of input data to determine its degree of membership to a fuzzy set, while all input variables are given a different number (from two to four) of membership functions of the type game (Gaussian function), which specifies a combination of membership functions in the form Gaussian curves and has the following form (Fig. 2) [29, 31, 33]:

$$\mu(P) = e^{-\frac{1}{2} \cdot \left(\frac{x-c}{\sigma}\right)^2}, \tag{1}$$

where σ is the coordinate of the maximum of the membership function, and c is the concentration coefficient of the membership function, which is determined based on the distribution of the characteristics in the general population.

The concentration coefficient determines the "transition point", that is, such a value of the universe of the function, for which the degree of confidence in belonging to the corresponding term will be equal to 0.5. This value reflects the maximum uncertainty in the belonging of the studied factor to the corresponding term. In the MatLab environment, the formulas of the membership functions are not written

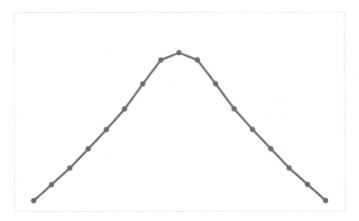

Fig. 2 Gaussian function

explicitly, instead, a concise record of the function parameters is provided, which for the Gaussian type are denoted as an ordered array of two numbers [σ c] [29, 31, 33].

The Anfis-editor add-on of the MatLab environment develops and tests fuzzy inference algorithms with rules in which the combination of terms of the input variables represents a complete set of all possible combinations of membership functions of the input variables in the designed system, while not all of them can be used for further analysis [34, 35]. On the basis of the selected rules, it is proposed to evaluate and form a change management strategy for the enterprise based on the creation of a system of rules that determine the degree of adaptability of the components of the enterprise strategy, namely:

- determination of flexibility and opportunities for transformation of the management system;
- level of competitiveness or market share of the enterprise;
- the efficiency of production activity as a basis for the implementation of changes and an indicator of their expediency;
- degree of adaptability.

The above list quite fully covers the system of "hard" factors that form the company's strategy, however, when implemented in the company's practical activities, it can be supplemented with additional factors that are important for the company and further determine the change management strategy. However, it should be taken into account that the introduction of additional factors can significantly complicate the model and make it cumbersome and unable to quickly take into account needs and the ability to change, while not significantly improving the obtained results. Therefore, the review of input factors requires a careful selection of factors and their additional justification.

3 Research Results

The study is based on the concept embedded in the McKinsey 7S model, where the company's readiness for change and the readiness to implement and implement changes is evaluated in seven interrelated dimensions, among which the structure, strategy, system, values, style, skills, personnel are important (Fig. 3).

All components of the McKinsey 7S model are interconnected, that is, when one element changes, the others will necessarily change. The named components are divided into "hard" elements (structure, strategy, system) and "soft" (values, style, skills, personnel). Solid elements are well amenable to measurement, formalization, and, accordingly, managerial influence, so very often management staff or leaders in change management try to transform these very elements. However, managing only one or several parts of an interconnected system either does not lead to changes, or the consequences and results are quite far from the intended ones. Therefore, when managing changes, it is necessary to consider other elements of the system.

Soft elements are much more difficult to assess and even describe, as they are highly dependent on the "human factor" and can sometimes even be invisible, but they have a significant impact on other component systems and are therefore important for change management.

Fig. 3 McKinsey 7S model
[36, p. 137]

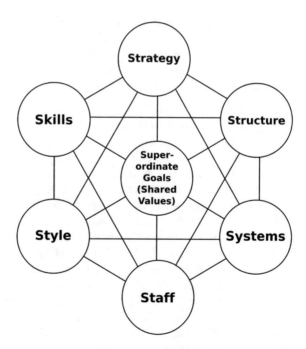

Usually, each element of the McKinsey 7S model is interpreted as follows [37, p. 137]:

"*Strategy* is a plan to support and build competitive advantages of the enterprise and measures aimed at responding to or predicting changes in the external environment and, accordingly, maintaining competitiveness.

Structure—represents the division of the enterprise into organizational units, and the relationships between them, also includes information connections and demonstrates the hierarchy of subordination and responsibility … The structure is considered one of the most obvious and visible elements of the enterprise and the one that is most easily subject to change.

Systems—daily actions and procedures in which employees are involved in order to achieve appropriate results at work; formal and informal procedures that support strategy and structure. Systems are more powerful than they think and should be top of mind for managers as they implement change.

Skills and abilities are the actual skills and competence of employees working at the enterprise. …

Personnel (status). Personnel management—processes used for the development of managers, socialization processes, ways of forming the main (nominal) values of managers, ways of introducing the company's young personnel, and assistance in managing the careers of employees, their motivation, remuneration, etc.

Style/culture—mainly represents the management style and culture of top-level managers. Organizational culture consists of two components:

- organizational culture: dominant values and beliefs, norms that develop and after a certain time become corresponding stable (lasting) features of organizational life—the atmosphere in the organization;
- management style: more attention to what managers do than what they say; how company managers spend their time; what they focus on.

Common values—represent the main values of the company, which are the basis of the general business culture and general professional ethics—these are the norms and standards of employee behavior and business conduct, therefore they are the basis (foundation) of every organization (enterprise)."

Among the "hard" factors in the development of construction enterprises used in the McKinsey 7S model, the element "strategy" occupies a special place, since it is from the construction of the strategy that the company's ability to manage changes depends. The direction of activity, the speed of transformations, and the financial, investment, and operational activities of the enterprise depend on the long-term planning of actions laid down in the strategy.

Under changing environmental conditions, when there is a great complexity of planning processes due to uncertainty, ambiguity, riskiness, variability, and volatility of many external and internal processes of the organization, the strategy acquires special importance, but at the same time it should not be rigid, but outline only the general direction and desired results changes, allowing to solve most issues regarding the organization of activities at the tactical or even operational levels.

This approach will allow the enterprise or organization not to lose its orientation due to the action of many factors and rapid changes, but at the same time to respond flexibly and quickly to new influences and challenges, while carrying out preventive management of changes in the desired direction for the enterprise.

Given the extreme importance of strategy for development and change management, this study evaluates the impact of parameters that serve as the economic basis for strategy formation (competitiveness and efficiency) and characterize the company's ability to change (the flexibility of the management system and the adaptability of the company) on the formation of the company's change management strategy.

Since all elements of the McKinsey 7S model must be considered as a whole, the models proposed in this paper must be used together with similar models developed for other elements of the system, for example, the system of fuzzy rules for the element "personnel" proposed in the article [33] or other similar solutions beside.

For training the hybrid network, the hybrid method was chosen with an error level of 0 and the number of cycles of 30. For the input factors S_1–S_4 and the integral S.

The system of fuzzy logical inference was also created using the Mamdani type system, which was implemented on the basis of [38].

As a result of training the network, the error was 0.312 percentage points (Fig. 4).

The complete set of all possible combinations of membership functions of the input variable in the designed Mamdani system is shown in Fig. 5.

The graphs (Fig. 6a, b) show Gaussian membership functions for the terms "low level of flexibility of the enterprise management system", and "high level of flexibility of the enterprise management system". As can be seen from Fig. 6, the flexible control system is marked as "0" and the rigid one as "1". The highest level of uncertainty is at a value of 0.5.

The graphs (Fig. 7a–c) show Gaussian membership functions for the terms "low competitiveness", "medium competitiveness", and "high level of competitiveness". As can be seen from Fig. 7, the maximum confidence that the level of competitiveness of the enterprise is low when the value of the competitiveness index is less, while the value of 0.3212 characterizes the maximum level of uncertainty.

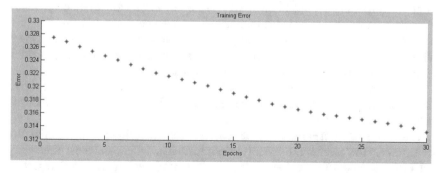

Fig. 4 Results of network training

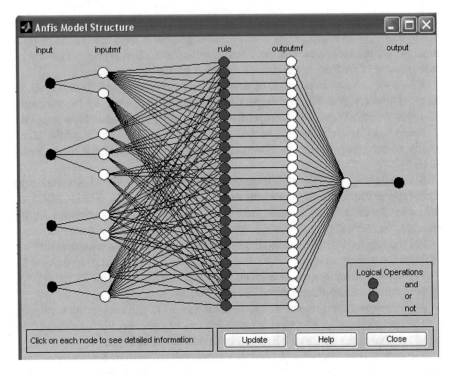

Fig. 5 The system of fuzzy inference rules

With maximum certainty, it can be stated that the enterprise has an average level of competitiveness when the competitiveness index has a value of 0.6523, and uncertainty about the average level of competitiveness is observed when the index values are less than 0.5212 and 0.086.

A high level of competitiveness in the enterprise is observed when the value of the index is more than one.

To assess the level of competitiveness as sufficient, for the formation of a strategy of development and not reduction or survival, which affects the level and planning of change management, the index of competitive advantages should be no less than 0.6523, which characterizes the average level of competitiveness or higher. Otherwise, the level of competitiveness may be assessed as insufficient for the formation of an effective strategy of preventive change management.

In order not to overload the model with an excessive number of rules, the other two parameters are evaluated on a dual linguistic scale in the format of "sufficient" or "insufficient" level.

As can be seen from Fig. 8, Ukrainian construction enterprises are characterized by an extremely low level of profitability of production activities, and most often carry out sales activities. This statement is indirectly confirmed by statistical reporting data, according to which most enterprises in the construction sector in the period from 2005 to 2019 operated at a loss. Therefore, the parameter "adequate level

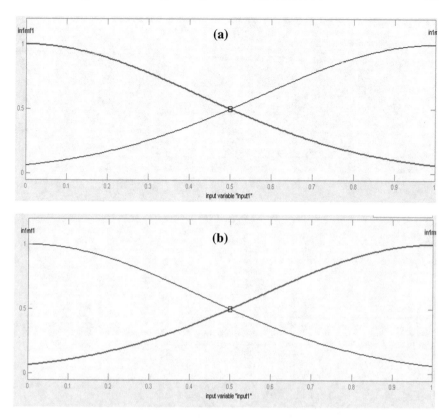

Fig. 6 Gaussian membership functions of the input parameter S_1 "level of flexibility of the enterprise management system" after entering the parameters for the terms "high" (**a**), "low" (**b**)

of efficiency of construction enterprises" has an average value, namely, more than 0.095. The maximum level of non-foaming is that the enterprise operates profitably at a value of 0.0557.

Based on the results of the analysis, it can be concluded that most enterprises have insufficient profitability to maintain development and increase or preserve the potential of change management since the profit received by them in the analyzed period is insufficient. Therefore, even those enterprises that operate profitably should work in the direction of increasing the efficiency of production activities, as well as financial and investment activities.

The factor "level of adaptability" of the enterprise was calculated as the ratio of the profitability of the enterprise's ordinary activities to the average level of profitability of enterprises in the construction sector for the corresponding year. This indicator varied from 0 to 8.5 (Fig. 9). However, in view of the extremely low profitability indicators, and more often the bankruptcy of the enterprises of the sector, it is more likely not about the high efficiency of the studied enterprises, but about normal

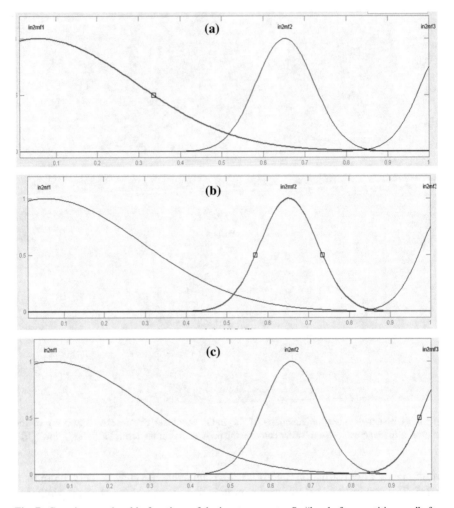

Fig. 7 Gaussian membership functions of the input parameter S_2 "level of competitiveness" after entering the parameters for the terms "low" (**a**), "medium" (**b**), and "high" (**c**)

activity against the background of insufficient profitability of the enterprises of the construction sector.

Finally, the membership functions for the four indicators that characterize the terms of the type "enough" to ensure deep changes and the formation of the company's change management strategy are given in Table 1.

According to the information provided in the Rule Editor dialog box (Table 2), the rules in the knowledge base have a weight of 1.

Since the system has three input variables, each of which has two terms and one variable has three terms, then the maximum number of rules in the knowledge base for formulating all possible dependencies between factors and consequences should

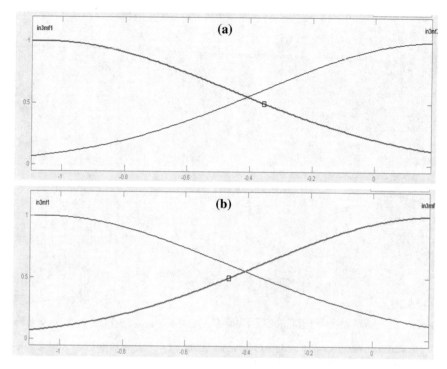

Fig. 8 Gaussian membership functions of the input parameter S_3 "level of efficiency" after entering the parameters for the terms "medium" (**a**) and "low" (**b**)

be $2^3 * 3 = 24$, each of which should have weight equal to 1. Ten rules were chosen in the final version, which reflects the impact of management flexibility, efficiency, competitiveness, and adaptability on the company's change management strategy.

The rules are formulated as follows:

1. If the flexibility of the management system is insufficient AND the level of competitiveness of the enterprise is low AND the efficiency of the enterprise's production activity is low AND the level of adaptability of the enterprise's production activity is sufficient, then the change management ability of the enterprise can have the following value: mf1 = 0.534.

2. If the flexibility of the management system is insufficient AND the level of competitiveness of the enterprise is low AND the efficiency of the enterprise's production activity is high AND the level of adaptability of the enterprise's production activity is sufficient, then the change management ability of the enterprise can have the following value: mf4 = 2.315, i.e. more than 2.3 times than in on average in enterprises of the sector.

3. If the flexibility of the enterprise's management system is insufficient AND the level of the enterprise's competitiveness is low AND the efficiency of the enterprise's production activity is high AND the level of adaptability of the enterprise's

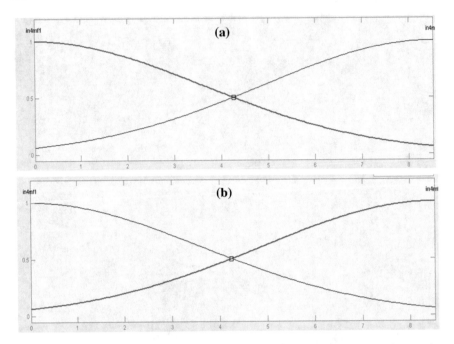

Fig. 9 Gaussian membership functions of the input parameter S_3 "adaptability level" after entering the parameters for the terms "high" (**a**) and "low" (**b**)

production activity is insufficient, then the enterprise's ability to manage changes can have the following value: mf5 $= 0.613$, i.e., reach only 61.3% of adaptive capabilities of sector enterprises. Such a striking difference is achieved due to the loss of adaptability of production activity, which causes a sharp decrease in potential and the inability to respond to changes with sufficient quality.

4. If the flexibility of the enterprise's management system is insufficient AND the level of the enterprise's competitiveness is average AND the efficiency of the enterprise's production activity is high AND the level of adaptability of the enterprise's production activity is insufficient, then the enterprise's ability to manage changes will be: mf7 $= 0.9855$, i.e., almost at the level of average adaptability of enterprises in the construction sector.

5. If the flexibility of the enterprise's management system is insufficient AND the level of the enterprise's competitiveness is high AND the efficiency of the enterprise's production activity is high AND the level of adaptability of the enterprise's production activity is insufficient, then the enterprise's ability to manage changes mf9 $= 1.306$.

6. If the flexibility of the enterprise's management system is insufficient AND the level of the enterprise's competitiveness is high AND the efficiency of the enterprise's production activity is high AND the level of adaptability of the enterprise's production activity is sufficient, then the enterprise's ability to manage changes mf12 $= 2.408$.

Table 1 Membership functions for factors S_1–S_4

Indicator	Membership function
S_1	$\mu_{heo6x}(S_1) = \begin{cases} e^{-\frac{1}{2}\left(\frac{S_1 - 0,4246}{0,99}\right)^2}, & \text{if } S_1 < 0,99 \\ 1, & \text{if } S_1 \geq 0,99 \end{cases}$
S_2	$\mu_{heo6x}(S_2) = \begin{cases} e^{-\frac{1}{2}\left(\frac{S_2 - 0,0076}{0,6523}\right)^2}, & \text{if } S_2 < 0,6523 \\ 1, & \text{if } S_2 \geq 0,6523 \end{cases}$
S_3	$\mu_{heo6x}(S_3) = \begin{cases} e^{-\frac{1}{2}\left(\frac{S_3 - 0,195}{0,557}\right)^2}, & \text{if } S_3 < 0,0557 \\ 1, & \text{if } S_3 \geq 0,0557 \end{cases}$
S_4	$\mu_{heo6x}(S_4) = \begin{cases} 1, & \text{if } S_4 \geq 8,504 \\ e^{-\frac{1}{2}\left(\frac{S_4 - 3,625}{8,504}\right)^2}, & \text{if } S_4 < 8,504 \end{cases}$

7. If the flexibility of the enterprise's management system is sufficient AND the level of the enterprise's competitiveness is low AND the efficiency of the enterprise's production activity is low AND the level of adaptability of the enterprise's production activity is sufficient AND the enterprise's ability to manage changes: mf15 = 0.6511.

8. If the flexibility of the enterprise's management system is sufficient AND the level of the enterprise's competitiveness is average AND the efficiency of the enterprise's production activity is low AND the level of adaptability of the enterprise's production activity is insufficient AND the enterprise's ability to manage changes: mf17 = 0.712.

9. If the flexibility of the enterprise's management system is sufficient AND the level of the enterprise's competitiveness is average AND the efficiency of the enterprise's production activity is high AND the level of adaptability of the enterprise's production activity is sufficient, then the enterprise's ability to manage changes: mf19 = 1.602.

If the flexibility of the enterprise's management system is sufficient AND the level of competitiveness of the enterprise is high AND the efficiency of the enterprise's production activity is high AND the level of adaptability of the enterprise's production activity is sufficient AND the ability to manage changes: mf23 = 2.683. The above rules reflect the main provisions of adapting the company's strategy to the condition of constantly acting profound changes and the development of its adaptability.

Table 2 A system of rules for displaying unclear dependencies between factors affecting strategy and the possibility of change management

№	Formula	Value (const)
1	If (x_1 is low1) and (x_2 is low1) and (x_3 is low1) and (x_4 is insuff1) then (x_5 is out1mf1)	0.5340
2	If (x_1 is low1) and (x_2 is low1) and (x_3 is hight2) and (x_4 is suf2) then (x_5 is out1mf4)	2.315
3	If (x_1 is low1) and (x_2 is low1) and (x_3 is hight2) and (x_4 is insuff1) then (x_5 is out1mf5)	0.613
4	If (x_1 is low1) and (x_2 is aver2) and (x_3 is hight2) and (x_4 is insuff1) then (x_5 is out1mf7)	0.9855
5	If (x_1 is low1) and (x_2 is hight3) and (x_3 is hight2) and (x_4 is insuff1) then (x_5 is out1mf9)	1.3060
6	If (x_1 is low1) and (x_2 is hight3) and (x_3 is hight2) and (x_4 is suf2) then (x_5 is out1mf12)	2.408
7	If (x_1 is enough2) and (x_2 is low1) and (x_3 is low1) and (x_4 is suff2) then (x_5 is out1mf15)	0.6511
8	If (x_1 is enough2) and (x_2 is aver2) and (x_3 is low1) and (x_4 is insuff1) then (x_5 is out1mf17)	0.7121
9	If (x_1 is enough2) and (x_2 is aver2) and (x_3 is hight2) and (x_4 is suf2) then (x_5 is out1mf19)	1.6027
10	If (x_1 is enough2) and (x_2 is hight3) and (x_3 is hight2) and (x_4 is suf2) then (x_5 is out1mf23)	2.6830

In Fig. 10 shows the influence of the factors "management system flexibility" and "level of competitiveness" on the ability to manage changes and form the company's strategy.

Figure 10 shows that a high degree of flexibility in the enterprise management system does not have a great impact on the ability to manage changes in the case of a low level of competitiveness, since the company in this case still cannot freely make changes, mainly practicing defensive strategies of protection and reduction of activities. This does not allow the enterprise to adapt to changes, in the best case it will react to already existing influences, and in the worst case—it will not be able to respond to changes and manage them at all. Significantly different results can be obtained at a high level of competitiveness. In this case, an increase in the flexibility of the strategy will lead to a significant increase in the possibilities of change management and strategy formation.

The flexibility of the management system provides an opportunity to manage changes at a higher level, if the production activity of the enterprise has low efficiency (Fig. 11), with a high level of efficiency of production activity, a rigid change management strategy provides the opportunity to achieve better results. The worst thing is when, at a high level of efficiency, the control system has insufficient rigidity

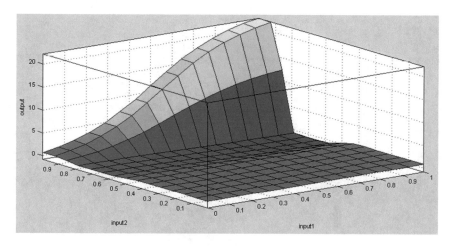

Fig. 10 Influence of parameters S_1 and S_2 on the resulting indicator

and reacts to the smallest external and internal influences. This situation leads to a low level of change management capability.

When combining a high level of adaptability and a low level of flexibility in the management system, the enterprise can achieve a sufficiently high ability to manage changes, but at the same time, the combination of high adaptability with the flexibility of the management system gives the opposite result (Fig. 12). Therefore, a stricter management system is recommended for enterprises that already have a high level of adaptability compared to enterprises that have an insufficient level of adaptability. Therefore, to compensate for insufficient adaptive capabilities. The enterprise can increase the flexibility of the management system.

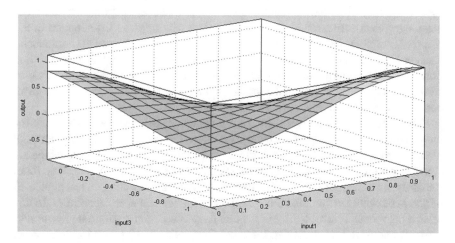

Fig. 11 Influence of parameters S_1 and S_3 on the resulting indicator

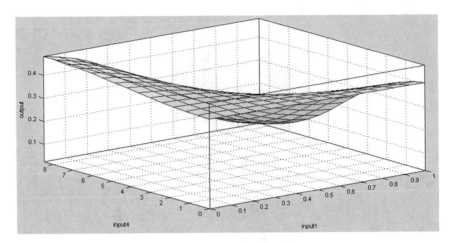

Fig. 12 Influence of parameters S_1 and S_4 on the resulting indicator

Figure 13 shows the combination of competitiveness and efficiency of the enterprise and their influence on the ability to change. From Fig. 13, it can be concluded that enterprises with a low level of production efficiency and a high level of competitiveness have the greatest motivation and ability to change, which serves as a basis for further development and allows not only planning but also resource support for the implementation of changes. At the same time, enterprises with insufficiently high market share compared to the main competitor, when achieving a high level of efficiency, will in any case try to improve their own market positions. In such cases, the potential and capacity of change management will be at a high level. If the level of efficiency of production activity is insufficient, or even unprofitable, enterprises are forced to implement an adaptive change management strategy. At the same time, they often have enough resources and financial opportunities to implement significant, structural changes that can radically improve the situation in the market. Therefore, the efficiency of production activity can be called the financial basis or basis for the implementation of changes. However, a high level of motivation regarding the implementation of changes is observed precisely in the case of strong competitive positions in the market, but a low level of efficiency. In this case, companies have an internal need to implement changes.

A high level of competitiveness and a strong market position serves as the basis for increasing the adaptability of the enterprise, as well as increase its ability to manage planned changes, as well as providing an opportunity to respond to spontaneous changes without losing the given development vector (Fig. 14). At the same time, the company achieves the highest degree of ability to manage changes at a high level of competitiveness in combination with the growth of adaptation capabilities. In this case, a high level of adaptability of the company ensures an appropriate reaction to negative and positive external and internal changes and will prevent a decrease in profits and a loss of market share of the company, and a strong market position will

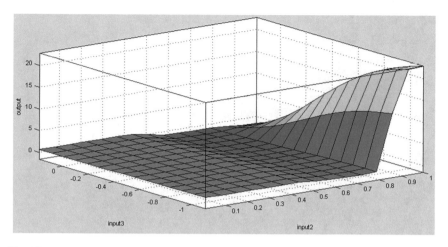

Fig. 13 Influence of parameters S_2 and S_3 on the resulting indicator

allow the introduction of planned changes and transformations, strengthening the level of competitiveness of the enterprise.

Figure 15 shows the influence of the factors "efficiency of production activity" and "level of adaptability" on the enterprise's ability to transform.

Figure 15 clearly shows that in the case of low efficiency of production activity, to continue successful functioning, enterprises need to increase the level of adaptability, and in case of high efficiency, the opposite is true, since the enterprise has chosen areas of activity that provide sufficient profitability, and the diversification of activities will reduce the overall level of efficiency. That is, the above equations are an effective tool for the company's management to determine the influence of strategy flexibility on

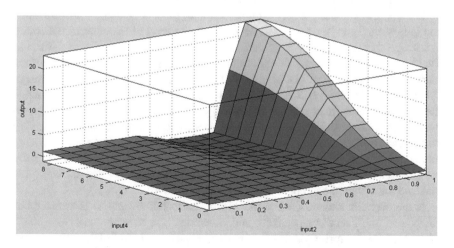

Fig. 14 Influence of parameters S_2 and S_4 on the resulting indicator

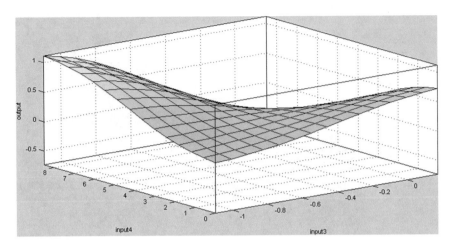

Fig. 15 Influence of parameters S_3 and S_4 on the resulting indicator

the adaptive capabilities of the enterprise, which will allow to make comprehensively justified and balanced decisions, balancing between the efficiency in the current time and the adaptive capabilities of the enterprise in the strategic period.

The proposed approach to assessing the capabilities of enterprise change management through strategy and other "hard" and "soft" factors of the McKinsey "7S" model will allow to comprehensively ensure the enterprise's adaptation to environmental changes.

According to the data in Table 2, the insufficient market and financial potential of the organization (factors S_2 and S_3 at a low or insufficient level) just as badly affects the ability of companies to implement deep changes as the insufficient degree of flexibility and adaptability of the enterprise (factors S_1 and S_4).

Thus, when evaluating the level of indicators S_1 as "low" and S_4—"insufficient", and indicators S_2 as "average" and S_3—"high", the level of the enterprise's ability to manage changes was 0.9855 of the average level among enterprises in the sector. That is, with potential financial opportunities for change management and a strong market position, enterprises lack the flexibility and adaptability to effectively manage change. For enterprises that belong to such a group, it is recommended to change the management system, management configuration, and the duties of workers in the direction of expanding the powers and opportunities to make decisions among managers of various management levels. Since all the elements of the model .... are interconnected, it is recommended to pay attention to the system, namely the daily decision-making procedures, business processes, work processes, and decisions that make up the standard operations in the organization.

In the opposite situation, when the level of indicators S_1 and S_4 is evaluated as "high" and "sufficient", respectively, and indicators S_2 as "average" and S_3—"low", the level of the enterprise's ability to manage changes was 0.6511 of the average level among enterprises of the sector. That is, despite the flexibility of the management system and adaptive capabilities, it is quite difficult to ensure sufficient depth and effectiveness of changes without financial support and strong competitive positions. Therefore, the enterprises included in this group should first build up their own financial potential, and maintain competitive positions, while implementing changes aimed at ensuring stable functioning on the market.

The given system of rules can be used both to assess the level of the company's change management capability and to identify factors that directly or indirectly hinder the implementation of changes, as well as stimulate or disincentivize the change management system.

Providing sufficient opportunities for implementing changes and building an effective change management system is possible provided that all four conditions of the rules listed in Table 1 are met, the simultaneous satisfaction of which is indicated by the minimum operation. However, these conditions are not always equivalent to each other.

The ability to manage changes and build a management strategy that will ensure smooth implementation of changes will correspond to the minimum value of the mathematical expectation of deviation of each condition.

The significance of each individual condition is determined by the concentration, which is calculated by raising each obtained result according to the corresponding membership function to the degree that determines the weight for the overall result of each of the rules (w_j). That is, the minimum value of each of the four functions listed in Table 1 determines the extent to which the organization's current change management potential corresponds to its strategy, goals, and change management intentions in the current period. At the same time, the level of ability to manage changes is calculated according to the following formulas:

$$\mu_{cms} = (\mu_1(s_1))^{\omega_1} \wedge (\mu_2(s_2))^{\omega_2} \wedge (\mu_3(s_3))^{\omega_3} \wedge (\mu_4(s_4))^{w_4}$$
$$\mu_{cms} = \min((\mu_1(s_1))^{\omega_1}; (\mu_2(s_2))^{\omega_2}; (\mu_3(s_3))^{\omega_3}; (\mu_4(s_4))^{w_4}) \tag{2}$$

To justify the weighting factors, the Fishburn weighting formula was used [39]:

$$\omega_j = \frac{2 \cdot (m - \bar{j} + 1)}{m \cdot (m + 1)}, \tag{3}$$

where m is the number of rules, m = 4; $\bar{j}$ is the rank of the jth indicator—a fuzzy set. Since the number of rules is 4, formula (3) has the form:

$$\omega_j = \frac{2 \cdot (4 - \bar{j} + 1)}{4 \cdot (4 + 1)} = \frac{2 \cdot (5 - \bar{j})}{20}, \tag{4}$$

The rules determining the weighting factors are as follows:

1. Since factors S_3 (efficiency of the enterprise's production activity) and S_4 (adaptability) have the greatest influence on the formation of the change management

Table 3 Calculation of weights for the factors of change management strategy

The indicator for which the term "necessary level for the formation of a change management strategy" is defined	Factor weight
S_1	$\omega_1 = \frac{2 \cdot (5-3)}{20} = 0.2$
S_2	$\omega_2 = \frac{2 \cdot (5-4)}{20} = 0.1$
S_3	$\omega_3 = \frac{2 \cdot (5-1.5)}{20} = 0.35$
S_4	$\omega_4 = \frac{2 \cdot (5-1.5)}{20} = 0.35$

strategy, each of them is assigned the same weight. That is the rank of the named components: $j_3 = j_4 = 1.5$.

2. Flexibility of the management system (S_1) is a necessary component for prompt implementation of changes. Therefore, according to preliminary estimates, the rank of this component will be $j_1 = 3$.
3. A strong market position is a necessary prerequisite for forming the potential for the future development of the enterprise while acting as a basis for change management. The rank of the S_2 indicator (competitiveness) will be 4.

Taking into account the mentioned remarks, weighting indicators were calculated for rules 1–4 (Table 3).

The result of the formula (4) reflects the enterprise's ability to implement and implement a change management strategy. The indicator is expressed in fractions of a unit and can be used to adjust the strategic and tactical goals of the company's development. Multiplying the target indicators of changes by the coefficient obtained according to formula (4) allows you to obtain a pessimistic development scenario, identifying possible deviations from the planned indicators and indicators that determine the effectiveness of the adopted strategy.

The proposed approach to assessing the capabilities of enterprise change management through strategy and other "hard" and "soft" factors of the McKinsey "7S" model will allow to comprehensively ensure the enterprise's adaptation to changes in the environment and internal environment, supplementing the enterprise management toolkit with additional models.

4 Conclusions

The proposed approach to assessing the capabilities of enterprise change management through strategy and other "hard" and "soft" factors of the McKinsey "7S" model will allow to comprehensively ensure the enterprise's adaptation to changes in the environment and internal environment, supplementing the enterprise management toolkit with additional models.

Identifying factors that will allow to support and development of positive targeted changes at enterprises is an important direction of scientific research. This study offers only one vector that will allow structuring and formalizing the change management process in the direction of forming, justifying, and supporting the company's strategy. It was found that the change management process is influenced by the factors "management system flexibility", "competitiveness", "operational efficiency", and "adaptability" of the enterprise. It is clear that the change management process is much more complex and involves a large number of factors that are not listed in this paper. The identification of such factors, their research, and the creation of appropriate models should become a new direction of scientific research.

In this work, the factors that affect the company's strategy are investigated, a formula is proposed for identifying the impact on the company's ability to implement changes, and a method of adjusting the target indicators of the development of construction companies, which is calculated taking into account the level of financial and market stability of the company and its ability to change.

In further research, it would be advisable to supplement the data sets with factors that take into account "soft" skill and success factors. Since the "strategy" element belongs to the "hard" factors of change management and is much easier to formalize than the "soft" factors, the toolkit of sparse set methods is the most suitable for this kind of research and can be used when identifying the impact on the possibility of implementing changes elements describing the organization's values, style, skills, and personnel.

Research on leadership and its impact on changes in organizations, internal readiness for personnel changes, organizational values and corporate culture, etc. deserves attention.

The study of change management is an interesting and promising field of research that will develop over time, revealing new facets and directions of scientific thought. Thus, the study of gender features in the implementation of changes, and the identification of reactions and interactions during changes in organizations of people with disabilities and vulnerable population groups is of great interest. Therefore, this work is one of the areas of research on a complex, multifaceted problem, the solution of which will be relevant for a long time.

References

1. Asikhia Olalekan, U., Nneji Ngozi, E., Olafenwa Abiodun, T., Owoeye Oladipo, A.: Change management and organizational performance: a review of literature. Int. J. Adv. Eng. Manag. 3(5), 67–79 (2021). https://doi.org/10.35629/5252-03056779
2. Ashkenas, R.: Change management needs to change. https://hbr.org/2013/04/change-management-needs-to-cha (2013)
3. Errida, A., Lotfi, B.: The determinants of organizational change management success: literature review and case study. Int. J. Eng. Bus. Manag. 13, 184797902110162, 1–15 (2021).https://doi.org/10.1177/18479790211016273

4. Harrison, R., Fischer, S., Walpola, R.L., Chauhan, A., Babalola, T., Mears, S., Le-Dao, H.: Where do models for change management, improvement and implementation meet? A systematic review of the applications of change management models in healthcare. J. Healthc. Leadership **13**, 85–108 (2021). https://doi.org/10.2147/JHL.S289176

5. Sujová, A., Simanová, Ľ.: Impacts of implemented changes on the business performance of Slovak enterprises. Central Eur. Bus. Rev. (2023). https://doi.org/10.18267/j.cebr.328

6. Jones, J., Firth, J., Hannibal, C., Ogunseyin, M.: Factors contributing to organizational change success or failure: a qualitative meta-analysis of 200 reflective case studies. In: Hamlin, R., Ellinger, A., Jones, J. (eds.) Evidence-Based Initiatives for Organizational Change and Development, pp. 155–178. IGI Global (2018). https://doi.org/10.4018/978-1-5225-6155-2.ch008

7. Junnaid, M.H., Miralam, M.S., Jeet, V.: Leadership and organizational change management in unpredictable situations in responding to Covid-19 pandemic. Int. Trans. J. Eng. Manag. Appl. Sci. Technol. **11**(16), 1–12 (2020). https://doi.org/10.14456/ITJEMAST.2020.322

8. Bushuyev, S., Bushuyeva, N., Bushuiev, D., Bushuieva, V.: Inspirational emotions as a driver of managing information-communication projects. In: International Scientific and Technical Conference on Computer Sciences and Information Technologies, 2022-November, pp. 438–441 (2022)

9. Beer, A., Sotarauta, M., Bailey, D.: Leading change in communities experiencing economic transition: place leadership, expectations, and industry closure. J. Chang. Manag. (2023). https://doi.org/10.1080/14697017.2023.2164936

10. Boorman, C., Jackson, B., Burkett, I.: SDG localization: mobilizing the potential of place leadership through collective impact and mission-oriented innovation methodologies. J. Chang. Manag. (2023). https://doi.org/10.1080/14697017.2023.2167226

11. Reznik, N., et al.: Systems Thinking to Investigate the Archetype of Globalization. Springer International Publishing (2022). https://doi.org/10.1007/978-3-031-08087-6_9

12. Kozyk, V., Novakivskyi, I., Mrykhina, O., Koleshchuk O.: Development of a strategy for improving organizational and production structures based on methods for non-convex programming. In: Conference NTI-UkrSURT 2019, Kharkiv, Ukraine (2019)

13. Chupryna, I. et al.: Designing a toolset for the formalized evaluation and selection of reengineering projects to be implemented at an enterprise. East.-Eur. J. Enterp. Technol. (2022). https://doi.org/10.15587/1729-4061.2022.251235

14. Ryzhakova, G., Pokolenko, V., Malykhina, O., Predun, K.: Structural regulation of methodological management approaches and applied reengineering tools for enterprises-developers in construction. Int. J. Emerg. Trends **8**(10), 7560–7567 (2020)

15. Ford, J., Ford, L., Polin, B.: Leadership in the implementation of change: functions, sources, and requisite variety. J. Chang. Manag. Reframing Leadership Org. Pract. **21**(1), 87–119 (2021). https://doi.org/10.1080/14697017.2021.1861697

16. Naslund, D., Norrman, A.: A conceptual framework for understanding the purpose of change initiatives. J. Chang. Manag. **22**(3), 292–320 (2022). https://doi.org/10.1080/14697017.2022.2040571

17. Oreg, S., Berson, Y.: Leaders' impact on organizational change: bridging theoretical and methodological chasms. Acad. Manag. Ann. **13**(1), 272–307 (2019). https://doi.org/10.5465/annals.2016.0138

18. TodnemBy, R.: Organizational change and leadership: out of the Quagmire. J. Chang. Manag. **20**(1), 1–6 (2020). https://doi.org/10.1080/14697017.2020.1716459

19. Walk, M.: Leaders as change executors: the impact of leader attitudes to change and change-specific support on followers. Eur. Manag. J. (2022). https://doi.org/10.1016/j.emj.2022.01.002

20. Ryzhakova, G.V. et al.: Modern structuring of project financing solutions in construction. In: 2022 IEEE International Conference on Smart Information Systems and Technologies (SIST) (2022)

21. Bielienkova, O., Novak, Y., Matsapura, O., Zapiechna, Y., Kalashnikov, D., Dubinin, D.: Improving the organization and financing of construction project by means of digitalization. Int. J. Emerg. Technol. Adv. Eng. **12**(8), 108–115 (2022)

22. Zeltser, R.Ya., Bielienkova, O.Yu., Novak, Ye., Dubinin, D.V.: Digital transformation of resource logistics and organizational and structural support of construction. Sci. Innov. **15**(5), 38–51 (2019)
23. Li, Y., Biloshchytskyi, A., Bronin, S., Liashchenko, T.: A conceptual model for diversification strategies choice. In: IEEE International Conference on Smart Information Systems and Technologies (SIST), 2021, pp. 1–4 (2021). https://doi.org/10.1109/SIST50301.2021.9465934
24. Li, Y., Biloshchytska, S.: Diversification of activity as a component of adaptive strategic management of construction enterprise. Manag. Dev. Complex Syst. **37**, 173–177 (2019). https://doi.org/10.6084/m9.figshare.9783233(2019)
25. Bielienkova, O., Stetsenko, S. Oliferuk, S., Sapiga, P., Horbach, M., Toxanov, S.: Conceptual model for assessing the competitiveness of the enterprise based on fuzzy logic: social and resource factors. In: IEEE International Conference on Smart Information Systems and Technologies (SIST), pp. 1–5 (2021). https://doi.org/10.1109/SIST50301.2021.9465923
26. Izmailova, K.V.: Enterprise Finance: Lecture Notes. KNUBA, Kyiv (2020)
27. Fedun, I., Stetsenko, S., Tsyfra, T, Valchuk, B., Valentyna, A.: Innovative Software Tools for Effective Management of Financial and Economic Activities of the Organization. Lecture Notes in Networks and Systemst, vol. 485, pp. 17–38 (2023)
28. Lehenchuk, S., Raboshuk, A., Valinkevych, N., Polishchuk, I., Khodakyvskyy, V.: Analysis of financial performance determinants: evidence from slovak agricultural companies. Agric. Resour. Econ. **8**(4), 66–85 (2022)
29. Bielienkova, O.Iu.: Strategy and Mechanisms for Ensuring the Competitiveness of Construction Enterprises Based on the Model of Sustainable Development: Monograph. Lira-K, Kyiv (2020)
30. Sorokina, L.V.: Capital cost management at banking institutions based on neuro-fuzzy modelling. Actual Probl. Econ. **154**(4), 506–515 (2014)
31. Sorokina, L.V.: Improving the procedure of forecasting changes in financial condition in construction works by means of two-stage model of fuzzy inference. Actual Probl. Econ. **120**(6), 285–293 (2011)
32. Gupta, N., Bari, A.: Fuzzy multi-objective optimization for optimum allocation in multivariate stratified sampling with quadratic cost and parabolic fuzzy numbers. J. Stat. Comput. Simul. **87**(12), 2372–2383 (2017). https://doi.org/10.1080/00949655.2017.1332195
33. Shpakov, A., Stetsenko, S., Shpakova, H., Sorokina, L., Akselrod, R.: Assessment of the influence of adaptability factors on the effectiveness of managing changes in enterprises by fuzzy logic. Sci. Horizons **24**(10), 72–82 (2021)
34. Al-Maitah, M., Semenova, O.O., Semenov, A.O., Kulakov, P.I., Kucheruk, V.Y.: A hybrid approach to call admission control in 5G networks. Adv. Fuzzy Syst. **2018**, 1–7 (2018). https://doi.org/10.1155/2018/2535127
35. Semenova, O., Semenov, A., Koval, K., Rudyk, A., Chuhov, V.: Access fuzzy controller for CDMA networks. In: 2013 International Siberian Conference on Control and Communications (SIBCON) (2013). https://doi.org/10.1109/SIBCON.2013.6693644
36. Hayes, J.: The Theory and Practice of Change Management. Palgrave Macmillan, London (2014)
37. Zapuhliak, I.B.: McKinsey 7S model as a tool for evaluation of readiness to change gas transmission companies. Sci. Bull. Uzhhorod Natl. Univ. **3**, 136–140 (2015)
38. Mamdani, E.H.: Application of fuzzy algorithms for the control of a simple dynamic plant. Proc. IEEE 121–159 (1974)
39. Fishburn, P.C.: Mathematics of Decision Theory. Methods and Models in the Social Sciences, vol. 3. The Hague, Mouton (1972)

Modeling the Level of Implementation of BIM by Enterprises as a Means of Optimizing the Cost

Lesya Sorokina⊙, Tetiana Tsyfra⊙, and Inna Vahovich⊙

Abstract Nowadays, construction companies need to obtain additional competitive advantages, which can be new ways of processing and modeling information such as BIM. Currently, existing models need to be improved, given that the situation for project implementation has become much more uncertain, flexible, and complex, and many processes are moving into the digital sphere. Therefore, the authors proposed an approach to assessing the feasibility of implementing BIM by subcontractors through simulation modeling. The article considers the main stages of the formation of BIM, features of construction cost management depending on the level of BIM used in designing, considers the main barriers to the implementation of BIM models in practical activities, including imperfect national standards or their absence, lack of demand from customers, lack of own experience in using these technologies, lack of training and its high cost, high cost of BIM technologies, as well as the small size of projects, which makes the use of BIM impractical. It is proposed to assess the feasibility of implementing BIM in the activities of subcontractors, taking into account barriers. The calculations are based on the universally recognized criterion of profitability of the capital invested in the construction project—ROI (Return On Investment), the assessment of changes as a result of the introduction of BIM is provided with the help of simulation modeling. At the same time, the simulation model takes into account the capabilities of domestic subcontractors regarding the intensity of use of such digital technology by constructing Markov chains.

Keywords BIM-modeling · Digital technology by enterprise · Return on investment · Markov chains · Simulation modeling

L. Sorokina · T. Tsyfra (✉)
Kyiv National University of Construction and Architecture, Kyiv, Ukraine
e-mail: tsyfra.tiu@knuba.edu.ua

L. Sorokina
e-mail: sorokina.lv@knuba.edu.ua

I. Vahovich
Department of Regional Development Priority Projects Implementation, Ministry for Communities and Territories Development of Ukraine, Kyiv, Ukraine

© The Author(s), under exclusive license to Springer Nature Switzerland AG 2024
A. Semenov et al. (eds.), *Data-Centric Business and Applications*, Lecture Notes on Data Engineering and Communications Technologies 195, https://doi.org/10.1007/978-3-031-54012-7_12

1 Introduction

The level of implementation of building information modeling (BIM) has been increasing over the years. Today, in developed countries, the priority is not to create new facilities but to efficiently maintain the existing buildings, which further stimulates the implementation of BIM at the state level.

Many countries have already received state support for BIM technologies. In Great Britain, in April 2016, the government decreed the use of BIM level 2 for all government projects [1]. In addition, the government obliged developers to use BIM in all residential projects, thus ensuring a high level of public health and safety [2]. The US pioneered the development and implementation of BIM starting in the 1970s. BIM adoption has accelerated over the past three years, but not all states are currently required to use it [1]. However, in 2003, the General Services Administration developed the "National 3D-4D-BIM Program," which established a policy requiring the adoption of BIM for all Public Building Service projects [3].

France intends to introduce BIM throughout the construction sector and has currently developed BIM standards for works in infrastructure projects. The government also allocated 20 million euros for the digital development of the construction industry [1]. Norway, Denmark, Finland, and Sweden were among the first to implement BIM. In these countries, the use of BIM in public projects is mandatory. In 2010, the "BIM Roadmap" was developed in Singapore with the goal that by 2015, at least 80% of construction industry entities would use BIM [3]. Singapore has implemented BIM on all government projects since 2015.

The UAE is a rapidly growing economy that significantly supports the construction industry sector for this tremendous growth. In 2013, the Dubai Municipality issued a circular that provided for the use of BIM for architectural works and works with the European Parliament on certain projects [3].

The introduction of BIM in Germany was slower than in other countries, but in 2015 the German government announced the creation of a digital construction platform—a BIM task force created by several industry organizations to develop a national BIM strategy [3].

In China, BIM adoption has grown significantly since 2016. BIM has become a crucial element and is now used for most projects. According to the five-year plan, which is designed for the period 2016–2020, by the end of 2020, the construction industry should master and implement these information technologies. Hong Kong, unlike China, has advanced much further in the implementation of Building Information Modeling (BIM). Many government departments are targeting BIM Level 2 standards, which are currently mandatory in the UK [3].

So, from the above, it can be concluded that the implementation of building information modeling is best carried out with state support.

It is estimated that BIM can eliminate unplanned changes by 40%, reduce project execution time by 7%, and also reduce the time for drawing up an estimate by 80% [4, p. 162].

Most often, BIM technologies are used [5, p. 17] for "repeat" customers in the construction of new buildings for the public sector (67%) and for the private sector (62%). For "one-time" customers, the indicator among the private sector is higher— 50%, and for the state—40%. As an example [5, p. 16], back in 2011, 43% of respondents did not know about BIM, and in 2020, 73% of respondents knew about BIM, 23% of respondents use BIM in all their projects, and 46% use BIM for most projects.

Therefore, developing a methodology for assessing the feasibility of implementing BIM in subcontractors' practice, keeping in mind barriers, is a relevant direction of research. The calculations are based on the universally recognized criterion of profitability of the capital invested in the construction project—ROI (Return On Investment), the changes which are assessed with the help of BIM using simulation modeling.

2 Literature Review and Methodology

The introduction of digital technologies is an inevitable process that changes the very basis of building business processes of enterprises [6–10]. Digitization today is one of the basic factors in the transformation of construction, so many scientists are engaged in researching the features of this process, and the advantages and disadvantages of introducing digital technologies into the work of enterprises [11–15]. In recent years, the attention of researchers has been focused on creating or supplementing existing digital models with new blocks, and functions, or creating new models [16–20]. Many papers are devoted to new ways of using robotics, unmanned aerial vehicles, robots, and augmented reality to update many processes in construction, beginning with planning, design, and organization, and ending with control over work progress and technical operation of buildings and structures [21–25].

Thus, digitization trends in various sectors of the economy were studied in the following works [1], certain aspects of improving BIM models were proposed in the articles [1], and other aspects of digitization such as modeling and automation of the construction process, robotics, modern methods of quality and reliability control, the use of drones and other means of observation to monitor the progress of construction works were analyzed [23, 26–29], etc.

It is worthy of note that this is the first study that analyses the digital transformation in the construction industry in the last decade. The study maps the construction industry research efforts on digital transformation and provides an assessment of the construction industry in terms of digital transformation [30–34]. It thus provides a theoretical and practical basis for researchers and practitioners alike [30].

Recent Ukrainian publications focus on the fact that, first of all, BIM requires the transition of the construction industry to the principles of life cycle management and market pricing. The latest attempts to develop the concept of the reformation of the construction industry have been confronted precisely with this first issue. Exploitation for renovation projects is considered in the article [38]. Tendencies in the development of BIM technologies throughout the globe and in eastern European countries for example Ukraine are shown. Examples of the exploitation of local building information systems in realizing renovation processes are given, and proposals for policymakers in terms of applying BIM technologies in housing renovation activities and facility management are formulated.

3   Research Results

Building Information Modeling (BIM) [31, 32] is a process of digital presentation and creation of graphic and non-graphic information and its exchange in a common data environment (CDE). In the future, this data is used to make decisions about the object during its life cycle.

There is a concept of "BIM levels" regarding a set of minimum generally accepted criteria, which are necessary to determine the degree of conformity of the project process with BIM technology. At BIM level 0, 2D drawings are used, most often to provide construction documentation, with practically no collaboration at this level. BIM level 1 is used by most organizations, where they use 3D concepts and 2D drawings for the development of regulatory documentation. BIM level 2 involves joint work on the project but with the use of its own 3D models. BIM level 3 involves the use of one common project model [33].

BIM dimensions are different from BIM levels. Dimensions determine the connection of specific types of data with the information model. By adding additional data types, you can get more detailed information about the construction project.

The following dimensions of BIM are currently known: 3D (object model), 4D (time), 5D (cost), 6D (operation), 7D (sustainability), and even 8D (safety). However, scientists determined [34, p. 477] that BIM has a multi-dimensional capability— "nD" modeling because the number of dimensions that can be added to BIM is nearly endless.

3D includes object representation, visualization, implementation—BIM object creation, and final documents with detailed and structural design, specifications, and sustainability.

4D is the process of planning construction work, presented in time schedules, with 3D models for the development of real-time graphical simulations [34, p. 477]. The planning data is added to the components, which will be formed in detail as the project progresses. This information can be used to obtain accurate information about the program and visualization, which will show how the project will consistently develop. The time-related information for each item may include information about the time required to fulfill the order, the time required to install/assemble and to start work/fix the sequence in which the components are to be installed. Over time, the information combined into a common information model will be developed into a precise project program. With this data, it is easy to get information about the project, and it is also possible to show how the project will develop and how each individual structure will look at each stage [35].

At this stage, cost optimization can be carried out due to clear planning [35], which ensures a safe, logical, and effective sequence of processes. The ability to create prototypes of how resources will be combined prevents wasteful and expensive rework of the project.

Integrating the 5th dimension "cost" into the BIM model generates a 5D model that allows you to instantly generate cost budgets. This reduces the time required to determine the quantity and estimate from weeks to minutes, improves the accuracy of estimates, minimizes cases of disputes due to ambiguities in the data [34, p. 477], and thus allows you to devote more time to optimizing the value of the object.

Data included in the cost may include capital costs (for example, the cost of purchasing and installing a component), the associated current costs, and the cost of upgrading/replacing in the future [35].

The advantages of this approach to cost estimation [35] are the ability to easily see costs in a three-dimensional form, to receive information about changes made, as well as automatic calculation of the cost of components/systems fixed to the project. When using BIM, costs are considered as part of the project and allow tracking of forecasted and actual costs during the course of the project. This technique allows you to effectively draw up a budget and not go beyond its scope in the process of project implementation.

6D BIM is information about the project life cycle. The construction industry has traditionally focused on initial capital costs for construction [35]. Dimension 6D enables a better understanding of the value of assets during the life of the facility and allows for more effective decisions regarding the facility, both in terms of cost and sustainability.

The data contained in 6D BIM contain [35] information to support the management and operation of objects. This information can be about the manufacturer of certain components, their installation, maintenance, configuration, and operation information, service life, and finally decommissioning. This approach helps to more effectively plan maintenance or current repairs and thus optimize the costs of operating the facility.

Adding sustainability components to a BIM model creates 7D models that allow designers to meet carbon targets for specific project elements and validate design decisions or test/compare different options [34, p. 477] (Figs. 1 and 2).

In practice, 4D BIM is used most often for task planning—86%, 5D BIM is used for cost estimation—84%, 6D BIM for sustainability—68% and less often for object management—28%, and 7D BIM is used for safety—56%.

Scientists call the main barriers to using BIM [4, p. 168, 5, p. 24] non-perfect national standards or their absence, lack of demand from customers, lack of experience in using these technologies, lack of training, and its high cost, high cost of BIM technologies, as well as the small size of projects, which makes the use of technologies such as BIM impractical.

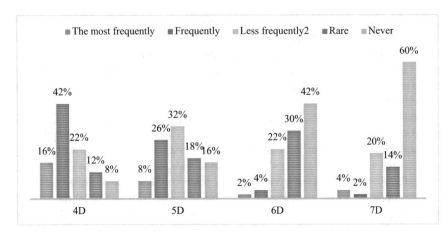

Fig. 1 Frequency of using BIM measurements [36, p. 21]

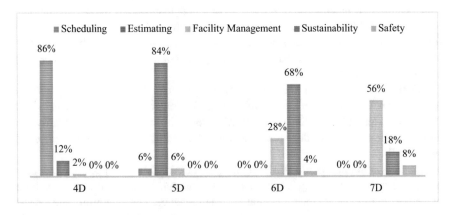

Fig. 2 Elements of BIM measurements [36, p. 21]

Barriers can be conditionally divided into certain categories [4, p. 163]:

1. Absence of a national standard that would define national priorities and give recommendations for the entire industry.
2. High cost of implementation, which includes both the high initial cost of the software and the high cost of the implementation process.
3. Lack of qualified personnel, which implies either a lack of specialists on the market or the costs of training/retraining existing personnel.
4. Organizational problems include professional responsibility, process problems, and lack of support from management.
5. Legal issues include BIM data ownership issues as well as licensing issues.

The above barriers to BIM implementation influence each other. The figure shows the relationship between the implementation of BIM technology and five categories of pre-code. The implementation of BIM is not just a technical problem, it affects the organizational structure and work processes in various companies in the construction industry. When the implementation of BIM is related to problems from outside the enterprise, such as suppliers, then economic and legal problems will be raised. The initial high cost of software security requires changes in the organizational structure of the enterprise or in the work process, which leads to organizational problems, and the high cost of human resources directly leads to a lack of professionals. Imperfect standards lead to conflicts in the exchange of information and data, which can cause legal problems regarding the ownership of data and access to it. At the same time, the absence of a national standard means the absence of a uniform level of personnel training, which increases the personnel deficit [4, p. 164].

Having analyzed the degree and nature of the main obstacles to the implementation of BIM, two main hypotheses can be identified [4, p. 165]:

1. A national strategy is needed to implement BIM in the construction industry.
2. The cost of implementation can be recognized as the biggest obstacle.

According to scientists, the national strategy for the implementation of building information modeling (BIM) is a vital factor in promoting the implementation of BIM [4, p. 166].

A key point is also the application of BIM standards, which can be both international and created by a separate country. If an international BIM standard is not used, then individuals in business and individual project teams will continue to repeatedly create and reproduce local solutions to the same problems. Businesses will continue to spend more time than necessary to bring new products and services to market because employees will need more time and effort to share their ideas and communicate concrete results. There will be more errors and omissions in structures that will be discovered during construction and where their correction will be expensive. Information will be re-collected rather than entered once and used continuously. Much more resources will be spent than necessary, both during construction and during the operation of the object [37, pp. 63–64].

Currently, there are no requirements for the use of BIM in the construction sector in Ukraine. However, on November 30, 2019, the Ministry of Development of Communities and Territories of Ukraine, the Office of Effective Regulation (BRDO), the UCSB Association, the Interstate Guild of Consulting Engineers, the Confederation of Builders of Ukraine, and the initiative group UA BIM Task Group signed the Memorandum "Road Map of Implementation information modeling of buildings (BIM) when creating construction objects, architectural objects" [38].

The main goal of the Memorandum [39] is the introduction of building information modeling when creating construction objects and architectural objects during 2019–2021. However, this Memorandum does not create mutual obligations and does not entail responsibility from the participants.

The Appendix to the Memorandum of Cooperation contains the "Road map of priority steps for creating conditions for the introduction of building information modeling (BIM) in the creation of construction objects and objects of architectural activity (2019–2021)". It contains 5 main tasks that must be completed during the specified period [15]:

1. Organization of works and support of measures for implementation of building information modeling (BIM). This stage involves the preparation and conclusion of a memorandum on the development and implementation of building information modeling. The deadline for the second half of 2019.

2. Implementation of regulatory and technical documents from building information modeling (BIM) into the system of technical regulation of construction. Harmonization and translation of international basic standards that determine the terminology, general principles of creating models, and their use at the stages of the object's life. The implementation period is the end of 2019—the beginning of 2020.

3. Making additions to the professional certification of responsible performers of certain types of work (services) in the field of architectural activity. Making additions regarding the use of building information modeling (BIM) to the qualification characteristics of an architect, design engineer, construction expert, and technical supervision engineer (in construction). The deadline is the second half of 2020.

4. Implementation of building information modeling (BIM) in construction. Practical application of building information modeling (BIM) at all stages of creation of real estate objects (including at the stage of design and construction of objects). The completion period is the I–IV quarter of 2021.

5. Activities on the use of information modeling (IM) in construction. Preparation of the draft Resolution of the CMU of Ukraine on the establishment of requirements for the mandatory use of building information modeling (BIM) in the construction of objects with budget funds. The completion date is the 4th quarter of 2021.

However, using exclusively administrative influence in order to achieve the goals of digitalization of real estate design, construction, and management is doomed to failure. The absence of clear economic benefits from the implementation of BIM technologies for any of the participants in the investment and construction process will only cause the search for various schemes to evade the fulfillment of too strict requirements. Instead, in the case of detailed substantiation of the benefits from investments in the product aimed at improving the quality of the processes of creating new and maintaining existing buildings, designers, developers, and contractors will more actively implement BIM technologies. At the same time, we should expect a reduction in cases of refusals to use information modeling tools in the construction industry, even in the case of designing and constructing relatively simple and inexpensive objects. This primarily concerns contractors, as there is a widespread opinion about the undeniable effectiveness of BIM technologies exclusively at the design stages. Therefore, we offer the authors a toolkit for evaluating the level of construction cost optimization by contractors due to the use of BIM tools, which, however, can be used by other participants in the investment and construction process.

Statistical methods are mainly used to model random processes. A special place among statistical methods takes a Markov chain method [4]. The idea of considering a time series as a Markov chain was expressed by the professor of the University of Bergamo, Sergio Ortobelli. To do this, it is necessary to discretize the value of profitability, that is, to divide the interval of variation of profitability into several subintervals, and according to statistical data to record events—falling of profitability into certain subintervals. An important prerequisite for the use of this approach is the verification of the markedness property of the obtained chain of events. Calculations are based on the generally recognized criterion of profitability of capital invested in a construction project—ROI (Return On Investment), the assessment of changes as a result of the introduction of BIM is provided using simulation modeling. At the same time, the simulation model takes into account the capabilities of domestic subcontractors regarding the intensity of use of such digital technology by constructing a Markov chain.

The choice of ROI as an efficiency criterion is due primarily to the availability of a public report on research on the results of the implementation of BIM in the world [39], which provides reasonable indicators of the impact of the level of involvement of BIM in the contractor's production processes on the return on investment. This document considers three options for the success of investment projects:

- low, when the return on invested capital is negative or does not reach 1%;
- average, limited to 25% return on investors' funds;
- high, for which the ROI exceeds 25%.

However, taking into account the profitability of construction enterprises and companies specializing in real estate operations, we specified ROI coefficients. In particular, according to the data of the State Statistics Service of Ukraine for 2010–2019 [5] there was only one case when companies specializing in real estate operations (they are the main consumers of BIM, taking into account the above overview of the use of various dimensions of these technologies in the world) managed to ensure the profitability of their main activity of more than 25% (Table 1).

That was in 2019 when the profitability of real estate management activities reached 34.7%. Taking into account the unfavorable changes in the economic environment in the world and Ukraine in 2020–2022, we assumed that the high level of ROI of domestic participants in the construction process should be limited to an interval of 25–34.7%.

Table 1 Justification of the average levels of ROI for construction projects in Ukraine (systematized by the authors)

Rate of return (ROI)	Number of observations 2010–2019	Periods and values of profitability	Average value ROI, %
High	1	2019: 34.7%	30
Medium	4	2012: 2.8%, 2013: 3.1%, 2017: 6.2%, 2018: 16.3%	8
Low	5	2010: 0.3%, 2011: −3.6%, 2014: −46.9%, 2015: − 33.4%, 2016: −8.1%	−18.3

The middle of this interval: 30%, thus, will reflect the maximum legalized level of construction profitability. Of course, the successful implementation of BIM by developers, designers, and contractors will contribute to achieving just such a level of profitability of capital investments in construction. According to the data of public state statistics [5], during the 4 periods of the last decade, the activity related to real estate operations was unprofitable (Table 1). In 2011, as well as in 2014–2016, according to estimates the average loss rate was 23%, primarily due to the decline of the industry in 2014–2015: then the losses were 46.9 and 33.4%, respectively. In 2010, the investigated type of activity showed a positive financial result, but the profitability of operations did not exceed 1%: it was only 0.3% (Table 1). Averaging the low profitability over half of the retrospective horizon, it is assumed that the low level of ROI of construction projects in Ukraine will be −18.3% on average. In the remaining years of the analyzed period, the profitability of real estate transactions was in the range of 2.8–16.3%, and the average level was (Table 1) 8%, which is less than the middle of the range of 1–25%. Since the introduction of BIM requires considerable expenditure of resources from contractors and a rather long period of reduced productivity, when evaluating the consequences of changes in information support, we took into account the level of average profitability of real estate transactions specified for Ukraine.

The average indicators of ROI for Ukrainian enterprises are shown in Fig. 4, where we also present the indicators of the probability of the appearance of each level of implementation of BIM at Ukrainian contracting enterprises. That is, Fig. 4 is a marked graph of states of the economic system of some contracting construction enterprises, built based on the theory of Markov chains. As is known [4], the Markov process is a random process, the specific values of which for any given time parameter t + 1 depend on the value at time t, but do not depend on its values at time t − 1, t − 2, etc. (discrete case of the Markov process). In other words, the "future" of the process depends only on the "present", but does not depend on the "past" (provided that the "present" of the process is known). In our opinion, the mentioned above quite objectively characterizes the effectiveness of the implementation of BIM technologies in the conditions of systematic disturbances in the economic environment of contracting enterprises. A discrete interval of the process can be considered a year or six months, during which 3 states of implementation of digital modeling of construction processes must be observed ([4], Fig. 4): low, medium, and high. These states can change primarily as a result of the 5 main barriers discussed above (Fig. 3).

In particular, a construction enterprise with a "low level" of using digital modeling tools, that is, one that did not use BIM at all, or used it very rarely, may or may not move to the "medium level" state. Of course, such a transition is natural, although it implies a significant intensification of the implementation of these technologies. The transition from the state of "low level" immediately to the state of "high level" is unlikely, since it requires the implementation of considerable costs. It will be necessary not only to invest in the purchase of software and training of personnel but also to spend almost 6 months of time to restore labor productivity on the new software. Therefore, a prerequisite for such a rapid update of information technology support

Fig. 3 Interrelationships
between the main barriers [4,
p. 165]

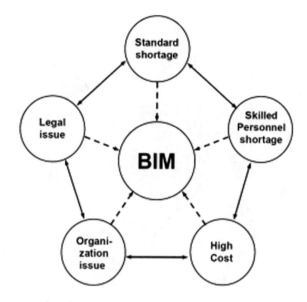

for the work of a subcontractor is a real opportunity to participate in the implementation of a technically complex, socially and economically significant construction project. Usually, in Ukraine, such projects are financed from the state budget or loans provided under state guarantees, and therefore only the winners of tenders are involved in the construction and ensuring the operational suitability of such buildings, in order to participate in which a number of strict conditions must be met. Anticipating in the medium-term perspective the reduction of corruption in construction, we thus do not exclude the transition of the contractor from the "low level" to the "high level" of BIM implementation, bypassing the "middle level".

However, in the process of developing the toolkit for assessing profitability due to the use of the latest information technologies, we did not adhere to positions of extreme optimism. Therefore, within 0.5–1 year, it is quite natural that there are three options for changing the state of implementation of BIM by subcontractors, which are currently characterized by an "average level". First, the negative impact of organizational, legal, and economic factors may lead to a change in the market segments of the provision of construction services, the activities of which will not require the use of BIM, which will cause a transition from "medium level" to "low". Secondly, when the conditions of the enterprise will not undergo significant changes, neither in terms of the technical complexity of the objects under construction, nor in terms of the requirements for construction participants, nor in terms of the profitability of operational activities, the contractor will remain at the "average level" of implementing BIM, without having significant incentives for investments in the development of information support for economic processes. Thirdly, the possibility of transition from "medium level" to "high" as a result of the emergence of economic and social incentives for the introduction of BIM by contractors is not excluded.

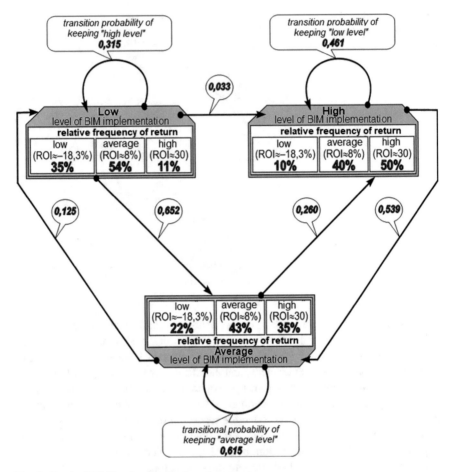

Fig. 4 Graph of BIM implementation by subcontractors

In our opinion, the complete digitization of information and documentary support of construction activity is an irreversible process, and therefore the transition from "high level" to "low" is impossible without reorganization of the construction enterprise through disintegration. For small and micro construction enterprises, which make up more than 50% of the industry, such forms of reorganization are not typical, as they are equivalent to liquidation. Therefore, an adverse environmental impact will, in the worst case, cause a transition from "high level" to "medium". However, the probability of maintaining the achieved "high level" of BIM implementation remains high, given global trends.

On the basis of the stated considerations regarding the change in the levels of implementation of BIM by subcontractors, we substantiated the quantitative values of transition probabilities, and it is these that are plotted on the graph of states (Fig. 4). Of course, the probability of a change in the level of saturation of the operational activity of the BIM contractor with technologies can be estimated quite conditionally, based on some a priori information. After all, the analyzed means of information support, firstly, have not yet gained proper distribution, and secondly, this direction is not covered by systematic statistical observations, the results of which are published for a wide range of users. In particular, the State Statistics Service does not provide such information. Therefore, we estimated the probability of the occurrence of one or another state of BIM use in Ukraine using data on the frequency of use of BIM measurements [36, p. 21] from Fig. 1, as the global trend of digitalization of construction processes is not slowing down, but only accelerating. Given the representativeness of the data [36, p. 21], it is considered possible to estimate the averaged shares for each of the 5 variants of the intensity of use of BIM, regardless of the measurement (w_j), as well as reliable intervals for such frequencies (Δw_j). The corresponding calculations are presented in Table 2.

As the calculations show (Table 2), the most variable was the relative frequency of respondents who never use BIM: in 95 cases out of 100, the volume weight of such contractors should be in the range from 12.5 to 50.5%, but the mathematical expectation of the lowest level digitization of the industry is 31.5%. It is this indicator, in our opinion, that will characterize the probability of permanent ignoring of BIM by contractors, and therefore this value is set as the transitional probability of maintaining the "low level" state in two adjacent time intervals (Fig. 4).

Optimistic forecasts regarding the digitalization of the construction industry should be made taking into account, at least, the lower limit of the share of enterprises that do not want to use BIM, i.e. 12.5% (Table 2). Therefore, we have made an assumption that at any given time, 12.5% of companies with an "average level" can move to a "low" level (Fig. 4).

"High level" of implementation of BIM by construction enterprises is characterized by indicators of intensity "often" and "very often". The mathematical expectation is the relative share of these two intensities (Table 2) 26% (= 18.5 + 7.5, underlined). It is worth assuming that the intensification of digital technologies in construction will be manifested in the fact that 26% of enterprises with an "average level" will move to the "high level" category within 0.5 years, which is shown next to

Table 2 Calculation of the averaged shares for the main variants of the intensity of BIM use (calculated by the authors based on the data [36, p. 21])

Intensity of BIM use	Average level of intensity (average share), w_j, %	The standard deviation of the proportion, σ_x, %	Limits of changes in the average level of intensity (p = 0.05, $t_{(0.05;3)}^* = 3.182$)
Very often	7.5	1.55	2.6–12.4
Often	18.5	4.77	3.3–33.7
Sometimes	24	1.35	19.7–28.3
Seldom	18.5	2.02	12.1–24.9
Never	**31.5**	5.98	**12.5–50.5**

the corresponding arrow in Fig. 4. As was discussed earlier, the probability of transition from a "low level" immediately to a "high level" is very low, but it exists, and we estimated it in accordance with the lower limit of the intensity of BIM application, marked as "frequent", i.e. 3.3% (Table 2, underlined).

Assuming that, similar to other countries of the world, domestic developers will positively evaluate the results of the application of BIM in Ukraine, and therefore the probability of retention by companies of a "high level" of the implementation of BIM was evaluated by taking into account the upper limit of relative frequencies for the intensities "often" and "very often ", which is 46.1% (= 12.4 + 33.7, highlighted with a double underline in Table 2).

We calculated the probability of reaching the "average level" state from other states, or its preservation (Fig. 4) based on the condition of a complete group of events, that is, by subtracting from 100% the percentage indicators of transitions to "high" and "low" levels.

Along with the use of placed transition graphs of the Markov chain (Fig. 4), the predictors of transition probabilities are presented in the form of matrices. Based on such a matrix, Table 3 was built, which systematizes information about possible changes in the situation of the use of BIM by a typical Ukrainian contracting enterprise.

For systems whose state transformations can be represented using Markov chains, there is a vector of the steady state of the system, i.e., a set of limit values of the probabilities of the appearance of each state, which will appear during the long time of the system's existence. It is such marginal probabilities that provide an assessment of the degree of optimization of the cost of construction by subcontractors in the case of the application of BIM within the framework of the proposed toolkit. The calculation of the marginal probabilities of each of the states, i.e., the levels of implementation of BIM (p_l, p_m, p_h) involves the solution of systems of equations based on the matrix of transition probabilities and the normalization condition. Taking into account the indicators of the Table 3, the system of equations has the form (see Eq. 1):

Table 3 Transient probabilities of changes in the state of implementation of BIM by subcontractors (Calculated by the authors)

Level of BIM use		The probability of transition in the next period to the level			
		Low	Medium	High	Total
In the current period	Low	0.315	0.652	0.033	100
	Medium	0.125	0.615	0.260	100
	High	0	0.539	0.461	100

$$\begin{cases} p_l = 0.315 \cdot p_l + 0.125 \cdot p_m \\ p_m = 0.652 \cdot p_l + 0.615 \cdot p_m + 0.539 \cdot p_h \\ p_h = 0.033 \cdot p_l + 0.260 \cdot p_m + 0.461 \cdot p_h \end{cases} \tag{1}$$
$$p_l + p_m + p_h = 1$$

where p_l, p_m, p_h are, respectively, the marginal probabilities of the construction enterprise achieving low, medium, and high levels of implementation of BIM in the long term.

Of course, the normalization condition consists in the equality of the sum of the marginal probabilities, which is reflected by the corresponding entry after system (1). Note that each equation of system (1) is formed by summing the columns of the matrix of transition probabilities (Table 3) taking into account the variant of the current state of the system.

As a result of solving system (1), it can be assumed that in the next few years the share of contracting enterprises with a low level of BIM penetration will be about 11% ($p_l \approx 0.11$), with medium and high—60%, respectively ($p_m \approx 0.60$) and 29 ($p_h \approx 0.29$).

Table 4 presents the calculation of the expected industry-wide efficiency from BIM introduction by enterprises of Ukraine. According to the calculations, at a low level of BIM use, the weighted average ROI indicator for construction projects in the country will be 1.2%, which still exceeds the lower limit of the average level of profitability. Average or high activity of construction enterprises in the use of BIM will increase the ROI of projects for the construction and management of real estate objects by 9.9 and 16.4 percentage points, respectively (Table 4). Taking into account the results of the calculations of the marginal probabilities of the intensity of the implementation of BIM in Ukraine, the ROI of construction projects will reach almost 11% > 8% on average, which is higher than the average level based on the statistical data of the last decade (Table 1).

Table 4 Calculation of industry-wide efficiency from BIM introduction by enterprises of Ukraine (Author's calculations)

Indicators		Level of BIM use		
		Low, $p_l = 0.11$	Medium, $p_m = 0.60$	High, $p_h = 0.29$
ROI of construction projects	Low $\approx -18.3\%$	35	22	10
	Medium $\approx 8\%$	54	43	40
	High $\approx 30\%$	11	35	50
Mathematical expectation of ROI under conditions of BIM use		**1.2%** $= (-18.3 \times 0.35 + 8 \times 0.54 + 30 \times 0.11)\%$	**9.9%** $= (-18.3 \times 0.22 + 8 \times 0.43 + 30 \times 0.35)\%$	**16.4%** $= (-18.3 \times 0.10 + 8 \times 0.40 + 30 \times 0.50)\%$
Mathematical expectation of ROI across the industry		**11%** $\approx (1.2 \times 0.11 + 9.9 \times 0.60 + 16.4 \times 0.29)\%$		

4   Conclusions

In the case of substantiating decisions regarding the introduction of BIM by each specific enterprise, our toolkit provides for the implementation of 5 series of 60 simulation experiments on the mathematical expectation of the ROI of projects in which a contractor could be involved. For each of the series, the average value of ROI is calculated, and based on the results of averaging the group averages, the expected effectiveness of the introduction of BIM is established.

For carrying out experiments, the well-known electronic spreadsheet processor Excel or similar programs are sufficient. The procedure of the simulation experiment is as follows:

1. A random number is generated in the range from 1 to 100. It is the value of the random number that will characterize the level of intensity of BIM use in the next 0.5–1 year. To implement this stage, it is advisable to use the built-in EXCEL function: = RANDBETWEEN(1;100), and 1 and 100 are function parameters that determine the upper and lower limits of random number variation. The results of using the toolkit will not be affected in any way by the use of other limits: from 0 to 99, as is traditionally noted in manuals on simulation modeling and risk management. An alternative version of the program implementation of this stage is the superposition of another random number generation function: = RAND()*100.

2. The interpretation of random numbers in preliminary experimental results is carried out, that is, the forecast level of the intensity of the use of BIM is determined. For this, the range to which a random number will fall is taken into account:

   – if the number does not exceed 11, then the intensity of BIM use will be set at a "low level", since pl = 0.11 according to the solution of system (1);
   – the value of a random number in the range from 12 to 71 is the basis for a conclusion about the "average level" of the intensity of use of BIM by the contractor, since the cumulative share of the average and low level is 0.71 = 0.11 + 0.60, since in accordance with system (1) pm = 0.60. moreover, the relative fractions after the 11th will fall on the "average level";
   – if the random number turns out to be greater than 71, the expected intensity of BIM use is high, since the percentiles above the cumulative share of "low" and "average levels" belong to the high level, and according to the system (1) ph = 0.29.

   If the parameters of the RANDBETWEEN function are set from 0 to 99, the limits of the intensity levels will be shifted one unit to the left: "low level": from 0 to 10, "medium level": from 11 to 70, "high"—from 71 to 99. Software implementation determining the forecast level of use of the BIM is a superposition of the IF function of the LOGICAL category with a reference to the corresponding cells of the previous stage and references or direct setting of the condition and the name of the level, for example: = IF(A22<=11;"low";IF(A22 > 71;"high";"medium")).

To simplify the input of functions, the sequence of conditions has undergone some inversion: the conditions of the "high" and "medium" levels are set in the form of single inequalities, and the double one for the "medium level" is set as a consequence of the failure to fulfill the remaining conditions.

3. Determination of ROI for a separate experiment taking into account Table 4: 1.2%, 9.9% and 16.4%, respectively, for "low", "medium" and "high" levels. The software implementation of this stage is similar to the previous one:

$$= IF(A22 <= 11; 1.2\%); IF(A22 > 71; 16.4\%; 9\%) \qquad (2)$$

This stage completes the simulation experiment, the obtained result will be averaged over the entire series.

We also conducted 5 series of 60 simulation experiments, the average ROI values for which were 6.92, 6.73, 7.43, 6.36, and 6.13%. In the end, it turned out that the profitability of projects at different levels of use of BIM is 6.71%. This result exceeds the average level of operational activity of construction enterprises for 2010–2019 [5], since the maximum value of 5.8% was recorded in 2014, and the achievement of this result is mainly connected with inflation and devaluation processes. Therefore, the implementation of BIM by each contractor is economically justified despite all the risks, barriers, and costs.

References

1. Leading Countries with BIM Adoption. https://www.united-bim.com/leading-countries-with-bim-adoption/. Accessed 24 Jan 2023
2. BIM Mandatory in UK High Rise Residential Projects: Implications for Builders & Designers. https://www.united-bim.com/bim-mandate-in-uk-high-rise-residential-projects-implications-for-builders-designers/. Accessed 24 Jan 2023
3. Smith, P.: BIM & the 5D project cost manager. Selected papers from the 27th IPMA (International Project Management Association) (2014)
4. Norris, J.R.: Markov Chains. Cambridge University Press (1997). ISBN 0-521-48181-3
5. Day of the sub'ektiv gospodaryuvannya 2019. Statistichny zbirnik/za redaktsieyu M.S. Kuznetsovoi: K.: Derzhanalitinform. http://ukrsat.gov.ua (2020). Accessed 24 Jan 2023
6. BIM Adoption Around the World: How Good are We? https://www.geospatialworld.net/article/bim-adoption-around-the-world-how-good-are-we/. Accessed 24 Jan 2023
7. Zeltser, R.Ya., Bielienkova, O.Yu., Novak, Ye., Dubinin, D.V.: Digital transformation of resource logistics and organizational and structural support of construction. Sci. Innov. **15**(5), 38–51 (2019)
8. Lee, C.-H., Liu, C.-L., Trappey, A.J.C., Mo, J.P.T., Desouza, K.C.: Understanding digital transformation in advanced manufacturing and engineering: a bibliometric analysis, topic modeling and research trend discovery. Adv. Eng. Inform. **50** (2021). https://doi.org/10.1016/j.aei.2021.101428
9. Klee, C.: Digitization of the property development industry: overview of current literature and research gaps. Espergesia **8**(1) (2021). https://doi.org/10.18050/esp.2014.v8i1.2692
10. Terentyev, O.O., Grigorovskiy, P.E., Tugaj, O.A., Dubynka, O.V.: Building a system of diagnosis technical condition of buildings on the example of floor beams using methods of fuzzy

sets. In: Proceedings of the 2nd International Conference on Building Innovations, vol. 73 (2020). https://doi.org/10.1007/978-3-030-42939-3_72

11. Honcharenko, T., Chupryna, Y., Ivakhnenko, I., Zinchenco, M., Tsyfra T.: Reengineering of the construction companies based on BIM-technology. Int. J. Emerg. Trends Eng. Res. **8**(8), 4166–4172 (2020). https://doi.org/10.30534/ijeter/2020/22882020

12. Akselrod, R., Shpakov, A., Ryzhakova, G., Honcharenko, T., Chupryna, I., Shpakova, H.: Integration of data flows of the construction project life cycle to create a digital enterprise based on building information modeling. Int. J. Emerg. Technol. Adv. Eng. **12**(01), 40–50 (2022). https://doi.org/10.46338/ijetae0122_02

13. Tytok, V., Bolila, N., Ryzhakov, D., Pokolenko, V., Fedun, I.: CALS–technology as a basis of creating modules for assessment of construction products quality, regulation of organizational, technological and business processes of stakeholders of construction industry under the conditions of cyclical and seasonal variations. Int. J. Adv. Trends Comput. Sci. Eng. **10**(1), 271–276 (2021). https://doi.org/10.30534/ijatcse/2021/381012021

14. Stetsenko, S., Tsyfra, T., Vahovich, I., Sichnyi, S., Lytvynenko, O.: Information and analytical tools for monitoring the prices of material and technical resources (MTR) of construction. Sci. J. Astana IT Univ. **7**, 63–76 (2021). https://doi.org/10.37943/AITU.2021.40.39.006

15. Yaser, G., Andrzej, C.: Digital transformation of concrete technology—a review. Front. Built Environ. **8**, 835236 (2022). https://doi.org/10.3389/fbuil.2022.835236

16. Nadiia, R., Yusuf, I., Yaroslav, K.S., Nataliia, B., Mykola, S., Olha, B.: Systems thinking to investigate the archetype of globalization. In: Alareeni, B., Hamdan, A. (eds) Financial Technology (FinTech), Entrepreneurship, and Business Development. ICBT 2021. Lecture Notes in Networks and Systems, vol 486. Springer, Cham (2022). https://doi.org/10.1007/978-3-031-08087-6_9

17. Franco, A.B., Jacqueline, Domingues, A.M., de Almeida Africano, N., Deus, R.M., Battistelle, R.A.G.: Sustainability in the civil construction sector supported by Industry 4.0 technologies: challenges and opportunities, infrastructures. 7 (2022). https://doi.org/10.3390/infrastructures 7030043

18. Goyko, A.F., Mikhels, V.O., Vakhovich, I.V., Pokrovsky, R.L., Hrytsenko, Yu.O.: Principles of Planning the Production Program of a Construction Enterprise and Methods of Standardizing Its Parameters. Kyiv National University of Civil Engineering and Architecture (2007)

19. Bozhanova, V., Korenyuk, P., Lozovskyi, O., Belous-Sergeeva, S., Bielienkova, O., Koval, V.: Green enterprise logistics management system in circular economy. Int. J. Math. Eng. Manag. Sci. **7**(3), 350–363 (2022). https://doi.org/10.33889/IJMEMS.2022.7.3.024

20. Shpakov, A., Stetsenko, S., Shpakova, H., Sorokina, L., Akselrod, R.: Assessment of the influence of adaptability factors on the effectiveness of managing changes in enterprises by fuzzy logic. Sci. Horizons **24**(10), 72–82 (2021)

21. Kozyk, V., Novakivskyi, I., Mrykhina, O., Koleshchuk, O.: Development of a strategy for improving organizational and production structures based on methods for non-convex programming. In: Conference NTI-UkrSURT 2019, Kharkiv, Ukraine (2019)

22. Izmailova, K.V., Parkhomenko, V.V.: Simulation modeling of financial indicators of investment activity of the enterprise. Ways to increase the efficiency of construction in the conditions of the formation of market relations, № 22, pp. 73–76 (1997)

23. Sorokina, L.V., Hoyko, A.F. (eds.): Econometric Toolkit for Financial Security Management of Construction Companies. Kyiv National University of Construction and Architecture (in Ukrainian), Kyiv (2017)

24. Vorobec, S., Voytsekhovska, V., Zahoretska, O., Kozyk, V.: The context of the circular economy model implementation, based on indicators of the European union in/for Ukraine by means of fuzzy methods. In: Kryvinska, N., Greguš, M. (eds.) Developments in Information & Knowledge Management for Business Applications. Studies in Systems, Decision and Control, vol 421. Springer, Cham (2022). https://doi.org/10.1007/978-3-030-97008-6_4

25. Chupryna, Ryzhakova, G., Chupryna, K., Tormosov, R., Gonchar, V.: Designing a toolset for the formalized evaluation and selection of reengineering projects to be implemented at an enterprise. East.-Eur. J. Enterp. Technol. (2022). https://doi.org/10.15587/1729-4061.2022.251235

26. Ryzhakova, G.V., et al.: Modern structuring of project financing solutions in construction. In: 2022 IEEE International Conference on Smart Information Systems and Technologies (SIST). (2022)

27. Ryzhakova, G., Pokolenko, V., Malykhina, O., Predun, K.: Structural regulation of method-ological management approaches and applied reengineering tools for enterprises-developers in construction. Int. J. Emerg. Trends **8**(10), 7560–7567 (2020)

28. Bielienkova, O.Iu.: Strategy and mechanisms for ensuring the competitiveness of construction enterprises based on the model of sustainable development: monograph. Lira-K, Kyiv (2020)

29. Novykova, I.V.: Marketing tools in stimulating innovative activity of enterprises. Int. J. Adv. Res. Eng. Technol. (IJARET) **11**(6), 241–251 (2020)

30. Adekunle, S.A., Aigbavboa, C., Ejohwomu, O., Thwala, W.D.: Digital transformation in the construction industry: a bibliometric review. J. Eng. Des. Technol. (2021). https://doi.org/10.1108/JEDT-08-2021-0442

31. Stanley, R., Thurnell, D.: The benefits of, and barriers to, implementation of 5D BIM for quantity surveying in New Zealand. (2014)

32. Charef, R., Alaka, H., Emmitt, S.: Beyond the third dimension of BIM: A systematic review of literature and assessment of professional views. J. Build. Eng. **19**. https://researchportal.bath.ac.uk/files/175603287/JOBE_2017_601_accepted_version.pdf (2018)

33. Liu, S., Xie, B., Tivendal, L., Liu, C.: Critical barriers to BIM implementation in the AEC industry. Int. J. Mark. Stud. **7**(6), 162–171 (2015)

34. Tsyfra, T.Yu.: BIM yak instrument reformuvannia systemy tsinoutvorennia (na prykladi dorozhno-budivelnykh pidpryiemstv Kazakhstanu. [BIM as a tool for reforming the pricing system in construction (on the example road construction companies in Kazakhstan)]. Shli-akhy pidvyshchennia efektyvnosti budivnytstva v umovakh formuvannia rynkovykh vidnosyn **47**(2), 167–178 (2021). https://doi.org/10.32347/2707-501x.2021.47(2)

35. BIM Levels explained. https://www.thenbs.com/knowledge/bim-levels-explained

36. McGrawHill Construction: Smart Market Report the Business Value of BIM for Construction in Major Global Markets. McGraw Hill Construction, New York. https://icn.nl/pdf/bim_construction.pdf (2014). Accessed 24 Jan 2023

37. Bielienkova, O., Novak, Y., Matsapura, O., Zapiechna, Y., Kalashnikov, D., Dubinin, D.: Improving the organization and financing of construction project by means of digitalization. Int. J. Emerg. Technol. Adv. Eng. **12**(8), 108–115 (2022)

38. Siniak, N., Zróbek, S., Nikolaiev, V., Shavrov, S.: Building information modeling for housing renovation—example for Ukraine. Real Estate Manag. Valuation **27**(2), 97–107 (2019). https://doi.org/10.2478/remav-2019-0018/. Accessed 2023/01/24

39. Perspectives of BIM in Ukraine and Implementation at the State Level. https://uscc.ua/ru/news/perspektivy-bim-v-ukraine-i-vnedrenie-na-gosudarstvennom-urovne. Accessed 24 Jan 2023

Digital Promotion as Innovative Business Management Technologies of Retail Chains

Olena Bilovodska⊙**, Kostiantyn Ivanchenko**⊙**, Ihor Ponomarenko**⊙**,
Zorina Shatskaya**⊙**, and Olena Budiakova**⊙

Abstract At the time of constant intensification of competition in domestic and international markets, the use of complex of marketing knowledge becomes the basis of competition. While entering new market, an important task is to form a promotion complex in order to spread information about the company and its products or services. The work is devoted to the development of a promotion complex with the use of digital tools in the retail market of grocery and consumer goods in Romania and the evaluation of its effectiveness. To do this, the authors considered statistical information on the use of the Internet, social networks and search engines in Romania, and provided recommendations for the development of a promotion complex in this market using advertising tools on the Facebook social network, the Google search engine and the use of advertising through opinion leaders on the Instagram social network and evaluated the effectiveness of the proposed promotion complex based on the calculation of the main marketing metrics. The results of the study can be used to form knowledge and decision-making experience regarding the development of a promotion complex when entering new markets in developing countries. During the work, the following methods were used: the method of analysis and synthesis, the method of comparison, the analytical method, the method of forecasting, etc.

Keywords Promotion · Retail · Digital · Digital promotion tools

O. Bilovodska (✉) · K. Ivanchenko
Taras Shevchenko National University of Kyiv, Volodymyrska Street, 60, Kyiv 01033, Ukraine
e-mail: o.bilovodska@knu.ua

I. Ponomarenko
State University of Trade and Economics, Kyoto Street, 19, Kyiv 02156, Ukraine
e-mail: i.ponomarenko@knute.edu.ua

Z. Shatskaya · O. Budiakova
Kyiv National University of Technologies and Design, Nemirovich-Danchenko Street, 2, Kyiv 01011, Ukraine
e-mail: shacka.zy@knutd.com.ua

O. Budiakova
e-mail: budyakova.oy@knutd.edu.ua

1 Introduction

Digital promotion has become an important tool for retail chains to achieve sustainable development. It has emerged as an innovative business management technology with potential to improve customer experience, reduce operating costs, and increase profitability. This article aims to discuss the advantages and challenges of digital promotion for retail chains, and explore the opportunities for this technology to help achieve sustainable development.

In today's digital age, the retail industry has seen tremendous growth over the last decade. This growth has been driven by the introduction of innovative business management technologies that allow retailers to better manage their operations and maximize profits. One of the most important of these technologies is digital promotion, which can provide retailers with an effective way to reach potential customers, increase brand awareness, and boost sales. This article will explore the various aspects of digital promotion, its advantages and disadvantages, and how it can be used as an innovative business management technology for retail chains. Additionally, this article will discuss the key considerations retail chains should keep in mind when implementing digital promotion strategies. Finally, this article will provide an overview of the current trends and best practices for using digital promotion as a business management technology. By understanding the various aspects of digital promotion and its potential benefits, retail chains can gain a competitive edge in the market and maximize their profits.

2 Literature Review

The use of digital promotion technologies has become increasingly important in recent years for retail chains and other businesses. This literature review explores the concept of digital promotion as an innovative business management technology and the related research that has been conducted in this area.

To sum up, it can be said that researchers mostly examine the various digital promotion techniques available to retailers, such as digital signage, loyalty programs, coupons, and promotions. The state is that retail chains must be able to effectively utilize these digital promotions to stay competitive in the market. By leveraging digital promotion, retailers can increase customer engagement, create stronger relationships with customers, and increase sales. Also, it's focused on how digital promotion can be used to improve the efficiency of retail chains and the impact of digital promotion on customer loyalty and how it can be used to improve customer retention.

Others investigate the relationship between digital promotion and customer loyalty. They discuss how digital promotion can increase customer loyalty by providing customers with personalized offers and discounts. Additionally, they examine how digital promotion can be used to improve customer satisfaction, increase brand awareness, increase sales, and impact of social media on the success of digital promotion strategies in retail. The authors discuss how social media can be used to target customers, increase customer engagement, and build customer loyalty.

Thus, O. Bilovodska et al. investigated the peculiarities of marketing in the digital environment [1], exploring the use of digital technologies for promotion and the impact of digital communication on consumer behaviour. F. Alsalim studied the role of the management information system for promotion and examined how digital technologies can be used to improve customer relationship management [2]. N. Muna et al. explored the influence of social networks on the work of small and medium-sized businesses [3], while A. Markhaeni et al. examined the features of advertising campaigns using opinion leaders [4]. L. Lang et al. investigated methodological approaches to the creation and evaluation of promotion measures [5], while A. Alaei et al. compared different models of sales via the Internet [6]. Y. Li and X. Gong explored the synergy between promotion channels [7], investigating the use of digital tools to improve customer engagement and loyalty, B. Melovic et al. studied the impact of digital transformation and digital marketing on the brand promotion and positioning [8]. Y. Dwivedi et al. examined the future of digital and social media marketing research [9].

Overall, this literature review has highlighted existing research related to digital promotion as an innovative business management technology for retail chains. The research has explored the use of digital tools for promotion, customer relationship management, and the impact of digital communication on consumer behavior. Furthermore, the research has examined the use of opinion leaders in advertising campaigns, methodological approaches to the creation and evaluation of promotion measures, and the synergy between different promotion channels. In conclusion, digital promotion is an important and innovative business management technology that can be used to improve customer engagement and loyalty.

3 Results

At the beginning, let's consider the number of people in Romania who are Internet users (Fig. 1).

According to the given data, the level of Internet penetration is sufficient to use promotion channels that are associated with the use of digital communications.

Consider how often the Internet is used in Romania (see Table 1).

So, the vast majority of almost all age groups (except 55–64) use the Internet every day.

At the first stage of promotion formation, it is necessary to determine the purpose of promotion. The goal is the formation of the brand in the Romanian market.

The second stage is the development of a promotion complex. For this, it is necessary to formulate the goals that must be achieved:

- To create high awareness of the brand in its target audiences.

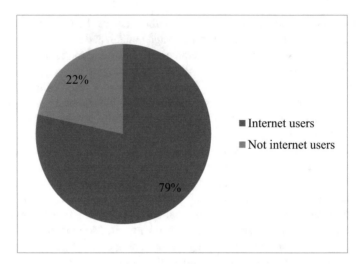

Fig. 1 The share of Internet users in the Romanian population in 2022. *Source* Compiled by the authors based on data [10]

Table 1 Share of daily Internet users among the population of Romania in 2020, by age group	No	Age	Share of daily Internet users (%)
	1	16–24	89
	2	25–34	80
	3	35–44	75
	4	45–54	62
	5	55–64	40

Source Compiled by the authors based on [11]

- To create positive associations and emotions that the brand evokes in target audiences.
- To create a stable loyal group of customers and supporters.

To develop an advertising strategy, consider the tools that should be used (see Table 2).

Search engine advertising works on a pay-per-click model, in which advertisers bid on keywords and pay for each click on their ad. Each time a search is performed, Google searches for advertisers and selects a set of winners to appear in the ad space on the search results page. Winners are selected based on a combination of factors, including the quality and relevance of their keywords and ad campaigns, as well as keyword bids.

More specifically, who appears on a page is based on the advertiser's rating and the advertiser's score, calculated by multiplying two key factors—the CPC rate (the highest amount an advertiser is willing to spend) and the quality score (a value that

Table 2 Advertising tools

Tool	Characteristic
Search engines	Allows advertisers to bid on search engine ads when someone searches for a keyword that is relevant to their business offerings The most popular PPC advertising system in the world is Google Ads
Social networks	Ad is placed on social networks such as Facebook, Instagram, TikTok, and so on Advertisement in social networks combines segmentation according to different groups of factors (for example, according to geographical, socio-psychological criteria, etc.) to make possible detailed identification of the target audience
Opinion leaders	At a fundamental level, influencer marketing is a type of social media marketing that leverages product endorsements and mentions from opinion leaders. Influencer marketing works through the trust that influencers have built, and whose endorsements serve as a form of social proof to potential customers about the quality of the advertised brand

Source Compiled by the authors based on sources [12–14]

Table 3 Market shares of search engines in Romania on March 2022

No	Search engine	Market share (%)
1	Google	97.37
2	Bing	1.69
3	Yahoo!	0.47
4	DuckDuckGo	0.18
5	Petal Search	0.16
6	Yandex	0.09

Source Compiled by the authors based on data [15]

takes into account the click-through rate, relevance and quality of the landing page This is a kind of auction [12].

In Romania, the dominant search engine is Google, which occupies almost 98% of the market (see Table 3).

Running an advertising company through Google Ads is especially important because, as the most popular search engine, Google receives a huge amount of traffic.

The next channel is advertising on social networks such as Facebook, Instagram, TikTok, etc. Consider the number of users of various social networks (see Table 4).

Table 4 Number of social media users in Romania in 2020

No	Social media	Users, million
1	Facebook	10.88
2	Instagram	3.898
3	Linkedin	2.833

Source Compiled by the authors based on [16]

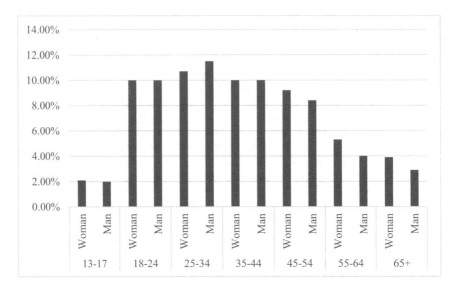

Fig. 2 Facebook users in Romania in March 2022. *Source* Compiled by the authors based on data [17]

As can be seen from the figure, Facebook is dominating among social networks. For completeness of information, let's consider the gender-age structure of users of the most popular social network—Facebook (see Fig. 2).

In March 2022, there were 13 mln. Facebook users in Romania, representing 67.8% of the total population. Most of them were women—51.1%. The largest group of users were people aged 25–34 (2.9 mln.). The biggest difference between men and women is in people aged 55–64, where women lead by 527,000.

Considering such data, it can be concluded that a large part of the target audience is users of this social network.

Considering advertising through opinion leaders, we can highlight that the main advantages of such advertising are: access to the target audience, celebrity effect, direct communication and feedback, trust and closeness to the audience [18, 19]. According to research, up to 92% of social network users trust the advice of influencers more than standard advertising [19]. An important role in establishing such a high level of trust is made by the possibility of direct communication with one's idol through social networks. In addition, the blogger's audience wants to get closer to their lifestyle by using the products and services that the influencer uses. Such marketing allows the brand to establish relations with a young audience [20].

The MediaKix study revealed how companies relate to influence marketing (see Table 5, Figs. 3 and 4) [19].

So, 80% of respondents consider advertising through opinion leaders to be better than other types of ads.

So, 89% of respondents consider advertising through opinion leaders no less profitable than other.

Table 5 Effectiveness of influence marketing, % of respondents

No	Performance evaluation	% of respondents (%)
1	Very effective	35
2	Effective	45
3	Neutral	15
4	Inefficient	5
5	Completely inefficient	0

Source Compiled by the authors based on data [20]

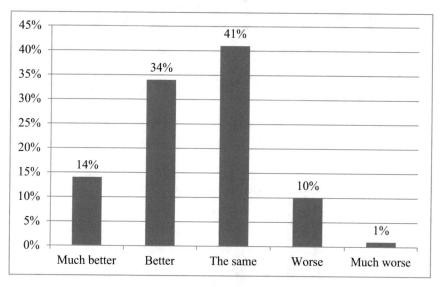

Fig. 3 Answers to the question, "How you evaluate influencer marketing ROI compared to ROI from other leads sources?". *Source* Compiled by the authors based on data [20]

As the main platforms for influence marketing, experts use Instagram and YouTube [21] due to their visibility, popularity and convenience.

Promotion payments are usually have three types: a fee, for discounts or barter.

Let's consider the gender and age structure of Instagram users in Romania (see Fig. 5).

In March 2022, Romania had 5.7 mln. Instagram users, representing 30% of its population. Most of them were women—51.6%. The largest group of users were people aged 18–24 (1.9 mln.). The biggest difference between men and women is seen in people aged 35–44, where women lead by 421,500 [22].

Let's consider possible options for such advertising (see Table 6).

Sales promotion is an important tactical task of international marketing. Within its framework, we propose to carry out measures regarding:

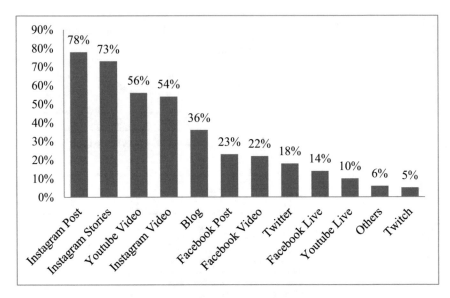

Fig. 4 Answers to the question: "Which content format is effective for influencer marketing?". *Source* Compiled by the authors based on data [20]

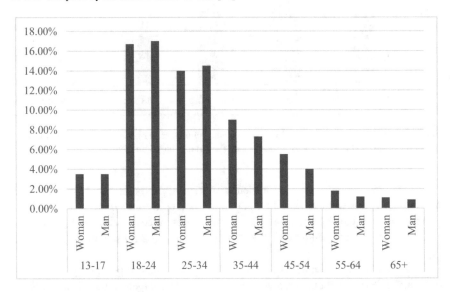

Fig. 5 Gender and age structure of Instagram users in Romania, March 2022. *Source* Compiled by the authors based on [22]

Table 6 List of opinion leaders in Bucharest

No	Account name	Account category	Subscribers	Involvement (%)	The cost of an advertising
With the greatest involvement					
1	@ianishagi	Sport	230 thousand	20.2	2100–3900$
2	@olga.das	–	38 thousand	19.9	420–790$
3	@taaooma	Motherhood, cinema, theater, dance	760 thousand	19.6	19,000–35,000$
With the most followers					
1	@inna	Music	2.8 million	1.5	2100–3900$
2	@andra	Music	2.2 million	1.8	1900–3600$
3	@salehgomaa	Sport	2.1 million	6.5	6300–12,000$

Source Compiled by the authors based on data [23]

1. stimulating own personnel to create ideas for sales improvement and sales promotion;
2. stimulating consumer.

Stimulating own staff. To improve the efficiency of our own staff, we offer measures such as awards, competitions for the best employee, training and various professional conferences for staff.

Stimulating consumers. To increase the demand for services, implement the following methods:

- providing promotional codes for a discount or free first delivery;
- a system of accumulating bonus points, for which you can get souvenir products or a discount on subsequent orders;
- creation of contests and raffles of gifts in the form of a lottery based on the number of ordered deliveries.

Digital and traditional public relations have one common goal—to help improve a company's reputation and increase its visibility among members of its target audience. The main difference between them is the choice of methods used to achieve this goal [24].

Digital PR aims to increase brand value through digital tools. Building a strong brand in the increasingly competitive digital age can be challenging, but there is a wide range of techniques that can be used to increase visibility and stay competitive. Examples include [25]:

- online press releases;
- on the Internet to create coverage on the Internet;
- business profiling;
- reviews and interviews on the Internet etc.

Next, it is necessary to choose a message that will be transmitted to consumers. Speaking about the structure of an advertising appeal, we can talk about its elementary (headline, text, slogan, illustration, address block) and compositional components [25].

The next stage is to draw up a budget for promotion. It is proposed to use the "top-down" approach and allocate $200,000 for promotion, of which $50,000 is for search engine advertising, $50,000 for social media advertising, $45,000 for Instagram opinion leader advertising, $5,000 for PR and $50,000 for sales promotion.

The next part of the process is the development of feedback channels. In this case, the main feedback channels are social networks, social media and blogs [26].

At the last stage, it is necessary to evaluate the effectiveness of the proposed promotion measures. To do this, we will evaluate the effectiveness of each proposal.

The efficiency evaluation is based on the suggestion that each consumer will, on average, give the company $3 of profit. The evaluation of the effectiveness of the advertising company is shown below (Table 7) and consists of the following components:

- Conversion rate—the percentage of website visitors who reach the desired goal (conversion) out of the total number of visitors [27].
- Daily budget—monthly budget divided by 30.
- Monthly budget.
- Impressions per month—the number of views of advertisements in 1 month.
- Clicks/month—the number of clicks per month.
- Conversions/month—the number of people who made purchases in a month.
- Average profit from each consumer.

Table 7 Analysis of the effectiveness of the advertising campaign through Google Ads

No	Indicator	Value
1	Conversion rate	10%
2	Daily budget	1,666,667
3	Monthly budget	50,000
4	Shows per month	3,500,000
5	Clicks/month	238,095.2
6	Conversions/month	23,809.52
7	Average profit from each consumer	3
8	Income per month	71,428.57
9	ROAS	1.43
10	CPA	2.1
11	CPC	0.21
12	CPM	14.28571
13	CTR	0.068027

Source Compiled by the authors based on [12]

- Monthly income—expected monthly income from consumers involved in the advertising campaign.
- ROAS is a metric that measures the effectiveness of a digital advertising campaign. Return on advertising spend (ROAS) shows how much revenue was generated from a particular advertisement or campaign [28, 29].
- CPA is a metric that determines how much it costs to generate action with advertising [30].
- CPC is a measure of how much advertisers pay for ads they place on websites or social media based on the number of clicks the ad receives [31].
- CPM (cost per mille or cost per millenium) is the cost of a thousand impressions of an advertisement or banner [32].
- CTR is the percentage of people who view a web page and then click on a specific ad that appears on that page [33].

So, in this advertising campaign, with a monthly budget of $50,000, about 3.5 million ads will be shown, about 238,000 will visit company's website, and about 24,000 will make a purchase. With calculating that each consumer for the entire time on average will bring the company a profit of $3, we have a profit of $71,428.57.

The next part is advertising through Facebook. The evaluation of the effectiveness of such advertising is shown in Table 8.

So, according to this advertising campaign, with a monthly budget of $50,000, about 5.8 million ads will be shown, about 263 thousand will go to the company's website, and about 31.5 thousand will make a purchase. When calculating that each consumer for the entire time on average will bring the company a profit of $3, we have a profit of $94,736.84.

Table 8 Analysis of the effectiveness of the advertising campaign through Facebook

No	Indicator	Value
1	Conversion factor	11%
2	Daily budget	1,666,667
3	Monthly budget	50,000
4	Shows per month	5,833,333
5	Clicks/month	263,157.9
6	Conversions/month	31,578.95
7	Average income	3
8	Income per month	94,736.84
9	ROAS	1.894737
10	CPA	1.727273
11	CPC	0.19
12	CPM	8.571429
13	CTR	0.045113

Source Compiled by the authors based on data [34]

The next advertising type is advertising through Instagram opinion leaders. The evaluation of the effectiveness of such advertising is shown in Table 9.

So, according to the advertising company, with a budget of $44,980, there will be about 8.13 million ad views, about 421,000 will go to the company's website, and about 38,000 will make a purchase. When calculating that each consumer for the entire time on average will bring the company a profit of $3, we have a profit of $113,692.1.

Public relations is an effective way to build consumer trust that contributes to long-term revenue. When shoppers are searching for a product online, they will be looking for different types of information about their future purchases [36]. Consumers can also view ratings from previous consumers. Maybe they'll go further by reading long testimonials about how the product performs over time. This information allows buyers to make informed purchasing decisions. Trust is important not only for increasing brand revenue, but also for strengthening the business as a whole [37].

Among PR researchers, there is no common opinion about the evaluation of effectiveness. There are at least three reasons for this [38]:

- PR is a multi-meaning concept closely related to advertising, propaganda, journalism, campaigning, and marketing. The lack of a clear frame makes it difficult to evaluate PR.
- PR activity is constantly under the influence of the "external environment", which is constantly changing and it is difficult to say exactly how much PR influenced a specific event or business indicator.
- Duration and cost of a detailed assessment.

Table 9 Analyzing the effectiveness of an advertising campaign through opinion leaders on Instagram

No	Account name	The price of an advertising post	The expected number of people who will go to the application	The price of the involved consumer	The expected number of people who will make a purchase	Aggregate advertising revenue	Profit
With the greatest involvement							
1	@ianishagi	3000	46,460	0.064572	4181.4	12,544.2	9544.2
2	@olga.das	580	7562	0.076699	680.58	2041.74	1461.74
3	@taaooma	27,000	148,960	0.181257	13,406.4	40,219.2	13,219.2
With the most followers							
1	@inna	3000	42,000	0.071429	3780	11,340	8340
2	@andra	2600	39,600	0.065657	3564	10,692	8092
3	@salehgomaa	8800	136,500	0.064469	12,285	36,855	28,055

Source Compiled by the authors based on data [21, 35]

Therefore, it is impossible to accurately evaluate the effectiveness of the PR company, since consumers could share information, influencing additional purchases or possible other actions. Even if we can prove that a single article generated a certain amount of revenue, there is an argument that it still generated much more value than that amount, which can be estimated through increased brand awareness, referrals, and other factors [39].

One of the ways to evaluate the effectiveness of a PR campaign is to evaluate the market value of links from various sources. This helps to understand the value of links and how much it would cost to attract the same number of consumers through advertising. For example, in 2017 Econsultancy built a total of 6,717 links with an estimated market value of $5,547,000. 20 specialists worked on it. Under such conditions, we can calculate the average amount of links created by one employee per month: 28 links for the amount of $23,112 [40].

Another example would be the international company Vimpelcom Ltd. In the study of the company Ex Libris [41], the value of the company's shares on NASDAQ was compared with the change in PRT and Media Presence indicators during six months in the leading and most influential international media.

PRt is an integral indicator that takes into account a number of simple qualitative parameters of the material: the emotional tone of the publication and its title, the transmission of the company's key message, the speaking activity of company representatives and third-party speakers (partners and experts), as well as the genre of the publication, the visibility of the material, and the density of mentions of the object of research.

MediaPresence is an indicator that allows you to judge the presence of the researched object in mass media. In its essence, this is an absolute indicator, which is equal to the number of publications published about the researched object in the total mass of mass media publications for the researched period.

In this study, a regression model was built. According to the confidence intervals for the data, when M-PRT increases by 1%, quotations increase by 3%, and when Media Presence increases by 1%, the value of Vimpelcom's assets strengthens by an average of 1.4%.

Taking into account the first method, it can be expected that each PR specialist should attract consumers in the amount of 23 thousand USD on average. Therefore, when hiring one specialist, an additional benefit of about $20,000 a month is expected.

At the expense of advertising and PR, taking into account the average profitability of retail companies at the level of 25% [42], we expect turnover at the level of $1.2 million per month, that will be 0.16% of the market [43].

4 Discussions and Conclusions

The study of digital promotion as an innovative business management technology of retail chains has shown that it can be an effective tool for increasing customer engagement, loyalty and sales. Digital promotion helps retailers to reach a wider audience, increase customer loyalty and attract new customers. It also enables retailers to track customer behavior and respond to their needs. However, it is important for retailers to use digital promotion properly. The use of digital promotion should be tailored to the needs of the particular retail chain and must be used strategically. Additionally, retailers should ensure that digital promotion strategies are integrated with their other marketing efforts, as this will help to ensure that the digital promotion initiatives are successful.

So, digital promotion can be an effective tool for increasing brand awareness, creating positive associations and emotions, and creating a stable loyal customer base in the Romanian market. It is necessary to take into account the specifics of the Romanian population, the structure of its Internet users and the popularity of various channels of communication. It is also important to consider the type of message that should be transmitted to the target audience and the budget for promotion. In order to evaluate the effectiveness of the proposed promotion measures, it is necessary to use conversion rates, daily and monthly budgets, shows per month, clicks per month, conversions per month, average profit from each consumer, income per month, ROAS, CPA, CPC, CPM and CTR.

Digital promotion should include search engine advertising, social media advertising, influencer advertising, public relations, sales promotion, and feedback channels. It is important to create a budget for promotion and to evaluate the effectiveness of each proposal. This can be done by calculating the number of ads shown, visits to the company's website, and purchases made. Additionally, market value and PRT and Media Presence indicators can be used to evaluate the effectiveness of a PR campaign. Finally, it is important to consider the cost of hiring a specialist and the expected turnover from promotion and PR activities.

Study shows that a digital promotional strategy is profitable and efficient for retail companies in Romania. Thus, using Google ads for promotion purposes has 142.86% profitability, Meta Ads—189.47% and advertising campaign through opinion leaders on Instagram—252.76% respectively.

References

1. Bilovodska, O., Boienko, O., Omelchenko, V., Kostynets, Iu., Ievseitseva, O., Omelchenko, H.: Marketing digital strategy for promoting brand of global retailer achieving sustainability. Rev. Econ. Finance **20**, 647–653 (2022). https://doi.org/10.55365/1923.x2022.20.75
2. Alsalim, F.: Influence of mis components on efficiency of e-marketing strategies: evidence from telecommunication organizations in Jordan. F.I.A. Int. J. Data Netw. Sci. (2022). https://www.scopus.com/record/display.uri?eid=2-s2.0-85121044746&origin=resultslist&sort=plfdt-f&lis tId=57414162&listTypeValue=Docs&src=s&imp=t&sid=e914d1217e31c9890d18a2ac0e1 f7bc8&sot=sl&sdt=sl&sl=0&relpos=0&citeCnt=0&searchTerm=. Accessed 30 Jan 2022

3. Muna, N., Yasa, N., Ekawati, N.: Market entry agility in the process of enhancing firm performance: a dynamic capability perspective. Int. J. Data Netw. Sci. (2022). https://www.scopus.com/record/display.uri?origin=recordpage&eid=2-s2.0-85121039462&citeCnt=0&noHighlight=false&sort=plfdt-f&listId=57414162&listTypeValue=Docs&src=s&imp=t&sid=e914d1217e31c9890d18a2ac0e1f7bc8&sot=sl&sdt=sl&sl=0&relpos=1. Accessed 31 Jan 2022

4. Marhaeni, A., Yasa, I., Fahlevi, M.: Gender and age in the language of social media: an easier way to build credibility. Int. J. Data Netw. Sci. (2022). https://www.scopus.com/record/display.uri?origin=recordpage&eid=2-s2.0-85121029130&citeCnt=0&noHighlight=false&sort=plfdt-f&listId=57414162&listTypeValue=Docs&src=s&imp=t&sid=e914d1217e31c9890d18a2ac0e1f7bc8&sot=sl&sdt=sl&sl=0&relpos=2. Accessed 30 Jan 2022

5. Lang, L., Lim, W., Guzmán, F.: How does promotion mix affect brand equity? Insights from a mixed-methods study of low involvement products. J. Bus. Res. (2021). https://www.scopus.com/record/display.uri?origin=recordpage&eid=2-s2.0-85121427159&citeCnt=0&noHighlight=false&sort=plfdt-f&listId=57414162&listTypeValue=Docs&src=s&imp=t&sid=e914d1217e31c9890d18a2ac0e1f7bc8&sot=sl&sdt=sl&sl=0&relpos=8. Accessed 30 Jan 2022

6. Alaei, A., Taleizadeh, A., Rabbani, M.: Marketplace, reseller, or web-store channel: the impact of return policy and cross-channel spillover from marketplace to web-store. J. Retail. Consum. Serv. (2020). https://www.scopus.com/record/display.uri?origin=recordpage&eid=2-s2.0-85090718999&citeCnt=0&noHighlight=false&sort=plfdt-f&listId=57414162&listTypeValue=Docs&src=s&imp=t&sid=e914d1217e31c9890d18a2ac0e1f7bc8&sot=sl&sdt=sl&sl=0&relpos=10. Accessed 30 Jan 2022

7. Li, Y., Gong, X.: What drives customer engagement in omnichannel retailing? The role of omnichannel integration, perceived fluency, and perceived flow. IEEE Trans. Eng. Manag. (2022). https://www.scopus.com/record/display.uri?origin=recordpage&eid=2-s2.0-85122875573&citeCnt=0&noHighlight=false&sort=plfdt-f&listId=57414162&listTypeValue=Docs&src=s&imp=t&sid=e914d1217e31c9890d18a2ac0e1f7bc8&sot=sl&sdt=sl&sl=0&relpos=15. Accessed 30 Jan 2022

8. Melović, B., Jocović, M., Dabić, M., Vulić, T.B., Dudic, B.: The impact of digital transformation and digital marketing on the brand promotion, positioning and electronic business in Montenegro. Technol. Soc. (2020). https://doi.org/10.1016/j.techsoc.2020.101425. Accessed 03 Feb 2022

9. Dwivedi, Y.K., Ismagilova, E., Laurie Hughes, D., Carlson, J., Filieri, R., Jacobson, J., Jain, V., Karjaluoto, H., Kefi, H., Krishen, A.S., Kumar, V., Rahman, M.M., Raman, R., Rauschnabel, P.A., Rowley, J., Salo, J., Tran, G.A., Wang, Y.: Setting the future of digital and social media marketing research: perspectives and research propositions. Int. J. Inf. Manag. (2021). https://doi.org/10.1016/j.ijinfomgt.2020.102168. Accessed 01 Feb 2022

10. World Development Indicators. https://databank.worldbank.org/source/world-development-indicators. Accessed 01 Feb 2022

11. Share of Daily Internet Users in Romania According to Age from 2014 to 2020 (2022). Statista. https://www.statista.com/statistics/1241861/romania-internet-users-use-accessed-internet-daily-age/. Accessed 11 April 2022

12. Google Ads. https://ads.google.com/intl/uk_ua/home/?subid=ua-uk-ha-aw-bk-c-bau!o3~Cj0KCQiA4NTxBRDxARIsAHyp6gCNGCV0_9hVgx7MPLpF5YeIX4aNoVN3f4hMAVehDA0TD4dPI0a0fAsaAnCmEALw_wcB~56890479749~aud-570778808110%3Akwd-19650782074~1485457199~412067010608. Accessed 30 Jan 2022

13. What Is Influencer Marketing: How to Develop Your Strategy (2019). https://sproutsocial.com/insights/influencer-marketing/. Accessed 01 Feb 2022

14. What Is Social Media. Advertising. ObelisMedia. http://obelismedia.com/social-media-advertising/. Accessed 01 Feb 2022

15. Search Engine Market Share Romania (2022). https://gs.statcounter.com/search-engine-market-share/all/romania. Accessed 25 April 2022

16. Social Media Users in Romania (2020). https://napoleoncat.com/stats/social-media-users-in-romania/2020/03. Accessed 17 Feb 2022

17. Facebook Users in Romania (2022). https://napoleoncat.com/stats/facebook-users-in-romania/2022/03. Accessed 29 Jan 2022
18. Criticism of Influencer Marketing: Disadvantages of Hypes—WDR (German TV) (2018). https://socialmediaagency.one/criticism-influencer-marketing-disadvantages-hypes/. Accessed 30 Jan 2022
19. Influencers Are the New Brands (2018). https://forbes.com/sites/deborahweinswig/2016/10/05/influencers-are-the-new-brands/. Accessed 30 Jan 2022
20. Influencer Marketing 2019 Industry Benchmarks. MediaKix (2019). https://mediakix.com/influencer-marketing-resources/influencer-marketing-industry-statistics-survey-benchmarks/. Accessed 30 Jan 2022
21. Instagram CPM, CPC, & CTR Benchmarks (2019). https://blog.adstage.io/instagram-cpm-cpc-ctr-benchmarks. Accessed 30 Jan 2022
22. Instagram Users in Romania (2020). https://napoleoncat.com/stats/instagram-users-in-romania/2020/03. Accessed 30 Jan 2022
23. Global Lite Report Quarter by Numbers Q1 2019 (2019). https://www.nielsen.com/wp-content/uploads/sites/3/2019/07/q1-2019-qbn-lite-report.pdf. Accessed 30 Jan 2022
24. How to Create the Perfect Digital PR Campaign Strategy. https://www.stateofdigital.com/create-perfect-digital-pr-campaign-strategy/. Accessed 30 Jan 2022
25. Poliakov, V.A.: Reklamne povidomlennia: protsesy rozrobky ta realizatsii tvorchykh kontseptsii (2015). https://stud.com.ua/34896/marketing/reklamne_povidomlennya_protsesi_rozrobki_realizatsiyi_tvorchih_kontseptsiy. Accessed 29 Jan 2022
26. Petrovska, I.P., Chernous, V.I.: DIGITAL PR yak potuzhnyi instrument suchasnoho marketynhu. Suchasni problemy ekonomiky ta pidpryiemnytstvo (2013). https://ela.kpi.ua/bitstream/123456789/19942/1/SPEP-12_21_Petrovska1.pdf. Accessed 29 Jan 2022
27. Conversion Rate. Optimization Glossary. Optipedia. https://www.optimizely.com/optimization-glossary/conversion-rate/. Accessed 10 April 2022
28. Return On Ad Spend (ROAS). https://www.appsflyer.com/glossary/roas/. Accessed 10 April 2022
29. What Is ROAS. Calculating Return On Ad Spend. https://www.bigcommerce.com/ecommerce-answers/what-is-roas-calculating-return-on-ad-spend/. Accessed 10 April 2022
30. Cost Per Action (CPA) Advertising. Grow Hack Scale. https://growhackscale.com/glossary/cost-per-action-cpa. Accessed 10 April 2022
31. CPC (Cost Per Click) Explained. Amazon Ads. https://advertising.amazon.com/library/guides/cost-per-click. Accessed 10 April 2022
32. CPM (Cost Per Millenium). Unisender. https://www.unisender.com/ru/support/about/glossary/cpm/. Accessed 10 April 2022
33. Click-Through Rate (CTR) Definition. Investopedia. https://www.investopedia.com/terms/c/clickthroughrates.asp. Accessed 10 April 2022
34. Facebook Business Manager (2022). https://business.facebook.com/home/accounts?business_id=153473009086365. Accessed 29 Jan 2022
35. Stoshikj, M., Kryvinska, N., Strauss, C.: Service systems and service innovation: two pillars of service science. Procedia Comput. Sci. **83**, 212–220 (2016). https://doi.org/10.1016/j.procs.2016.04.118
36. Kryvinska, N., Zinterhof, P., van Thanh, D.: An analytical approach to the efficient real-time events/services handling in converged network environment. Netw. Based Inf. Syst. 308–316. https://doi.org/10.1007/978-3-540-74573-0_32
37. Explore New Influencers (2022). https://www.heepsy.com/influencers. Accessed 29 Jan 2022
38. Chy mozhna vymiriaty efektyvnist PR i otsinyty yoho vplyv na biznes? Asotsiatsiia kompanii konsultantiv v sferi zviazkiv z hromadskistiu. http://www.akospr.ru/12623. Accessed 31 Jan 2022
39. Burek, M.: How to Measure PR in Revenue Marketing. Shift Communication (2020). https://www.shiftcomm.com/blog/how-to-measure-pr-in-revenue-marketing/. Accessed 30 Jan 2022
40. Digital PR: A Guide to Choosing & Reporting on KPIs. Econsultancy (2018). https://econsultancy.com/digital-pr-a-guide-to-choosing-reporting-on-kpis/. Accessed 30 Jan 2022

41. Otsinka biznes-efektyvnosti PR ta prohnoznyi rozrakhunok vartosti kotyruvan na prykladi keisu kompanii. Vimpelcom Ltd. (2020). https://exlibris.ru/news/otsenka-biznes-effektivn osti-pr-i-prognoznyj-raschet stoimosti-kotirovok-na-primere-kejsa-kompanii-vimpelcom-ltd/. Accessed 31 Jan 2022
42. FMCG Outlook 2020 (2020). https://www.iriworldwide.com/getattachment/News/News/ 2020-FMCG-Outlook/2020-FMCG-Outlook-Report.pdf?lang=en-AU. Accessed 29 Jan 2022
43. Best Practices to Generate Backlinks to Your Website. Digital Marketing Institute (2020). https://digitalmarketinginstitute.com/blog/9-best-practices-to-generate-backlinks-to-your-website. Accessed 30 Jan 2022

Support for the Development of Educational Programs with Graph Database Technology

Iryna Zinovieva⬤, **Nina Sytnyk**⬤, **Olha Denisova**⬤, and **Volodymyr Artemchuk**⬤

Abstract The paper discusses the challenges in training students as an qualified specialists in the education system, particularly in higher education. It emphasizes the need for modern approaches to address the challenges and the importance of aligning educational programs with national and European standards as well as dynamic updating of programs and standards in accordance with the changing labor market and technological progress. To support the development of high-quality educational programs and ensure compliance with the standards, the paper proposes using modern information technologies, such as graph database technology. Specifically, the paper describes the development and implementation of a graph database in the DBMS environment Neo4j, which models the connections between educational components (disciplines or courses), competencies and learning results by the national academic standards. The graph database helps to analyze the degree of compliance between educational programs and the standards, allows controlling the semantic connection between educational components, competencies and learning results, and, on this basis, provides support for decisions regarding the educational program's adoption or improvement. The paper concludes that the use of graph database technology is essential to support the development of educational programs and ensure their quality and effectiveness in preparing qualified specialists.

I. Zinovieva · N. Sytnyk · O. Denisova
Kyiv National Economic University Named After Vadym Hetman, Kyiv, Ukraine
e-mail: ira.zinovyeva@kneu.edu.ua

O. Denisova
e-mail: denisova@kneu.edu.ua

V. Artemchuk (✉)
G.E. Pukhov Institute for Modelling in Energy Engineering of The National Academy of Sciences of Ukraine, Kyiv, Ukraine
e-mail: volodymyr.artemchuk@npp.nau.edu.ua

State Institution "The Institute of Environmental Geochemistry of The National Academy of Sciences of Ukraine", Kyiv, Ukraine

National Aviation University, Kyiv, Ukraine

© The Author(s), under exclusive license to Springer Nature Switzerland AG 2024
A. Semenov et al. (eds.), *Data-Centric Business and Applications*, Lecture Notes on Data Engineering and Communications Technologies 195, https://doi.org/10.1007/978-3-031-54012-7_14

315

Keywords Graph database technology · Educational program · Curriculum · Competence · Neo4j

1 Introduction

The sphere of education is important for the stable functioning and development of any country in the world, the development of its socio-economic sphere, and the level of competitiveness in the international arena. A high level of education in the country affects the indicators of the quality of life and well-being of the population, the state of security and the degree of protection of national interests, provision of resource needs of the economy, volumes of social production, intellectual and innovative potential, processes of globalization and digitalization. To ensure a given level of education, the governments of various countries invest significant funds in its development and strengthening.

One of the ways to achieve the desired level of education, provided there is a sufficient level of funding and state support, is to meet, on the one hand, the needs of the labor market for a qualified workforce, and on the other hand, the needs of individuals seeking to acquire a sought-after profession and a highly paid job.

1.1 Relevance of the Research to Current Issues

The relevance of the research is determined by the requirements of the modern economy, which is characterized by instability and disparities. In particular, there is a professional imbalance in the labor market, as well as an imbalance of demand and supply regarding the qualifications of employees. Employment in the stratum of middle-skilled workers is decreasing due to demographic changes and the revision of career patterns among young people [1]. In a significant number of countries, including India, Great Britain, and Germany, there is a discrepancy between the skills of graduates of educational institutions of all levels and the requirements of employers [1, 2]. Accordingly, the problem of finding qualified workers is common in both developed countries and developing economies [1].

The latest information and communication technologies are causing significant structural changes in the labor market [3]. According to experts' forecasts, in the near future, there will be a mass layoff of a large number of workers and the disappearance of certain professions due to automation, robotics, and the introduction of artificial intelligence systems against the background of the trend of increasing demand for workers with training in accordance with the requirements of the technical frameworks of Industry 4.0 and Industry 5.0 and the emergence of so-called "professions of the future". These processes strengthen the tendency to increase the number of persons who, by their own will or at the demand of the market, change their profession or undergo retraining, often multiple times, throughout their lives [3, 4].

For the labor markets of a significant number of countries in Europe and the world, the next bifurcation factor after the COVID-19 pandemic was the consequences of the Russian military invasion of Ukraine. Among the many problems caused by the Russian military aggression, are the issues of migration of a significant part of the population of Ukraine, socialization and integration of forced migrants, their adaptation to new living conditions, access to the labor market, etc. According to the research results, the majority of Ukrainians fleeing the war, according to the UNHCR—almost 5 million people [5], faced employment problems, the difficulty in confirming their level of qualifications, experience, and knowledge. The OECD notes that, as of January 10, 2023, up to 40% of forced migrants are employed in EU countries, mainly in low-skilled jobs, and a mismatch of qualifications is a common phenomenon [6]. The authors see the root cause of this situation in the insufficient level of harmonization of the education system in Ukraine with the requirements, educational norms, and regulations of Europe and the world and, as a result, inconsistency of qualification levels and the lack of methods for evaluating foreign qualifications.

In the conditions of the identified trends, the education sector should focus on both the current rapid changes in requirements for graduates and the future needs of the market and the education seekers themselves, taking into account that quality education for all is one of the Sustainable Development Goals of the UN member states [7]. Taking into account the acceleration of all processes in the modern world and the scale of the consequences of events and threats, educational standards must be quickly updated, while ensuring the training of qualified personnel ready for future challenges, taking into account potential threats and being unified with the standards and qualification levels of developed countries of the world.

1.2 Defining the Research Problem

The training of specialists in Ukraine involves a multi-iteration process of learning throughout life: obtaining a professional pre-higher or higher education, self-education at the workplace, periodic training courses and mastering new skills, and obtaining additional professional education. As a result of this process, the specialist receives knowledge, experience, and skills—learning results that allow comparing the acquired competencies with the level of qualification.

The entire educational process is regulated by a significant number of normative legal acts, in particular, the laws of Ukraine "On education" [8], "On professional pre-university education" [9], "On higher education" [10], resolutions of the Cabinet of Ministers of Ukraine "On approval of the National Framework of Qualifications" [11], "On approval of the list of fields of knowledge and specialties for which higher education students are trained" [12], etc. On the basis of these and other normative documents and with the aim of harmonizing with the existing education systems of the developed countries of the world, in particular, the member countries of the European Higher Education Area, the National Framework of Qualifications [11] and education standards were developed and implemented, for example, for the field of higher education, these are standards of the first (bachelor's), second (master's) and third (educational-scientific/educational-creative) levels in terms of specialties [13].

National standards have a framework nature and act as a tool for determining and classifying competencies and structuring qualifications, determining the list of competencies and learning results of graduates, and summarizing the requirements for professional knowledge, abilities, skills, and other components of competence for each educational and qualification level and correspond to tasks and responsibilities (jobs) of a certain level of professional activity. Qualification levels are structured by competencies and systematically described in the National Qualifications Framework, which meets the criteria and procedures: The European Qualifications Framework [14], The Framework for Qualifications of the European Higher Education Area [15]. In particular, the similarity of 6–8 levels is determined by their descriptors—knowledge, skills and abilities, communications, responsibility, and autonomy. Further harmonization of national education standards with international standards requires their coordination with:

- The European Qualifications Framework for lifelong learning [16], which covers all levels of education and uses a similar set of descriptors to describe the eight qualification levels;
- International Standard Classification of Education [17], in terms of harmonizing the requirements for the level of education and learning results, the list of mandatory competencies of the graduate, the normative content of the training of those seeking higher education, formulated in terms of learning results;
- European Skills, Competences, Qualifications, and Occupations or ESCO Classification [18], which contains a list of qualification requirements (Skills and Competences) that must be met by a specialist in a certain specialty in EU countries;
- other similar documents.

On the basis of national standards, which include educational and qualification characteristics of graduates, educational training programs, and means of diagnosing the quality of higher education, educational institutions develop variable parts of relevant documents, as well as curricula and programs of academic disciplines.

Educational programs (abbreviated as EP) are a multifunctional tool for ensuring the quality of education, which, among other things, performs the following functions:

- informing: EP contains descriptive information about the specifics and features of training in a given specialty at a specific educational institution during a specified period of time and the amount of ECTS credits;
- organizations: EP contains a list of educational components (mandatory/normative and optional educational disciplines) and regulates the structural and

logical connection between them; determine the competencies (integral, general, professional) that the student must master and correlate them with the expected learning results;

- monitoring and control: thanks to the information contained in the EP, in particular regarding the system and principles of evaluation, the available resource provision of the educational process (personnel, material, and technical, educational, and methodological support), the identification of the purpose and goals of training, and others, the competent control bodies can compare, for example, during the accreditation of the EP, the declared characteristics with the actual ones. On the basis of the conducted analysis, recommendations can be made to improve such EP.

Thoughtfully and qualitatively developed EP contributes, on the one hand, to the formation of a qualified specialist in accordance with the declared competencies and learning results, which are relevant and in demand in the labor market, on the other hand, it allows the student to understand how the learning process will be structured, its purpose and goals.

The Ministry of Education and Science of Ukraine, in the Regulation on the Accreditation of Educational Programs for the Training of Higher Education Candidates [19], defined criteria for evaluating the quality of EP (see Fig. 1), according to which criteria 1–2, 4, 8 require: compliance with the objectives of EP the results of training, trends in the development of the specialty, the needs of the labor market, the experience of similar domestic and foreign EP, the requirements of domestic legislation and the institution's internal policy; the presence of a structural and logical connection between educational components, their balance, gradual deepening and consolidation of the content; consistency between educational components and the competencies that form them, the possibility of achieving the expected learning results according to the selected set of educational components; conformity of the scope of EP and educational components to the requirements of the standards and the curriculum in ECTS credits, hours, percentage ratio. Without clear compliance of the developed EP with the specified criteria, it cannot be considered high-quality, and therefore, the quality of the educational process in the educational institution that developed, approved, and implemented such EP is called into question.

The complexity, labor- and resource-intensiveness of EP development are caused, in particular, by the problems faced by the heads of project groups and their members, experts, and internal and external stakeholders:

- EP development works are performed without proper automated support, which means complete dependence on the human factor, and therefore, the probability of errors, inaccuracies, inconsistencies, and contradictions increases;
- the various information used to form and/or update the EP is, in the vast majority, weakly structured and insufficiently formalized, requiring high professional skills and analytical thinking, a deep understanding of the mechanism and regulatory framework, knowledge of international, industry and regional requirements from the persons involved, awareness of the needs of potential students, students, external stakeholders;

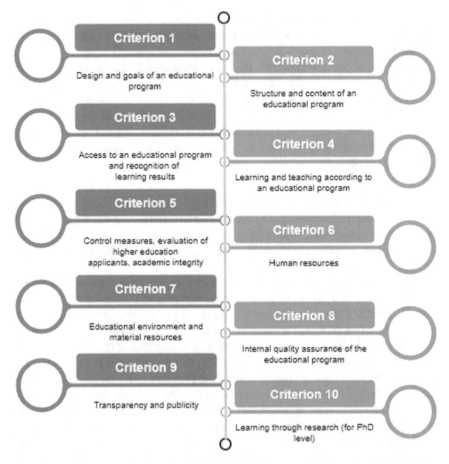

Fig. 1 Criteria for evaluating the quality of the educational program of higher education institutions

- due to the large amount of information and the number of documents and structural elements that are processed during the work with EP, it is difficult to check the completeness, consistency, and connections between the requirements of the labor market, goals, competencies, learning results, the content of educational plans and educational components, results training of specialists, as well as to assess the necessary resources on this basis, manage the development of the EP and monitor its implementation;
- working with documentation, regulatory and legal acts, and other sources of data necessary for the development of the EP, the need to be aware of the latest changes and innovations increases the complexity and duration of the process and makes it impossible to quickly modernize documents.

As a result, the educational sector needs means of supporting the dynamic formation and timely updating of the EP, ensuring its compliance with the content and learning results, analyzing the level of compliance of the EP project with quality criteria, assessing the balance of the relationship between educational components, competencies, and learning results.

It is also worth noting that until recently, the issue of the implementation of education standards was neglected by specialists in the informatization of educational processes. Among the reasons that caused such a situation are the generally low level of digitization and the absence of a single long-term national strategy of digitization in the education system, the low level of investments in this direction, and the non-triviality of processes that require automation [20]. According to the authors, the incompatibility of relational databases with the task of operating text information with a large number of semantic connections can also be determined by other reasons. The emergence and development of non-relational NoSQL databases gave an impetus to solving some of the problems in the field of digitization of the education system, in particular, in the direction of developing computer systems to support the development of educational programs.

The purpose of this study is to reveal the possibilities of using graph database technology as a tool to support the processes of developing and updating educational programs on the example of an educational and professional training program for first (bachelor) level higher education students, specialty 122 "Computer Science".

2 Theoretical Foundations and Materials

2.1 Stages of Educational Program Development

The Law of Ukraine "On Higher Education" defines an educational program as "a single complex of educational components (learning disciplines, individual tasks, practices, control measures, etc.) aimed at achieving the learning results provided for by such a program, which gives the right to receive a specified educational or educational and professional (professional) qualification(s)" [10]. In the context of this study, we are talking about the educational program as a local (university) normative act that defines the goals, tasks, methods, tools, and resources directed by the educational institution (faculty, department) to the training of specialists in a certain specialty. Depending on the level of education, the name of the educational program may change:

- for the first (bachelor's) and second (master's) levels, they are called educational and professional programs;
- for the third (educational-scientific/educational-creative) level—educational-scientific or educational-creative program.

The basis for the development of the EP for the specialty 122 "Computer Science" is the current legal framework in the field of higher education:

- national level: Law of Ukraine "On Higher Education" [10], Resolution of the Cabinet of Ministers of Ukraine "On Approval of the National Framework of Qualifications" [11], Resolution of the Cabinet "On Approval of the List of Fields of Knowledge and Specialties for which Higher Education Candidates are Trained" [12], Standard of higher education in specialty 122 "Computer science" for the first (bachelor's) level of higher education [21] (hereinafter—Standard), etc.;
- local (university) level: Regulations on educational programs [22]; Regulations on the organization of the educational process [23], etc.

The educational program is a central element in the system of implementation of educational standards, as it links the competencies contained in the framework of qualifications and directly the content of educational disciplines (educational components) in the context of formal or non-formal education.

According to [24], the introduction of competency-based education is possible with an understanding of the holistic concept, the limitations of a specific educational institution, as well as the availability of models and methods of curriculum management at two levels [24]. At the curriculum level, according to the authors, this concept most corresponds to the content of the document, which in the national system is called EP, the profile of the graduate with target competencies, the composition of academic disciplines, their scope and sequence of teaching, as well as the mechanism for achieving educational goals should be defined contribution of each academic discipline. The last task is the most difficult since the measurement of skills is not as unambiguous as the assessment of knowledge. At a lower level, the development of the syllabus should take place with the continuous coordination of their content and goals with other programs, with the verification of compliance with the set program results, as well as the analysis and control of the entire set of competencies that the student should receive as a result of training. This is a complex iterative process, in which the following main steps can be distinguished: determination of program learning results and general competencies, specification and structuring of competencies, distribution of competencies by specific topics, determination of criteria for measuring competencies, development of curriculum components (practices, tasks, etc.), determination of educational learning environment and methods, adjusting the curriculum based on the results of the analysis of graduates' competencies and their self-assessment. The result of these steps is a set of artifacts that correspond to the levels of implementation of educational standards and are managed by the relevant information systems (Fig. 2).

IEEE standards (for Learning Object Metadata [25], for Learning Technology [26]) and specifications, in particular TinCan API and xAPI, ensured the achievement of compatibility and multiple uses of learning objects in distance learning systems. At the same time, there are no generally accepted approaches for operating with artifacts of higher levels.

In the context of the mentioned study, the authors proposed their own view on the process of content formation and development of EP, which can be presented in the form of an interconnected set of stages described below.

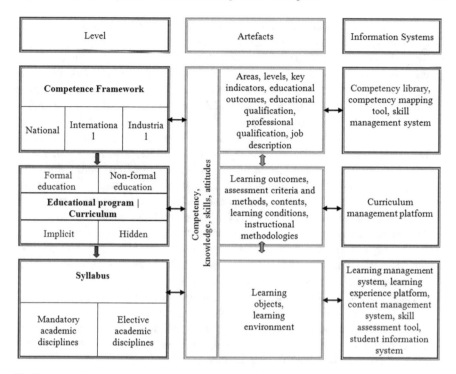

Fig. 2 Levels of implementation of education standards in the educational process

- Stage 1—coordination of documentary formalities and creation of a project group, selection of the head of this group. This stage is organizational and is designed to record the initiative and obtain approval from the university administration for the start of the development procedure.
- Stage 2—analysis of the regulatory framework; holding meetings, surveys, and discussions with students, graduates, future students, employers, and other interested parties, working with stakeholders, researching international experience. This is a communication and analytical stage, the purpose of which is to collect information for the purpose of harmonizing the purpose and goals of the EP with international, regional, and industry requirements, ideas, and wishes of interested parties.
- Stage 3—development of the EP project, formation of its concept (purpose, goals, competencies, program results), structure (educational components, their logical connection), and context (personnel, material and technical, educational and methodological, information support)—stage intense intellectual work. This is the most difficult and responsible stage, because the information with which the project group and its manager have to work is contained in the form of text files with a minimal amount of metadata, and is weakly structured and interconnected. This makes the task of automating this part of the work quite difficult, and, at the same time, makes the EP development procedure dependent on the

knowledge, experience, degree of involvement, responsibility, and assessments (perhaps subjective) of the specialists who are part of the project group.

- Stage 4—discussion of the EP project, preliminary assessment of the quality of the formed project, checking for compliance with quality criteria (see Fig. 1) in order to identify inconsistencies, inconsistencies, redundancy (insufficiency) of data, as well as balancing the purpose and goals of the EP with competencies, learning results, educational components. Based on the results of the described activities, a final document is prepared. According to the authors, this stage is critically important for the preparation of a high-quality EP, however, as experience shows, sometimes project groups approach the described procedures formally, which ultimately affects the quality of the educational process as a whole. As with the previous stage, the assessment and analysis procedures are currently not automated, so the problems described for Stage 3 are relevant at this step as well. Within the framework of this research, the authors propose to use graph database technology as a tool to support the work at this stage, the feasibility of using it for problems of a similar class is described in Sect. 2.3, and the development result and possibilities are described in Sect. 4 of the work.
- Stage 5—approval of EP at all levels, bringing information about EP to a wide range of people. This final stage is also organizational.
- Stage 6—internal monitoring and audit (survey, self-analysis, taking into account the suggestions of participants in the educational process and stakeholders), external evaluation of the EP (accreditation).

2.2 Approaches to Simplification and Optimization of Educational Program Development Processes

As repeatedly noted above, the EP development procedure is non-trivial, requires a creative approach, and requires knowledge, experience, and understanding of the purpose, tasks, and goals of the process from the persons involved in the work. These and the problems described above contributed to the fact that the EP development procedure is still not automated, although there are a number of studies and software solutions aimed at correcting the situation.

In general, the problem of automating the processes of development of documents important from the point of view of the organization of the learning process, such as curricula, is relevant for most countries of the world, in particular, the USA, Canada, Great Britain, Australia, Malaysia, China, Ukraine, etc. [27–29] and, together with which is still relevant due to the fact that different countries have their own approaches to design curricula, their content, and implementation.

One of the common ways to solve the problem of developing curriculums is to create centralized databases that contain educational content and enable developers to analyze, compare, compare, find best practices, etc. Curriculum mapping built on their basis contribute to the simplification of the procedure for developing such programs [20, 30, 31].

In [32], the process of developing an educational program is proposed to be simplified using a designer system, where the main component of the process management is the Curriculum Design Module. He is responsible for: curriculum maintenance, assessment, and analysis. The work of the module is based on the relation database and is integrated with external resources that are responsible for the control and monitoring of these documents [32].

One of the important moments in the process of developing an educational program is the process of communication with stakeholders, in particular, employers and graduates who graduated from the university in previous years [33]. In particular, a number of studies emphasize the need to maintain a single database of university graduates for the purpose of tracking employability. Along with the analysis of profiles on Facebook or LinkedIn, it is suggested to use the so-called student E-Portfolio - a web application that is formed at the stage of a student's studies at an educational institution and can be used in the future both as a resume and as a tool for tracking a graduate's career by the university [34]. Another study [35] suggests using a Mobile App—GES App (Global Employability Skills), which helps students to self-analyze acquired competencies and develop them, compare them with employers' requests, and build their own careers [35].

In the paper [36] it is noted that the ability to evaluate and improve educational programs for the training of specialists depends on the conceptual foundations, methods, and tools available to (existing) scientists and experts, which will allow determining the content of competences oriented to the current standards of training and the expected results of such training [36]. The author made an attempt to critically assess the possibility of using the content-focused curriculum mapping tool to analyze training plans for specialists in the field of administrative management. In other papers, it is proposed to use the mapping method as a tool for developing and improving curricula [37], statistical significance whit method relative importance index (RII) [38], LISREL (linear structural relation) model [39], graph matching [40] to solve the problems of curriculum development; to automate the procedures of data mining and analysis of educational plans, oriented to the result, use RDBS [41] and graph models [42, 43]. In the study [44] an attempt was made to formalize the curriculum development procedure and evaluate the quality of such a program using the social network analysis (SNA) approach, in particular, bipartite graphs to represent the relationships between professions and their qualification requirements [44]. The authors proposed the use of special metrics (evaluation and centrality) for curriculum evaluation, which are ranked in accordance with the SNA rules.

2.3 The Use of Graph Database Technology for the Purposes of Developing an Educational Program

For the development of a database model and the selection of tools for digitalization of the process of supporting the development of educational programs, taking into account the need to display numerous connections between loosely structured, mostly textual data, the authors suggest choosing graph database technology.

Graph database technology is a collective concept that, according to modern ideas about NoSQL databases, is based on graph theory [45], includes components [46] (Fig. 3), focused on the presentation of data and the relationships between them in the form of a graph [47–49].

In the classical sense [46], a graph is an ordered data structure consisting of nodes (entities) and arcs connecting these nodes and defining the relationship between them. Arcs can be directed (that is, have a direction) and have labels (attributes) characterizing or quantifying the connection [46]. They are able to present large volumes of data in a human-understandable form through visualization, which helps to identify patterns, identify discrepancies, or anomalies [50].

Graph database technology is widely used for researching linked data on Google, Facebook, and LinkedIn, as well as for solving innovative problems in industry, business, banking and financial institutions, medicine, biology, education, and other fields of activity [51–53].

The study [54] substantiates the idea of the expediency of working with graphical models of data presentation in the identification of qualification frameworks in the assessment of various educational programs for their compliance with existing standards and desired sets of competencies [54]. The authors prove that the use of a graph data model allows for to improvement of the quality of perception of a large array of weakly structured data by means of visualization, to perform an in-depth analysis of educational programs, and form a basis for their further improvement [54]. Authors of a number of other studies [55–57] came to similar conclusions regarding the ease of use of graph databases in modeling the processes of educational activity.

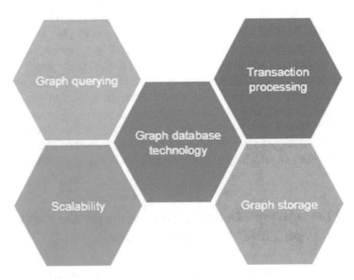

Fig. 3 Basic components of graph database technology

The perspective of graph database technology is evidenced by the data of the IT and database market review, in particular, according to Gartner forecasts, by 2025 the technology will be used in 80% of innovative solutions in the field of data and analytics (for comparison, in 2021—only 10%), which will contribute to faster and more effective decision-making [58]. According to [59], research on knowledge graph and graph embedding technology is identified as one of the most promising research areas in 2022. The tendencies are explained by the low level of efficiency of RDBMS when working with unstructured, but closely related, data.

According to the DB-Engines web resource, at the beginning of 2023, there are 39 DBMSs on the software market that support graph databases, among which Neo4j has been the leader in popularity for the last decade [60].

Neo4j is an aggregate-free Graph DBMS that supports Property Graph and can be used in the development of OLTP and OLAP systems, supports transactions, fulfilling ACID (atomic, consistent, isolated, durable) requirements, and is characterized by high data search performance [60]. Neo4j has a number of advantages over RDBMS, which makes this graph database technology a useful tool within the scope of this research: support for loosely structured and unstructured interconnected information; availability of graph visualization tools; support for horizontal scaling; the ability to remove or supplement the graph model with new nodes and edges, while maintaining data consistency; availability of the Neo4j Graph Data Science software platform, which allows data analysis using the GDS library of graph algorithms and machine learning; the ability to work autonomously or in the cloud using the Neo4j AuraDB cloud service; support for the most common programming languages; high search and filtering performance; availability of an open source version that does not require a license, which will reduce additional costs [61–64].

As we can see, graph databases have been used in recent years to help manage and solve various educational management problems. The use of graph databases allows for the representation and analysis of complex relationships between different educational entities, such as courses, learning results, and competencies.

3   Results

3.1   Nodes and Edges Description

Data processing and graphical modeling of the database in the Neo4j environment were performed using the materials of the educational and professional bachelor's training program in the specialty 122 "Computer Science". A graph database is implemented as a set of nodes characterizing the entities of the chosen specialty and edges (relationships) that connect them to each other. The nodes of the graph contain the characteristics of the following main entities:

- normative (mandatory) educational components, which are represented by two groups—educational disciplines of the general cycle of training and the cycle of professional (professional) training;
- learning results, representing all the knowledge, skills, and abilities acquired and mastered by the student after completing the educational component.

Edges reflect relationships (connections) between described entities. The simulated graph contains two types of edges:

1. edges reflecting the relationship (connection) of general and special competencies with normative educational components (general and professional training cycles). This group of ribs defines a set of competencies that should form the disciplines of specialty 122 "Computer Science". That is, the initial node of the edge of this type is a certain competence, and the final one is the corresponding educational discipline, which must take into account this competence and ensure its implementation in the EP. The relationship is built based on the fact that the study of each specific academic discipline should ensure and form certain general and professional competencies from the set defined by national educational standards in the acquirers;

2. edges reflecting the relationship (connections) of normative educational components (educational disciplines of general and professional cycles of training) with the corresponding learning results. These edges indicate a set of final learning results that learners should obtain in the form of knowledge, skills, and abilities. The starting node of such an edge is the discipline, and the final node is the corresponding learning result, the achievement of which is provided by this educational discipline. The list of learning results that should form the chosen specialty is also regulated by national educational standards.

The projected graph database in the Neo4j environment contains 92 nodes and 369 edges, its fragment is presented in Fig. 4.

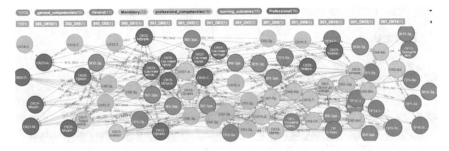

Fig. 4 A fragment of the projected graph database. Nodes and edges are described in Ukrainian terms, in accordance with the requirements of current Ukrainian national legislation

3.2 Queries, Visualization and Data Analysis

The graph of the educational program, developed in Neo4j, provides users with extensive opportunities to analyze its components (educational components, competencies, learning results), provides verification of links between educational disciplines with competencies and learning results for semantic correspondence; serves as a convenient tool for conducting an internal audit of an educational program at the stage of its development and/or revision. Identified shortcomings and semantic inconsistencies can be eliminated by making appropriate modifications.

The database will allow end users, first of all, members of the project group, as well as the management of the department and university administration, employees of educational and methodical units, etc., to perform various types of samples from the database for the purpose of analysis and making appropriate decisions regarding the improvement of educational programs in accordance with their standards of higher education.

Let's consider examples of requests to this graph, which will allow you to view individual components of the database and carry out their corresponding analysis. Examples of queries are described in the declarative language Cypher, which is a component of Neo4j.

Example #1
Purpose: to display a list of educational components that ensure the formation of specific competence in the acquirers.

Description of the content of the request: to display the connections of the general competence "ZK1-Ability to abstract thinking, analysis, and synthesis" with all educational components of the educational and professional program for the specialty 122 "Computer Science".

Query syntax:

```
MATCH (k: general_competencies {name: "ЗК1-Здатність до
абстрактного мислення, аналізу та синтезу"}) <-[]->(d)
RETURN k,d
```

The result of the request execution is shown in Fig. 5.

Example #2
Purpose: to review the complete list of competencies that form a specific educational component.

Description of the content of the request: to display a complete list of competencies (including general and professional), which forms the educational discipline "System Programming".

Query syntax:

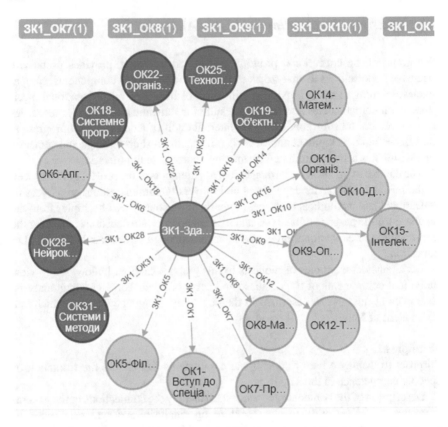

Fig. 5 Connections of general competence of ZK1 with educational components of cycles of general and professional training (OKn). Nodes and edges are described in Ukrainian terms, in accordance with the requirements of current Ukrainian national legislation

MATCH (ok18: Mandatory: Professional {name: "OK18-System programming", hours: 150, credits:5, semester:2, ekz: "1 kur_pr"})<-[]-(zk) RETURN ok18 AS Code_and_name_of discipline, zk AS Competencies

Visualization of the result of the request is shown in Fig. 6.

Example #3

Purpose: an overview of the list of learning results formed by the student as a result of studying a specific academic discipline.

Description of the content of the request: to display the full learning results that are formed by the student as a result of studying the educational discipline "Intellectual data analysis".

Query syntax:

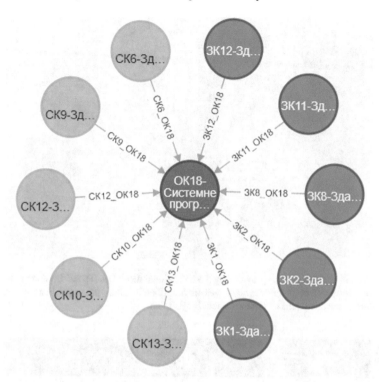

Fig. 6 List of general (SK) and professional (SK) competencies formed by the educational component (OK18-System programming). Nodes and edges are described in Ukrainian terms, in accordance with the requirements of current Ukrainian national legislation

MATCH(ok15: Mandatory: General {name: "ОК15-Інтелектуальний аналіз даних", hours: 120, credits:4, semester:4, ekz:1})-[]->(pr)
RETURN ok15 AS Дисципліна, pr AS Програмні_результати

The visual display that the end user will receive (Fig. 7) after completing this request provides an opportunity to review and semantically analyze the compliance of the essence of the educational discipline with the results announced as a result of its study.

The results of the request from Example 3 can also be presented in tabular form (Fig. 8).

Example #4

Purpose: to determine how many ECTS credits belong to educational components of general or professional training cycles.

Description of the content of the request: to display the educational components of the general training cycle and express them in ECTS credits.

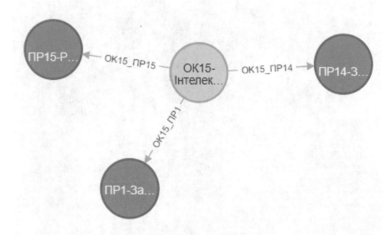

Fig. 7 Learning results (PRn) provided by the educational discipline (OK15-Intelligent data analysis) in the form of a graph. Nodes and edges are described in Ukrainian terms, in accordance with the requirements of current Ukrainian national legislation

"Дисципліна"	"Програмні_результати"
{"name":"OK15- Інтелектуальний аналіз даних ","hours":120,"semester":4 ,"ekz":1,"credits":4}	{"name":"ПР1-Застосовувати знання основних форм і законів абстрактно-л огічного мислення, основ методології наукового пізнання, форм і методі в вивчення, аналізу, обробки та синтезу інформації в предметній облас ті комп'ютерних наук"}
{"name":"OK15- Інтелектуальний аналіз даних ","hours":120,"semester":4 ,"ekz":1,"credits":4}	{"name":" ПР14-Застосовувати знання методології та CASE-засобів проект ування складних систем, методів структурного аналізу систем, об'єктно- орієнтованої методології проектування при розробці і дослідженні функ ціональних моделей організаційно-економічних і виробничо- технічних си стем"}
{"name":"OK15- Інтелектуальний аналіз даних ","hours":120,"semester":4 ,"ekz":1,"credits":4}	{"name":" ПР15-Розуміти концепцію інформаційної безпеки, принципи без печного проектування програмного забезпечення, забезпечувати безпеку ко мп'ютерних мереж в умовах неповноти та невизначеності вихідних даних"}

Fig. 8 Learning results (PRn) provided by the educational discipline (OK15-Intelligent data analysis) in a tabular presentation. Nodes and edges are described in Ukrainian terms, in accordance with the requirements of current Ukrainian national legislation

Query syntax:

```
MATCH (ok: Mandatory: General) RETURN ok.name AS Загальні_
дисципліни, ok.credits AS кількість_кредитів
```

A fragment of the query result is presented in Fig. 9.

Example #5

Purpose: to calculate the number of competencies formed by the educational component.

"Загальні_дисципліни"	"кількість_кредитів"
"ОК3-Фізичне виховання"	4
"ОК1- Вступ до спеціальності"	5
"ОК2-Українознавство"	4
"ОК4-Іноземна мова"	10
"ОК5-Філософія"	4
"ОК6-Алгебра і геометрія"	4
"ОК7-Програмування та алгоритмічні мови"	4
"ОК8-Математичний аналіз"	10
"ОК9-Операційні системи"	4
"ОК10-Дискретна математика"	5

Fig. 9 Tabular presentation of educational components of the general cycle of training in ECTS credits. Nodes and edges are described in Ukrainian terms, in accordance with the requirements of current Ukrainian national legislation

Description of the content of the request: calculate how many competencies are formed by the following educational components: "Intelligent data analysis", "Information and communication technologies", "Foreign language".

Query syntax:

```
MATCH (ok15: Mandatory: General {name: "ОК15-Інтелектуальний
аналіз даних"}) <-[zk]-()
RETURN ok15.name AS Назва_дисципліни,
COUNT (*) AS Кількість_компетентностей
UNION
MATCH (OK23:Mandatory: Professional {name: "ОК23-Інформаційно-
комунікаційні технології"}) <-[zk]-()
RETURN OK23.name AS Назва_дисципліни,
COUNT (*) AS Кількість_компетентностей
UNION
MATCH (OK4: Mandatory: General {name: "ОК4-Іноземна мова"}) <-[zk]-
()
RETURN OK4.name AS Назва_дисципліни,
```

"Назва_дисципліни"	"Кількість_компетентностей"
"ОК15- Інтелектуальний аналіз даних "	8
"ОК23-Інформаційно-комунікаційні технології"	7
"ОК4-Іноземна мова"	2

Fig. 10 Tabular presentation of the calculation of the number of competencies forming educational components (OK15-Intellectual data analysis, OK23-Information, and communication technologies, OK4-Foreign language). Nodes and edges are described in Ukrainian terms, in accordance with the requirements of current Ukrainian national legislation

COUNT (*) AS Кількість_компетентностей

The result of the request is shown in Fig. 10.

4 Discussions and Conclusion

Well-balanced and organized educational programs make it possible to improve the quality of the educational process and to ensure compliance of EP with national educational standards and requirements for accreditation.

The conducted research showed that the data in the EP, in the vast majority, are loosely structured and closely related to each other, requiring frequent modification. These and other circumstances made it possible to use graph database technology and Neo4j as the basic toolkit. The developed database can be used in the field of education management for educational program design and curriculum mapping. The database uses graph structures for semantic queries, where nodes and edges represent entities and their relationships, respectively. Graph database Neo4j is intuitive, clear due to visualization, productive when implementing queries, and meets all modern requirements for the development of modern software for educational purposes. The Cypher query language allows you to implement requests for mass updates of labels, and modification of properties of entities and relations (when changing national educational standards), while maintaining data integrity. Graph database Neo4j allows you to support any evolutionary changes of the EP without loss of productivity, and the resulting information model will be flexible and adaptive, which will allow users (head and members of project groups, department management, and university administration, employees of educational and methodological units, etc.) to perform in-depth analysis EP and make informed decisions about their improvement, and as a result, to improve the quality of the educational process and the training of highly qualified specialists.

The simplicity and convenience of manipulating the data contained in the EP with the help of graph database technology confirm the appropriateness of using Neo4j not only for solving the problem in the higher education system but also for the formation of various educational programs and advanced training courses, retraining of specialists, etc. Having defined the clear purpose, goals, and tasks of such courses, it is possible to easily supplement the database with new sets of competencies and learning results, modify existing ones and establish new relations without violating data consistency.

At the same time, the authors understand that the approach proposed in the work is not devoid of shortcomings. In particular, the research presents a database model that solves the outlined problem, rather than a ready-made tool in the form of a software product. To access the database, the user must have knowledge of query languages and programming fundamentals in order to master Cypher and use it effectively to work with data. The solution proposed in the work is not comprehensive, as it provides support for decisions at only one of the stages of the process of developing educational programs, and therefore requires further research and development.

The use of graph database technology to solve the problems of educational activity opens wide prospects for their further application in the field of educational digitization. We consider the use of graphs as a basis for modeling knowledge bases of educational programs (knowledge graph) and their analysis (text mining) for compliance with declared competencies, for the presence of certain elements of duplication, as one of the further directions of research. An important issue that requires further research is the integration of the developed graph database with the system of intellectual processing of texts and the system for supporting the formation of educational programs.

References

1. Education at a Glance 2022: OECD Indicators. OECD Publishing, Paris (2022). https://doi.org/10.1787/3197152b-en
2. Levels, M., van der Velden, R., Di Stasio, V.: From school to fitting work. Acta Sociologica 57, 341–361 (2014). https://doi.org/10.1177/0001699314552807
3. Autor, D.: The Labor Market Impacts of Technological Change: From Unbridled Enthusiasm to Qualified Optimism to Vast Uncertainty. National Bureau of Economic Research (2022). https://doi.org/10.3386/w30074
4. Bast, G.: The Future of Education and Labor, pp. 9–19 (2019). https://doi.org/10.1007/978-3-030-26068-2_2
5. Operational Data Portal UNHCR. https://data.unhcr.org/en/situations/ukraine. Accessed 18 Jan 2023
6. What We Know About the Skills and Early Labour Market Outcomes of Refugees from Ukraine: OECD. https://www.oecd.org/ukraine-hub/policy-responses/what-we-know-about-the-skills-and-early-labour-market-outcomes-of-refugees-from-ukraine-c7e694aa/. Accessed 12 Jan 2023
7. Gindler, P.: United Nations Office for Outer Space Affairs, UNOOSA. https://www.unoosa.org/oosa/en/ourwork/space4sdgs/sdg4.html. Accessed 17 Jan 2023

8. On Education. Law of Ukraine (2017). https://zakon.rada.gov.ua/laws/show/2145-19#Text. Accessed 26 Jan 2023
9. On Professional Pre-university Education. Law of Ukraine (2019). https://zakon.rada.gov.ua/laws/show/2745-19#Text. Accessed 26 Jan 2023
10. On Higher Education. Law of Ukraine (2014). https://zakon.rada.gov.ua/laws/show/1556-18#Text. Accessed 26 Jan 2023
11. On Approval of the National Framework of Qualifications. Resolutions of the Cabinet of Ministers of Ukraine (2011). https://zakon.rada.gov.ua/laws/show/1341-2011-%D0%BF#Text. Accessed 28 Jan 2023
12. On Approval of the List of Fields of Knowledge and Specialties for Which Higher Education Students Are Trained. Resolutions of the Cabinet of Ministers of Ukraine (2015). https://zakon.rada.gov.ua/laws/show/266-2015-%D0%BF#Text. Accessed 28 Jan 2023
13. Approved Standards of Higher Education in Ukraine, Ministry of Education and Science of Ukraine. https://mon.gov.ua/ua/osvita/visha-osvita/naukovo-metodichna-rada-ministerstva-osviti-i-nauki-ukrayini/zatverdzheni-standarti-vishoyi-osviti. Accessed 20 Dec 2022
14. European Qualifications Framework (EQF). https://www.cedefop.europa.eu/en/projects/european-qualifications-framework-eqf. Accessed 02 Feb 2023
15. European Higher Education Area and Bologna Process, The Framework of Qualifications for the European Higher Education Area. http://ehea.info/media.ehea.info/file/2018_Paris/77/1/EHEAParis2018_Communique_final_952771.pdf. Accessed 09 Dec 2022
16. Council Recommendation of 22 May 2017 on the European Qualifications Framework for Lifelong Learning and Repealing the Recommendation of the European Parliament and of the Council of 23 April 2008 on the Establishment of the European Qualifications Framework for Lifelong Learning (2017/C 189/03). https://eur-lex.europa.eu/legal-content/EN/TXT/HTML/?uri=CELEX:32017H0615(01)&from=EN. Accessed 19 Nov 2022
17. ISCED Levels. https://iqa.international/isced-levels/. Accessed 19 Nov 2022
18. ESCO Classification. https://esco.ec.europa.eu/en/classification. Accessed 22 Nov 2022
19. On Approval of the Regulation on Accreditation of Educational Programs for the Training of Higher Education Applicants. Order of the Ministry of Education and Science of Ukraine (2019). https://zakon.rada.gov.ua/laws/show/z0880-19#Text. Accessed 29 Jan 2023
20. Majerník, J., Kacmarikova, A., Komenda, M., Kononowicz, A.A., Kocurek, A., Stalmach-Przygoda, A., Balcerzak, Ł, Hege, I., Ciureanu, A.: Development and implementation of an online platform for curriculum mapping in medical education. Bio-Algorithms Med-Syst. **18**, 1–11 (2021). https://doi.org/10.1515/bams-2021-0143
21. On Approval of the Standard of Higher Education in the Specialty 122 "Computer Science" for the First (Bachelor's) Level of Higher Education. Order of the Ministry of Education and Science of Ukraine (2019). https://mon.gov.ua/ua/npa/pro-zatverdzhennya-standartu-vishoyi-osviti-za-specialnistyu-122-kompyuterni-nauki-dlya-pershogo-bakalavrskogo-rivnya-vishoyi-osviti. Accessed 29 Jan 2023
22. Regulations on Educational Programs at the Kyiv National Economic University Named After Vadym Hetman. Order by Kyiv National Economic University Named After Vadym Hetman (2022). https://drive.google.com/file/d/1eP2ZTzjbwDmFSIDol9MupfHYBWBvEVcx/view. Accessed 11 Jan 2023
23. Regulations on the Organization of the Educational Process at the Kyiv National Economic University Named After Vadym Hetman. Order by Kyiv National Economic University Named After Vadym Hetman (2022). https://drive.google.com/file/d/1FcfjeLJpSi2b6M8B6V0DIMKTCLzFL2Kv/view. Accessed 11 Jan 2023
24. Carvajal-Ortiz, L., Florian-Gaviria, B., Diaz, J.F.: Models, methods and software prototype to support the design, evaluation, and analysis in the curriculum management of competency-based for higher education. In: 2019 XLV Latin American Computing Conference (CLEI) (2019). https://doi.org/10.1109/CLEI47609.2019.235114
25. 1484.12.1–2020—IEEE Standard for Learning Object Metadata. https://ieeexplore.ieee.org/document/9262118. Accessed 14 Dec 2022

26. 1484.11.1–2022—IEEE Standard for Learning Technology—Data Model for Content Object Communication. https://ieeexplore.ieee.org/document/9756410. Accessed 14 Dec 2022

27. Walker, A., White, G.: Teachers using technology. In: Technology Enhanced Language Learning, pp. 137–151. Oxford University Press, Oxford (2013)

28. Dafoulas, G., Barn, B., Zheng, Y.: Curriculum design tools: using information modelling for course transformation and mapping. In: 2012 International Conference on Information Technology Based Higher Education and Training (ITHET) (2012). https://doi.org/10.1109/ithet.2012.6246064

29. Loeis, M., Hubeis, M., Suroso, A.I., Dirdjosuparto, S.: A strategy for reducing skills gap for the game development sector of the Indonesian creative industries (2023). https://doi.org/10.5267/j.dsl12:97-106

30. Curtis, D.M., Moss, D.M.: Curriculum Mapping: Bringing Evidence-Based Frameworks to Legal Education (2010). https://core.ac.uk/download/pdf/51074056.pdf. Accessed 02 Dec 2022

31. Treadwell, I., Botha, G., Ahlers, O.: Initiating curriculum mapping on the web-based, interactive learning opportunities, objectives and outcome platform (LOOOP). AJHPE **11**, 27 (2019)

32. Ling, T., et al.: Validation of Curriculum Design Information System Model (2018). https://zenodo.org/record/4320362

33. Asefer, M., Abidin, N.: University program as tools for graduates' employability from employers' perspectives: a review of academic literature. Southeast Asia J. Contemp. Bus. Econ. Law **25**(2), 1–10 (2021). https://seajbel.com/wp-content/uploads/2022/01/SEAJBEL25.ISU-2_65.pdf. Accessed 18 November 2022

34. Gutiérrez-Santiuste, E., García-Segura, S., Olivares-García, M.Á., González-Alfaya, E.: Higher education students' perception of the E-portfolio as a tool for improving their employability: weaknesses and strengths. Educ. Sci. **12**, 321 (2022)

35. Iqbal, M., Abbas, A., Petersen, S.A.: Development of graduate employability skills with regards to dream job using mobile app. EDULEARN Proc. (2022). https://doi.org/10.21125/edulearn.2022.1776

36. Arafeh, S.: Curriculum mapping in higher education: a case study and proposed content scope and sequence mapping tool. J. Further High. Educ. (2015). https://eric.ed.gov/?id=EJ1104503. Accessed 25 Dec 2022

37. Al-Eyd, G., Achike, F., Agarwal, M., Atamna, H., Atapattu, D.N., Castro, L., Estrada, J., Ettarh, R., Hassan, S., Lakhan, S.E., Nausheen, F., Seki, T., Stegeman, M., Suskind, R., Velji, A., Yakub, M., Tenore, A.: Curriculum mapping as a tool to facilitate curriculum development: a new school of medicine experience. BMC Med. Educ. **18** (2018). https://doi.org/10.1186/s12909-018-1289-9

38. Manoharan, K., Dissanayake, P.B.G., Pathirana, C., Deegahawature, D., Silva, K.D.R.R.: A curriculum guide model to the next normal in developing construction supervisory training programmes. BEPAM **12**, 792–822 (2021)

39. Chen, C., Chen, C.W.K., Shieh, C.-J.: A study on correlations between computer-aided instructions integrated environmental education and students' learning outcome and environmental literacy. EURASIA J. Math. Sci. Tech. Ed. **16**, em1858 (2020)

40. Limongelli, C., Marani, A., Sciarrone, F., Temperini, M.: Measuring the similarity of concept maps according to pedagogical criteria. IEEE Access **10**, 27655–27669 (2022)

41. Kang, Y.-B., Nadarajah, D., Nadarajah, V.: Building a database for curriculum mapping and analytics. MedEdPublish **8**, 38 (2019)

42. Yu, X., Stahr, M., Chen, H., Yan, R.: Design and implementation of curriculum system based on knowledge graph. In: 2021 IEEE International Conference on Consumer Electronics and Computer Engineering (ICCECE) (2021). https://doi.org/10.1109/ICCECE51280.2021.9342370

43. Bai, J., Che, L.: Construction and application of database micro-course knowledge graph based on Neo4j. In: The 2nd International Conference on Computing and Data Science (2021). https://doi.org/10.1145/3448734.3450798

44. Chen, C.-W., Huang, N.-T., Hsiao, H.-S.: The construction and application of E-learning curricula evaluation metrics for competency-based teacher professional development. Sustainability **14**, 8538 (2022)
45. Robinson, I., Webber, J., Eifrem, E.: Graph Databases, 2nd edn. In: O'Reilly Online Learning. https://www.oreilly.com/library/view/graph-databases-2nd/9781491930885/. Accessed 17 Dec 2022
46. Pokorný, J.: Graph databases: their power and limitations. Comput. Inf. Syst. Ind. Manag. 58–69 (2015). https://doi.org/10.1007/978-3-319-24369-6_5
47. A Brief Introduction to Graph Data Platforms—NEO4J (2020) VP, Product Marketing, Neo4j. https://go.neo4j.com/rs/710-RRC-335/images/Neo4j-Brief-Introduction-Graph-Data-Platforms-EN-US.pdf. Accessed 10 Nov 2022
48. Sytnyk, N., Zinovieva, I.: Modern NoSQL databases for training bachelors of "computer science" specialty. Inf. Technol. Learn. Tools **81**(1), 255–271 (2021). https://doi.org/10.33407/itlt.v81i1.3098
49. Zinovieva, I.S., Artemchuk, V.O., Iatsyshyn, A.V., Romanenko, Y.O., Popov, O.O., Kovach, V.O., Taraduda, D.V., Iatsyshyn, A.V.: The use of MOOCs as additional tools for teaching NoSQL in blended and distance learning mode. J. Phys.: Conf. Ser. **1946**, 012011 (2021). https://doi.org/10.1088/1742-6596/1946/1/012011
50. Friel, S.N., Curcio, F.R., Bright, G.W.: Making sense of graphs: critical factors influencing comprehension and instructional implications. J. Res. Math. Educ. **32**, 124 (2001)
51. Pingali, K.: An introduction to graph technology. In: Embedded.com (2022). https://www.embedded.com/an-introduction-to-graph-technology/. Accessed 25 Jan 2023
52. Timón-Reina, S., Rincón, M., Martínez-Tomás, R.: An overview of graph databases and their applications in the biomedical domain. Database (2021). https://doi.org/10.1093/database/baab026
53. Donnelly-Hermosillo, D.F., Gerard, L.F., Linn, M.C.: Impact of graph technologies in K-12 science and mathematics education. Comput. Educ. **146**, 103748 (2020)
54. Pasterk, S., Bollin, A.: Graph-based analysis of computer science curricula for primary education. In: 2017 IEEE Frontiers in Education Conference (FIE) (2017). https://doi.org/10.1109/FIE.2017.8190610
55. Hadzhikoleva, S., Borisova, M., Hadzhikolev, E., Hristov, H.: Model for assessing higher order thinking skills in training in graph databases. ICERI Proc. (2020). https://doi.org/10.21125/iceri.2020.1409
56. Rizun, M.: Knowledge graph application in education: a literature review. Folia Oeconomica **3**, 7–19 (2019)
57. Gartner identifies top 10 data and analytics technology trends for 2021. In: Gartner. https://www.gartner.com/en/newsroom/press-releases/2021-03-16-gartner-identifies-top-10-data-and-analytics-technologies-trends-for-2021. Accessed 09 Jan 2023
58. 2022 Research Fronts, English.casisd.cn. http://english.casisd.cn/research/rp/202212/P02022 1228378217394333.pdf. Accessed 09 Jan 2023
59. DB-Engines Ranking of Graph DBMS. https://db-engines.com/en/ranking/graph+dbms. Accessed 18 Jan 2023
60. Neo4j. https://neo4j.com. Accessed 29 Oct 2022
61. Bechberger, D., Perryman, J.: Graph databases in action: examples in Gremlin. In: Amazon (2020). https://www.amazon.com/Graph-Databases-Action-Dave-Bechberger/dp/1617296376. Accessed 15 Jan 2023
62. NoSQL distilled: a brief guide to the emerging world of polyglot persistence. In: Guide Books. https://doi.org/10.5555/2381014. Accessed 17 Jan 2023
63. Ian, R.: Graph databases: new opportunities for connected data. In: Amazon (2016). https://web4.ensiie.fr/~stefania.dumbrava/OReilly_Graph_Databases.pdf. Accessed 17 Jan 2023
64. Gosnell, D.K., Broecheler, M.: The Practitioner's Guide to Graph Data: Applying Graph Thinking and Graph Technologies to Solve Complex Problems O. 'Reilly Media Inc., Sebastopol, CA (2020)

Information Technology for Identifying Hate Speech in Online Communication Based on Machine Learning

Oleksiy Tverdokhlib⬤, Victoria Vysotska⬤, Petro Pukach⬤, and Myroslava Vovk⬤

Abstract Proposed in this paper information technology for identifying hate speech in online communication via machine learning methods is realized through the next steps: collecting data from reliable sources and forming datasets, data preprocessing (noise removal, text normalization, stop words removal, tokenization), labeling and data marking (hate, offensive or no hate), extracting significant linguistic features (using Bag-of-Words, TF-IDF, Word2Vec, GloVe, BERT), machine learning method choice, model realization, study and training of the model, estimation of classifier model accuracy. The basis of the considered model is implementation of the data cleaning, dataset partitioning, model training, fasttext using and prognostication. Machine learning method is selected taking into account its suitability for the tasks of text classification into categories of hate and hate speech and the previous efficiency evaluations. Different classifiers from the set of options as KNN (K-Nearest Neighbors), Naive Bayes, Decision Tree, Logistic Regression and Random Forest are estimated. According to metrics Accuracy KNN classifier achieves the accuracy of hate identification 0.832, Naive Bayes 0.315, Decision Tree 0.878, Logistic Regression 0.904 and Random Forest 0.879. According to metrics ROC AUC (Receiver Operating Characteristic Area Under the Curve) KNN classifier achieves performance of hate identification 0.82, Naive Bayes 0.61, Decision Tree 0.80, Logistic Regression 0.92 and Random Forest 0.89. Overall, random forest and logistic regression stand out as the most effective classifiers due to their high values of ROC AUC

O. Tverdokhlib · V. Vysotska · P. Pukach (✉) · M. Vovk
Lviv Polytechnic National University, 12 Bandera Str., Lviv 79013, Ukraine
e-mail: petro.y.pukach@lpnu.ua

O. Tverdokhlib
e-mail: oleksii.tverdokhlib.sa.2019@lpnu.ua

V. Vysotska
e-mail: victoria.A.Vysotska@lpnu.ua

M. Vovk
e-mail: Myroslava.I.Vovk@lpnu.ua

V. Vysotska
Osnabrück University, 1 Friedrich-Janssen-Str., 49076 Osnabrück, Germany

for all classes and general AUC. Decision Tree and KNN also demonstrate sufficient performance, whereas Naive Bayes lags behind other classifiers in terms of discriminative power and overall classification performance.

Keywords Hate identification · Hate classification · NLP · Twitter · Machine learning · KNN · K-Nearest Neighbors · Naive Bayes · Decision Tree · Logistic Regression · Random Forest

1 Introduction

The relevance of the study to identify hate speech is explained by necessity to regulate and reduce the harmful effects of the abuse of hate speech in online communities [1]. Hate speech has become a widespread problem in modern digital environment, that's why individuals, social/political/community groups and society as a whole are at risk [2]. A few key points that emphasize the relevance of research on detecting hate speech are the next [3–5]:

1. Maintaining progress while natural language processing based in machine learning.
2. Ensuring and monitoring the level social media responsibility.
3. Ensuring an adequate self-awareness education level of cultural communication in society and therefore providing education with respect for the interlocutor's opinion without using hate speech.
4. Difficulty in identifying and managing legal and ethical implications.
5. Negative impact on individuals and communities in the absence of online communication sources responsibility.
6. The growing dynamics of online hate speech using, its subsequent prompt identification and even blocking in necessary cases.

A striking example of hate speech support in social spaces are the so-called trolls n Russia's information war against Ukraine. This is one of the powerful reasons for the activation and support of the war in the center of Europe. Large media corporations through various social platforms support this war daily by publishing hostile hate speech and images of various thematic orientations, including those based on fakes and propaganda. That is the reason of chaos in the social space of ordinary social media users and further supporting information targeted attacks in this war of the twenty-first century at a new level. The war for public, political, cultural and social opinion of social media users is waged daily in the information space. As a result, their worldview is shaped distorting the truthfulness of the real events of the war at various levels of its manifestation. Hate speech, insults and hostility, combined with fake news and propaganda, have become a powerful tool for managing public opinion and consciousness through the Internet and various mass-media for today.

The prospect of implementing projects to identify hateful or offensive speech follows from the urgent need to stop the use of hate speech, to protect the well-being of individuals, to promote inclusive online communities and to ensure that social media platforms fulfill their responsibility to create safe and respectful communication environments [6]. The implementation of effective detection systems will contribute to a more inclusive and harmonious development, as well as to the online eco-environment support. Aim of creating an information technology for detecting hate speech is to identify and analyze its content and dynamics. Also, to reduce cases of hate speech in online communication by blocking posts or regulating mechanisms for checking future publications, in particular, to realize several key tasks [7–9]:

1. Improvement of the user online communication experience;
2. Facilitating the moderation of online platforms with a large number of users;
3. Ensuring the human rights for free communication in respect for the discourse participants;
4. Promotion the norms of civility and cultured communication in the online environment;
5. Protecting users from the harmful effects of hate speech.

The goal of the hate speech detection system creating is to protect users, to promote civil Internet communication, to enforce policies and regulations, to moderate content, to improve user experience and to raise awareness of the harmful effects from the hate speech [10]. Due to accomplishing these tasks the system will help to create safer and more respectful online environment for all users. The processes of identifying and classifying cases of offensive language in text data are the research object while designing of an offensive language detection system. The emphasis is on the development of algorithms, models and techniques that can effectively analyze and detect the content of offensive language with high accuracy. The methods and tools for processing natural language based on machine learning to identify and classify hate and insults in a textual content stream are the research subject while designing of hate speech detection system. These methods and means include the following elements [11, 12]:

1. Patterns and characteristics of hate speech;
2. Data collection and annotation;
3. Extraction and representation of linguistically significant features;
4. Machine learning and deep learning algorithms;
5. Evaluation of models and performance metrics;
6. Ethical and social implications of the abuse/control of hate speech in online communications;
7. Integrating and deploying of trained model for detecting hate speech.

2 Goal, Scientific and Practical Value of the Study

Research focuses on the development of identifying and classifying hate speech in text system. This system aims to improve online safety and to promote respectful communication by identifying and flagging content that violates community rules.

Due to the textual data automatic analysis, the system can provide timely and accurate identification of hate speech, allowing for appropriate action to be taken [13–15]. Scientific novelty can be described in several of the following aspects [16–18].

1. Models and algorithms are improved. They are specially designed to detect hate speech. This includes research into advanced machine learning and deep learning techniques, as well as the use of natural language processing and sentiment analysis methodologies. Accent is to improve the accuracy, efficiency and reliability of hate speech detection systems.

2. Linguistic analysis and contextual understanding. Proposed study involves in-depth linguistic analyzing to identify patterns, linguistic features and contextual factors leading to the identification of hate speech. That's why the linguistic dynamics of hate in the different social groups, regions and platforms must be considered. To improve the effectiveness of detection algorithms it is necessary to capture all the nuances of hate.

3. Data collection and annotation. Diverse and representative datasets to identify hate must be collected. So new data collection methodologies including different languages, cultures and demographics must be developed. The study also focuses on manual annotation and labeling of the data, solving questions related to subjective interpretation and inter-rater reliability.

3　Literature Review

Existing systems for detecting hateful/offensive speech in text include a number of functionalities and processes, in particular:

1. Twitter introduced various measures to identify hate speech and to promote safer environment for users [19–21]. One of them is the development of algorithms and tools for detecting hate speech [3, 16, 22].

2. Perspective from Google is API, developed by Jigsaw company (Google's technology incubator), with aim to analyze and moderate online content [23–26]. It uses machine learning models to evaluate quality and potential toxicity of textual content, such as comments, reviews or forum posts.

3. Content Policy from OpenAI is a set of guidelines and principles that govern the acceptable use of OpenAI models and services [27–29]. As an organization, OpenAI, endeavors for the responsible development and deployment of technology, focusing on security, ethical considerations and respect for user rights.

4. HateSonar is a machine learning-based system designed to detect hate speech and offensive content in text [30–32]. It's developed and supported by HateSonar Technologies. HateSonar Technologies creates new technologies for content moderation and online security.

5. Perspective API is a machine learning-based system designed by Jigsaw, being subsidiary company Alphabet Inc. (main company Google) [33–35]. It aims to simplify the detection and analysis of toxic language and harmful content in text.

API proposes a scalable content moderation solution, integrating its capabilities into platforms, apps or websites. These systems use Natural Language Processing (NLP) technics, machine learning algorithms and big annotated datasets to detect hate speech and offensive content in the text [36–38]. It's useful to notice that effectiveness of these systems can vary and updates are constantly needed to adapt to the ever-changing forms of hate speech.

It is significant to note that hate speech detection systems are not perfect and may sometimes make mistakes or omit some cases of hate speech. That's why Twitter encourages users to report any content that they believe violates the platform's policies, as user feedback plays an important role in improving and enhancing the system.

4 Methods and Technologies

As mentioned above research aims to help create a safer online environment, promote inclusivity and reduce the negative impact of hate speech on individuals and communities. Possible stakeholders of the information system development may be:

- Users are the individuals who interact with online platforms and fall under influence of hate speech. They benefit from a safer and respectful online environment.
- Platforms providers are companies and organizations that host online platforms and are responsible for moderating content using tools that can help identify and combat hate speech.
- Content moderators are individuals or groups responsible for reviewing and moderating user-generated content. They benefit from efficient and accurate systems that identify and mark hate speech, reducing their workload.
- Regulatory authorities are government agencies or regulatory bodies that are responsible for maintaining ethical and safe online spaces. They benefit from tools that comply with norms and guidelines on hate and offensive language.
- Advocacy groups are organizations that fight with hate and promote online safety. They benefit from technology solutions that assist in their advocacy efforts.
- Scientists and researchers who study online behavior, linguistics and social dynamics. Their benefit is in access to the data and insights provided by the system for research purposes.
- Society as a whole will benefit from reducing the impact of hate speech and promoting more respectful and inclusive online communication.

Some potential risks associated with an information system project are the next:

- Data preconception. Biased data being used to train models can lead to prejudiced predictions and inaccurate identification of hate speech or offensive language.

- False positive and false negative reviews. The risk is that system incorrectly classifies content as hateful or offensive causing the removal of innocuous content or the inability to identify problematic content.
- Evolution of language and context. The risk is that it will be complicate for the system to keep up with the ever-changing nature of language and the changing contexts where hate speech and offensive language can occur.
- Legal and ethical considerations. Confidentiality law can be broken or freedom of speech can be violated while detecting and mitigating hostile speech.
- User perception and acceptance. The risk is that users will perceive system as aggressive and overly restrictive leading to resistance or even avoidance of the platform.
- Scalability and productivity. The risk is that the system disables to process a large amount of data or to provide results in real or near real time affecting usability and efficiency.
- Cyber attacks. System's algorithms and mechanisms are used or bypassed by computer attackers potentially leading to the spread of hate and offensive content.
- Maintenance and upgrading. Risks associated with updating the system to keep up with new trends and user expectations.

It is important to identify and remove these risks throughout the research life cycle to reduce their potential impact on the effectiveness and ethical implications of the system.

Anxiety about the spread of hate speech, which can lead to harmful consequences such as discrimination, harassment and incitement to violence grows with the increasing use of online platforms and social media. So, it makes sense to develop effective automated detective system to satisfy timely intervention and moderation of safe and respectful online environment. The technical task of this study is to design and develop a hate speech detection system being able to analyze text datasets and accurately classify them as hateful/offensive speech or non-hostile speech via machine learning algorithms. Let's formulate several main steps for designing a hate speech detection system.

1. *Data collection.* Obtaining the different and representative textual dataset containing examples of hate speech from tagged text samples and datasets that cover a wide range of hate speech and non-hostile speech [35–37]. Text samples must be taken from the different sources, for example, social media and networks, online forums and public datasets. The dataset should cover the thematically focused information flow of textual content with different types of hate speech to ensure the system effectiveness.

2. *Preliminary processing of data,* containing hate speech cases. That means their cleaning and preparing for analysis, removing irrelevant information, handling special characters and applying methods like lemmatization and/or stemming to standardize the text [38–41]. This step includes also removing unnecessary information and stop words, normalizing the text (converting to lowercase), working out variations in spelling and applying techniques such as tokenization.

The process converts input information into acceptable data format for further analysis.

3. *Labeling and marking* means the manual process of marking each text sample as hate or no hate. To do this domain experts or human annotators review text samples and label them respectively. This task requires clear directives and well-defined criteria to ensure consistency in the annotation process.

4. *Extraction* of meaningful features from textual data is principal process for training a hate speech detection model. This means conducting exploratory data analysis and developing functions to remove relevant features for hate detection. In particular, it is removal and presentation of linguistically significant negatively features of the text as well as positively emotionally colored ones. Also, it's necessary to extract relevant features from pre-processed text data, for example, word frequency, N-grams and sentiment indicators to present text samples in a numerical format acceptable for the machine learning algorithms via FastText and Python [42–48]. Text samples must be converted into numerical representations that display important information. Commonly used feature extraction techniques contain "Bag-of-Words", TF-IDF (term frequency- inverse document frequency), words vectors (for example, Word2Vec, GloVe) or more complicated approaches as BERT (Bidirectional Encoder Representations from Transformers).

5. *Selecting a machine learning algorithm*, as KNN (K-Nearest Neighbors), Naive Bayes, Decision Tree, Logistic Regression and Random Forest, Support Vector Method, Recurrent Neural Networks, based on their suitability for the task of classifying text into hate and hate speech categories and preliminary performance evaluations [49–53].

6. *Model training*. Developing and training models of hate speech detection involves machine learning or deep learning model on annotated and pre-processed data. This problem is to optimize model parameters using cross-validation method. The aim is to develop the model that can precisely distinguish hate speech from non-hate speech. Machine learning models must be trained using pre-processed and extracted data.

7. *Evaluate the performance and effectiveness of the trained model* using appropriate evaluation metrics and fine-tune the model based on the results. These evaluation metrics can be accuracy, sensitivity, specificity, recall and F1-scores. Meanwhile evaluation methods can be applied cross-validation, confusion matrix analysis, and ROC curves. The main task is to identify strength and weakness of each model and compare their effectiveness, then to optimize model parameters and to estimate their effectiveness via the corresponding estimating indicators.

8. *Improvement and optimization*. After evaluating the initial model, it may be need to be refined and optimized to improve performance. It means tuning of the hyper-parameters, exploring different algorithms or using ensemble methods to improve the accuracy and reliability of the model.

9. *Model selection*. Task is to select the most effective model based on the results of the estimation and to deploy it as a hate detection system.

10. *Implementation of a scalable and efficient system architecture* for processing large volumes of data and providing real-time or near-real-time text processing.

11. *System implementation.* Deploying a trained hate speech detection model to an application or platform includes the development of APIs or interfaces for easy deployment and accessibility that allow for real-time or batch processing of user content. The system should be integrated into existing infrastructure to ensure effective and scalable hate detection.

12. *Integration of a trained hate detection model.* The task is to integrate the selected model into user-friendly system that can accept text input and provide real-time predictions on hate classification.

13. *Testing.* This means to test system with a separate validation dataset to ensure its efficiency, accuracy and reliability.

14. *Verification.* System must be tested on reliable data. Successful completion of this task enables robust and effective hate speech detection system that can contribute to migrating hate speech and promoting a safer online environment.

15. *Regular monitoring and updating of the system* to adapt to new forms of hate speech and improve its effectiveness over time.

Product proposed in this paper can be compared with existing analogues in terms of various factors: ethical considerations; user experience and interface; adaptability and flexibility; scalability and efficiency; accuracy and performance. Comparison helps to identify unique features and advantages that differ the developed system from existing solutions, highlighting its potential value in the field of hate speech detection.

Advantages of implementing an information system for detecting hate speech over analogues.

1. A comprehensive approach. The project uses several machine learning algorithms such as KNN, Naive Bayes, decision tree, logistic regression and random forest [49–53]. It ensures a more reliable and accurate hate detection system than projects that rely on a single algorithm. Therefore, the chances of accurate detection and classification of hate speech increase.

2. Interpretable results. Some of the algorithms, in particular, decision tree and logistic regression, provide results that can be interpreted. It means that one can get a sense of the decision-making process and understand the importance of different characteristics in the classification of hate speech. Interpretable results can be valuable for identifying key indicators of hate speech and for explaining the results of the system to stakeholders.

3. Flexibility and customization. The possibility to experiment with the different algorithms, adjust their parameters and explore ensemble methods aiming to improve system effectiveness. The hate speech detection system can be adapted to identify specific requirements and datasets.

4. Various function presentation. Project can use different types of functions and views due to the algorithm's combination. Each algorithm excels at capturing different aspects of hate speech, such as specific words, patterns or contextual

information. This diversity in function representation can improve the system's ability to detect and classify hate speech in different contexts and languages.

Disadvantages of the project in comparison with analogues.

1. Increased complexity. Using multiple algorithms complicates the project. This requires a deep understanding of each algorithm, parameters setting and potential problem associated with ensemble methods. Managing and maintaining of multiple algorithms can take more time and resources than working with a single algorithm.
2. Potential performance variations. Different algorithms can work differently depending on the dataset and the specific characteristics of the hate speech being analyzed. In spite of the fact that a variety of algorithms can improve system performance, it also makes it possible to vary performance across different datasets or text samples. Ensuring stable and reliable performance in different scenarios may require additional efforts.
3. Selecting and optimizing the algorithm. Choosing the right combination of algorithms and optimizing their parameters can be a difficult task. Determination the most appropriate algorithms for a particular hate speech detection problem requires experience and experimentation. Choosing inappropriate algorithms or incorrectly configuring their parameters can lead to non-optimal performance or increased computational complexity.
4. Model maintenance and scalability. Working with multiple algorithms can complicate model maintenance and scalability. In case when new data appear or the system needs to be updated, several models may need to be adapted and retrained simultaneously. Ensuring consistency and compatibility between implementations of different algorithms can be a potential challenge, especially as the system evolves over time. Overall, while this study offers advantages in terms of completeness, representation of various functions, flexibility and interpretable results, it also poses questions in terms of complexity, performance variations, algorithm selection and model support. Theses factor need to be carefully considered to get the most out of using multiple algorithms in a hate speech detection system.

Project challenges are formulated as:

- Challenge: to identify hate speech in the text accurately and effectively. Solution: to develop hate speech detection system using machine leaning, that effectively classifies text into categories of hostile and non-hostile statements.
- Challenge: dealing with evolving challenges of various forms of hate speech. Solution: to develop system that can adapt to different types and variations of hate speech, including new patterns and linguistic nuances.
- Challenge: large volumes of data for efficient processing and classification. Solution: to develop an efficient data processing pipeline and optimize algorithms to process large-scale datasets and complete classification tasks in a timely manner.

- Challenge: ensuring with generalizability across languages and cultural contexts. Solution: to develop functions and models that can capture the main characteristics of hate speech regardless of the specific language or cultural context.
- Challenge: problems of interpretation and explanation of the hate detection system. Solution: to include methods of interpretation and provide explanations for the system's classification decisions to increase transparency and accountability.
- Challenge: handling false-positive and false-negative results when detecting hate speech. Solution: to apply methods for reducing false positives (misclassification of hate speech as hate speech) and false negatives (absence of cases of genuine hate speech) by improving the accuracy and precision of classification algorithms.
- Challenge: ensuring ethical considerations and mitigating prejudice in identifying hate speech. Solution: to introduce measures to minimize bias, take into account cultural sensitivities and address potential ethical issues related to automatic hate speech detection.

Solving of these problems and fulfilling the corresponding tasks allow to develop a robust, accurate and scalable hate detection system that can effectively identify and classify hate speech in text.

5 Experiments

The following information technologies and methods were used to implement the hate speech detection system.

- Programming languages. Python was used for data preprocessing, model training and evaluation as the main programming language. HTML, CSS and JavaScript were used for the development of Google's expansion.
- Machine learning libraries. Python libraries, such as scikit-learn and fasttext were used for implementing and training machine learning models. These libraries provide efficient and reliable tools for text classification tasks.
- Data pre-processing. Text pre-processing methods including such tasks as text cleaning, tokenization and feature extraction were applied to the dataset. Python libraries, such as NLTK and scikit-learn, were used for these pre-processing steps.
- Machine learning algorithms. To detect hate speech there were used several machine learning algorithms, in particular, Naive Bayes, Decision Tree, Random Forest, Logistic Regression and KNN. These algorithms were chosen based on their effectiveness in text classification tasks and their availability in popular machine learning libraries.
- Development of Google extensions. There were used HTML, CSS and JavaScript. These web technologies created an interactive and user-friendly interface for the text classification.

- Model evaluation indicators. Various indicators including precision, accuracy, recall and F1-score were used to evaluate the performance of machine learning models. These indicators quantified the effectiveness of the models in detecting hate speech.
- Testing and setting up. To ensure the proper functioning of the Google extension and the accuracy of the classification results, extensive testing and setting up was carried out. Test cases are created to verify the results, identify and resolve problems.

The combination of these technologies and methods allowed for the successful implementation of the hate speech detection project, which resulted in an accurate and convenient solution for text classification and hate speech detection.

A dataset using Twitter data investigated the detection of hate speech [35–37]. Text is classified as: hate speech, offensive speech and no hate speech (Figs. 1, 2, 3, 4 and 5). Given the nature of the research, it is important to note that this dataset contains text that could be considered racist, sexist, homophobic or generally offensive.

As can be seen from Fig. 6, the majority of offensive language is used, followed by neutral and hateful language. Data is heavily clogged with links, nicknames, symbols and so on, that's why it's necessary to clean the text. WordNetLemmatizer is tool, used to lemmatize words in the Python programming language. Lemmatization is

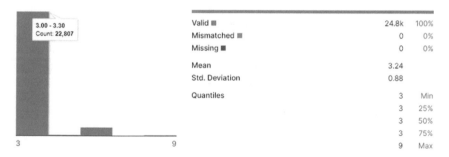

Fig. 1 The dynamics of tweet marking (≥3 users marked tweet)

Valid ■		24.8k	100%
Mismatched ■		0	0%
Missing ■		0	0%
Mean		3.24	
Std. Deviation		0.88	
Quantiles		3	Min
		3	25%
		3	50%
		3	75%
		9	Max

3.00 - 3.30
Count: 22,807

Valid ■		24.8k	100%
Mismatched ■		0	0%
Missing ■		0	0%
Mean		0.28	
Std. Deviation		0.63	
Quantiles		0	Min
		0	25%
		0	50%
		0	75%
		7	Max

0.00 - 0.35
Count: 19,790

Fig. 2 The dynamics of tweet marking as hate speech (hate speech)

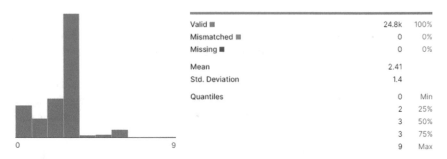

Fig. 3 The dynamics of tweet marking as offensive (offensive speech)

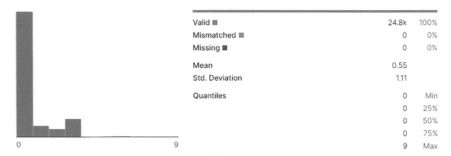

Fig. 4 The dynamics of tweet marking as neutral (no hate speech)

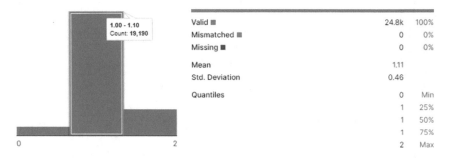

Fig. 5 Class marking, where 0 indicates hate speech, 1indicates offensive speech, 2 indicates no hate speech

the process of transforming a word to its basic form, known as a lemma. Word-NetLemmatizer uses WordNet English lexical database containing syntactic and semantic relations between words. It uses this information to determine the basic form of a word called a lemma which represents its semantic meaning. For example, if input word is "running", WordNetLemmatizer will turns it into the lemma "run". This is useful for standardizing words and reducing the size of the dictionary while processing text, searching for similar words, or performing other language operations. WordNetLemmatizer is part of the library NLTK (Natural Language Toolkit) in Python, which provides a variety of tools for word processing and language work. Next using regex we will define patterns and replace unnecessary data in the dataset, that is, we will clean the data (Fig. 7).

Vectorizing data. Countvectorize simplifies the use of text data directly in machine/deep learning models, such as text classification [54]. Countvectorizer is a method to converting text to numeric data. To demonstrate how it works, let's look at an example:

text = ['Hi, I am Victoria, she is my daughter Sofia and she is LPNU student']
The text is converted to a sparse matrix as shown below.

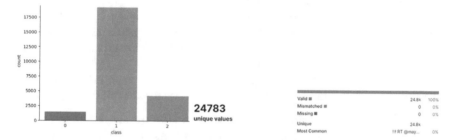

Fig. 6 Histogram of tweet vocabulary analysis, where 0 indicates hate speech, 1 indicates offensive speech, 2 indicates no hate speech

```
['rt USER a woman you shouldn complain about cleaning up your house amp a man you should always take the trash out ',
 'rt USER boy dat cold tyga dwn bad for cuffin dat hoe in the 1st place ',
 'rt USER dawg rt USER you ever fuck bitch and she start to cry you be confused a shit ',
 'rt USER USER she look like tranny ',
 'rt USER the shit you hear about me might be true or it might be faker than the bitch who told it to ya 57361 ',
 'USER the shit just blow me claim you so faithful and down for somebody but still fucking with hoe 128514 128514 128514 ',
 'USER can not just sit up and hate on another bitch got too much shit going on ',
 '8220 USER cause tired of you big bitch coming for u skinny girl 8221 ',
 'amp you might not get ya bitch back amp thats that ',
 'USER hobby include fighting mariam bitch ',
 'keeks is bitch she curve everyone lol walked into conversation like this smh ',
 'murda gang bitch it gang land ',
 'so hoe that smoke are loser yea go on ig ',
 'bad bitch is the only thing that like ',
 'bitch get up off me ',
 'bitch nigga miss me with it ',
 'bitch plz whatever ',
 'bitch who do you love ',
 'bitch get cut off everyday ',
 'black bottle amp bad bitch ',
 'broke bitch cant tell me nothing ',
 'cancel that bitch like nino ',
 'cant you see these hoe wont change ',
 'fuck no that bitch dont even suck dick 128514 128514 128514 the kermit video bout to fuck ig up ',
 'got ya bitch tip toeing on my hardwood floor 128514 URL ',
 'her pussy lip like heaven door 128524 ',
 'hoe what it hitting for ',
 'met that pussy on ocean dr gave that pussy pill 128524 ',
 'need trippy bitch who fuck on hennessy ',
 'spend my money how want bitch it my business ',
 'txt my old bitch my new bitch pussy wetter ',
 'say im back to the old me but my old bitch would get excited 128524 ',
 'if you aint bout that murder game pussy nigga shut up ',
 'if you re toe ain done you pussy stink ',
```

Fig. 7 Result of data cleaning

	hi	I	am	Victoria	she	is	my	daughter	Sofia	and	LPNU	student
0	1	1	1	1	2	2	1	1	1	1	1	1

There are 12 unique words in the text and therefore, 12 different columns, each representing a unique word in the matrix. Raw represents the number of words. Since words "she" and "is" were repeated twice we have the number of these particular words as 2 and 1 for the rest. Let's create bag of words using CountVectorizer with parameter max_features = 2000 in Python and scikit-learn library. The parameter max_features indicates the maximum number of features (words) that will be taken into account in bag of words. In this case, the limit is set to 2000 most frequent words. Calling the fit_transform on CountVectorizer object using the corpus variable (a list of text documents or sentences that will be used to create a bag of words). The result of the fit_transform call is a matrix X, where each raw represents one document and each column corresponds to one word from the bag of words. The values in the matrix reflect the number of occurrences of each word in each document. (toarray) converts resulting object into a regular Numpy array. So, this code performs the process of creating a bag of words from the text data and presents it as a matrix that can be used for further data analysis or modeling.

Splitting a dataset into a training set and a testing set is an important step in machine learning and data analysis. The main reasons for this splitting are the next:

1. Evaluation of model performance. The test set is used to evaluate the performance of the model that was trained on the training set. It allows to estimate how well the model fits new previously unseen data. This gives an idea of the model's overall generalizability.
2. Detecting overtraining. Splitting dataset into the training and test sets helps to detect model overtraining. Overtraining occurs when the model is too well fitted to the training data, but cannot effectively fit the new data. Using a separate test set allows to identify this phenomenon and take measures to improve the model.
3. Setting up hyper-parameters. Splitting dataset can also be used to set up the model hyper-parameters. Hyper-parameters are the parameters, that are not learned by the model during training but affect its behavior. Using of textual dataset allows to estimate different hyper-parameters combinations and to choose the optimal ones.
4. Maintaining independence. Splitting dataset into the training and test sets helps to preserve independence between process of the model training and evaluation of its effectiveness.

Splitting dataset into the training and test sets also allows to preserve the objectivity of the model's evaluation. Evaluation of the effectiveness on the new independent on training data enables to obtain realistic evaluation of the model performance.

For example, while classifying text, the model is trained on the training dataset, where it has access to class labels. After training, the model is evaluated on a test dataset, where class labels are used to compare model predictions with actual values. Hence, one can determine the precision, F-measure, recall or other metrics that

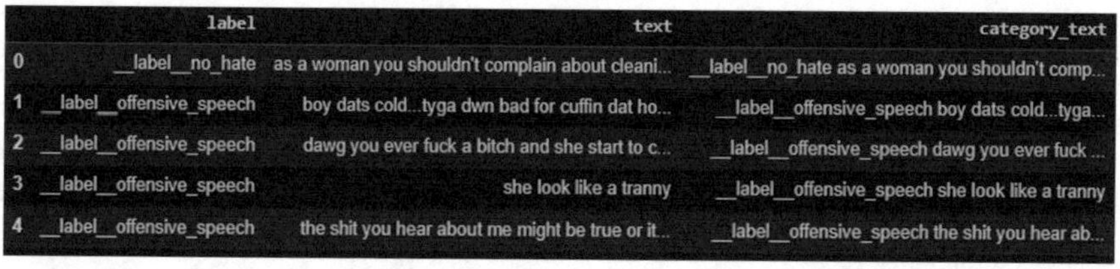

	label	text	category_text
0	__label__no_hate	as a woman you shouldn't complain about cleani...	__label__no_hate as a woman you shouldn't comp...
1	__label__offensive_speech	boy dats cold...tyga dwn bad for cuffin dat ho...	__label__offensive_speech boy dats cold...tyga...
2	__label__offensive_speech	dawg you ever fuck a bitch and she start to c...	__label__offensive_speech dawg you ever fuck ...
3	__label__offensive_speech	she look like a tranny	__label__offensive_speech she look like a tranny
4	__label__offensive_speech	the shit you hear about me might be true or it...	__label__offensive_speech the shit you hear ab...

Fig. 8 Labeling examples

indicate the model's performance. In general, splitting dataset into the training and test sets is an important step for evaluating and improving machine learning models. It identifies problems with overtraining, sets up hyper-parameters and ensures objective evaluation of the model's performance on new data.

Program code splits dataset into the training set (X_train, y_train) and test set (X_test, y_test) via function (train_test_split). The parameter test_size = 0.20 indicates that 20% of the data will be allocated for testing, i.e. will be used to evaluate the model's effectiveness. The parameter random_state = 0 is used to fix random values while splitting data. This ensures reproducible splitting, it means that every time the code is realized with the same random_state value, the same splitting dataset will be obtained. The code results in four variables: X_train (training features), X_test (test features), y_train (training goal values) and y_test (test goal values). These datasets can be used to train the model on training data and evaluate its effectiveness on test data. At first, you need to prepare dataset by labeling the data (Fig. 8). Then let's split test and training data.

Program code saves data from two DataFrames (train and test) as CSV-files. Specifically, it saves the column "category_text" from every DataFrame to the separate file. Format of saving data in CSV files makes it easy to read and process them in the future.

- test.to_csv("speech.test", columns = ["category_text"], index = False, header = False) saves data from DataFrame test to file named "speech.test" with column "category_text".
- train.to_csv("speech.train", columns = ["category_text"], index = False, header = False) saves data from DataFrame train to file named "speech.train". Parameter columns = ["category_text"] indicates which column you want to save. Parameters index = False and header = False indicate that it's necessary to miss saving the raw indices and column header.

Import fasttext to train the text classification model based on the training dataset "speech.train" and test its performance on the test set "speech.test". The model will be trained to classify the text based on the provided categories. Method (test) calculates the model effectiveness metrics such as precision, recall and F-measure and displays test results. Model accuracy is 89% on 3716 text dataset records. Let's classify artificially entered text (Table 1).

Table 1 Examples of text content classification results

N	Text	Label	Result
1	"nigga"	__label__hate_speech	0.95959681
2	"I hate that black person"	__label__hate_speech	0.70769596
3	"just brought some bread and went home"	__label__no_hate	0.81684661
4	"you ugly bustard don't show your face to anybody"	__label__hate_speech	0.60036236
5	"you are not that intelligent machine"	__label__offensive_speech	0.45037791
6	"I can not just sit up and HATE on another bitch.. I got too much shit going on"	__label__offensive_speech	0.99653608

Next, the developed extension will be checked on real examples. Let's take Twitter (Fig. 9) and YouTube (Fig. 10) as examples. As a result, plugin has highlighted different results for different texts depending on the content—hate speech not detected and offensive speech detected.

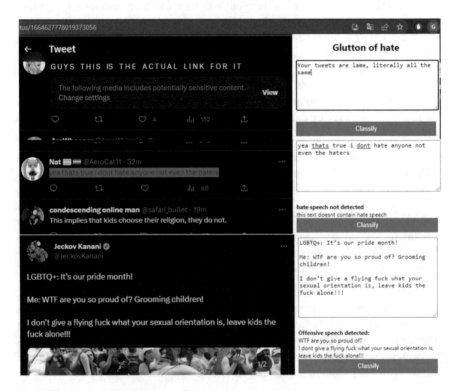

Fig. 9 The result of the extension on Twitter

Fig. 10 The result of the extension on YouTube

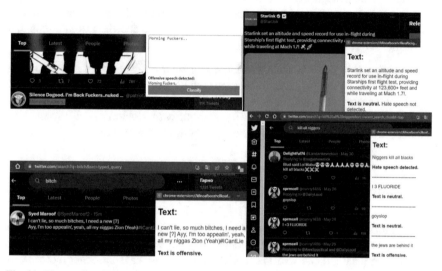

Fig. 11 The results of extension

There is option to go to a web page and activate extension (Fig. 11). In this case a window with all visible classified text will appear.

Some conclusions on the project:

1. Comfortable interface. Google extension provides an intuitive and user-friendly interface for text classification. Users can easily enter text or select existing text to get instant classification results. That's why the convenience and accessibility of the tool increase.
2. Universal classification methods. Using the pre-trained rapid text model, the extension offers accurate classification of hate, offensive language and neutral text. High accuracy rate of 92% indicates the effectiveness of the classification algorithm in distinguishing between different classes.
3. Real-time classification. The possibility to classify text in real time by simply selecting or pasting text makes extension efficient and responsive. Users can receive classification results instantly without delays, enabling quick analysis and decision-making.

4. Potential to identify and eliminate hate. The extension's focus on hate speech classification demonstrates its relevance in combating online harassment and promoting a safer digital environment. Due to accurately identifying hate speech, users can take appropriate action to combat it and create more inclusive online community.
5. Practical application. The functionality of the extension makes it suitable for a number of scenarios, from analyzing single texts to classifying entire web pages. Its flexibility allows users to adapt it to their specific needs, whether for personal use, content moderation or research purposes.
6. Constant improvement: in spite of high accuracy level of 92%, there is always room for improvement. Continuous efforts to fine-tune the classification model, eliminate false positives and false negatives, and expand the training data can further improve the performance and accuracy of the extension.

To sum up, Google's text classification extension is a valuable tool for identifying hate speech, offensive language, and neutral text. Its user-friendly interface, real-time classification capability and high level of accuracy contribute to its efficiency and practicality. The project demonstrates the possibility to develop a useful solution to combat online harassment and promote a safer online environment.

6   Results and Discussion

Classification_report is a report that provides comprehensive information on the effectiveness of the classification model for each class, where 0 indicates hate speech, 1 indicates offensive speech, 2 indicates no hate speech (Table 2). It contains the following evaluation metrics: precision, recall, F1-score and support. The classification report displays the next indicators for each class 0–1:

- Precision: it is the ratio of the number of correctly classified examples for a given class to the total number of examples that were predicted to be of that class. High precision means a low number of false-positive results.
- Recall: it is a ratio of the number of correctly classified examples for a given class to the total number of examples in this class. High recall means a low number of false-positive results.
- F1-score: it is a harmonic mean between precision and recall. It provides a balanced measure of model performance that takes into account both precision and recall.
- Support: it is number of examples in every class (class 0–1, where 0 indicates hate speech, 1 indicates offensive speech, 2 indicates no hate speech).

Accuracy is a metrics that defines the ratio of the number of correctly classified examples to the total number of examples. It indicates the general accuracy of the model in predicting classes. Accuracy is calculated as: (the number of correctly classified examples)/(the total number of examples). Classification_report helps to

Table 2 Values of classification_report for the different methods

Method		Precision	Recall	F1-score	Support
KNN	Class 0	0.36	0.16	0.22	279
	Class 1	0.87	0.95	0.90	3852
	Class 2	0.70	0.52	0.60	826
	Accuracy			0.83	4957
	Macro avg	0.64	0.54	0.57	
	Weighted avg	0.81	0.83	0.81	
Naive Bayes	Class 0	0.07	0.69	0.12	279
	Class 1	0.90	0.24	0.38	3852
	Class 2	0.44	0.54	0.48	826
	Accuracy			0.31	4957
	Macro avg	0.47	0.49	0.33	
	Weighted avg	0.77	0.31	0.38	
Decision Tree	Class 0	0.30	0.32	0.31	279
	Class 1	0.93	0.93	0.93	3852
	Class 2	0.84	0.85	0.84	826
	Accuracy			0.88	4957
	Macro avg	0.69	0.70	0.69	
	Weighted avg	0.88	0.88	0.88	
Logistic Regression	Class 0	0.45	0.25	0.32	279
	Class 1	0.94	0.95	0.94	3852
	Class 2	0.84	0.89	0.87	826
	Accuracy			0.90	4957
	Macro avg	0.74	0.70	0.71	
	Weighted avg	0.89	0.90	0.90	
Random Forest	Class 0	0.40	0.19	0.25	279
	Class 1	0.90	0.95	0.93	3852
	Class 2	0.83	0.76	0.79	826
	Accuracy			0.88	4957
	Macro avg	0.71	0.63	0.66	
	Weighted avg	0.86	0.88	0.87	

evaluate the effectiveness of the model for each class separately and make comparisons between the different classes. It defines classes with the best accuracy, recall and F1-sccore, and besides that detects problems that arise while classifying certain classes. Precision-recall is based on the understanding and degree of relevance, accuracy is the proportion of relevant samples among those found and refers only to positive results (Figs. 12, 13, 14, 15 and 16). Completeness is the proportion of the total number of positive samples that are actually found. These metrics are used to identify hypotheses in statistics.

The KNN classifier achieves an accuracy of 0.83, so it correctly classifies approximately 83% of the instances in the dataset (Fig. 12). Accuracy is relatively good to consider that the KNN model works quite well on this particular dataset. Analyzing the accuracy, recall, and F1-score values, one can see that the classifier performs differently for each class. For KNN classifier has an accuracy of 0.36, that means 36% of cases classified as class 0 are correctly classified. Class 1 (offensive speech) has an accuracy of 0.87, that indicates high percentage of correct predictions. Class 2 (no hate speech) has also an accuracy of 0.70. Recall values, representing the percentage of actual positive cases correctly classified vary for different classes. Class 0 has recall of 0.16, class 1 has recall of 0.95 and class 2 has recall of 0.52. F1-score combining precision and recall into a single indicator demonstrates the overall performance of the classifier. F1-score for classes 0, 1 and 2 are 0.22, 0.90 and 0.60, respectively. In summary, the KNN classifier achieves acceptable accuracy and performs well for class 1, but has relatively lower performance for classes 0 and 2. Accuracy and recall values vary for the different classes, indicating that the classifier performance depends on the class. F1-score values provide an overall estimation of classifier performance.

In the Naive Bayesian classifier evaluation, the accuracy for class 0 (hate speech) is 0.07, indicating a low proportion of correctly classified cases among the predicted positive results. Recall for class 0 is 0.69, indicating that a relatively high proportion of actual positive results were correctly identified (Fig. 13). For class 1 (offensive speech) the accuracy is 0.90, indicating a high proportion of correctly classified cases among the predicted positive results. Recall is 0.24, indicating that a low proportion of actual positive results were correctly identified. For class 2 (no hate speech) the accuracy is 0.44, indicating a moderate proportion of correctly classified cases among the predicted positive results. Recall is 0.54, indicating that a half of actual positive results were correctly identified. The accuracy of a Naive Bayesian classifier is 0.31, that means approximately 31% of the instances in the dataset are correctly classified.

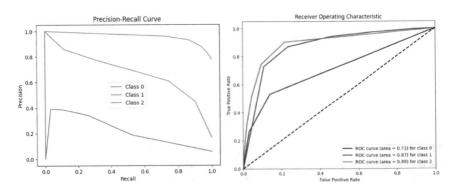

Fig. 12 Precision recall curve and ROC curve for KNN

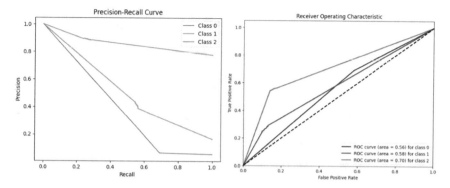

Fig. 13 Precision recall curve and ROC curve for Naive Bayes

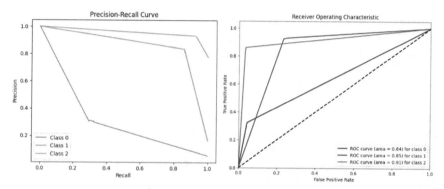

Fig. 14 Precision recall curve and ROC curve for Decision Tree

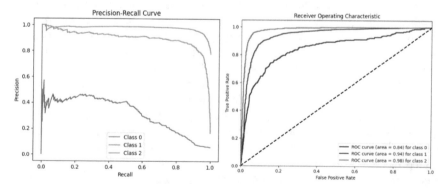

Fig. 15 Precision recall curve and ROC curve for Logistic Regression

It's very poor result. In terms of the macro average F1-score, which provides a balanced measure of accuracy and recall for all classes, the value is 0.33. In general, the Naive Bayes classifier shows different performance in different classes, with

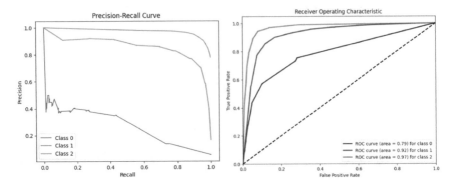

Fig. 16 Precision recall curve and ROC curve for Random Forest

higher precision and lower recall for class 1 (offensive words) and lower precision and higher recall for class 0 (hate speech). Accuracy and F1-score prompt that the classification characteristics of the Naive Bayes model can be improved.

The decision tree classifier shows an accuracy of 0.88, indicating that it correctly classified approximately 88% of the instances in the dataset (Fig. 14). This is relatively high accuracy and indicates that the decision tree model works well on this particular dataset. Looking at the accuracy, recall, and F1-score values, one can notice that the classifier performs well in all classes. For class 0 accuracy is of 0.30, that means, 30% cases classified as class 0, are correctly classified. For class 1 accuracy is of 0.93, indicating high percentage of correct predictions. For class 2 accuracy is also high of 0.84. Recall values representing percentage of actual positive correctly classified cases are high for all classes also. Class 0 has recall 0.32, class 1 has recall 0.93, and class 2 has recall 0.85. F1-score being a harmonic average of accuracy and recall, provides a balanced measure of classifier performance. F1 scores for all classes are relatively high, with Class 1 having the highest F1 score of 0.93. To summarize, the Decision Tree classifier demonstrates high performance with high precision, accuracy, recall and F1-score on this dataset. It effectively classifies instances into all classes and can be proposed as a good model for the task at hand.

The logistic regression classifier achieves an accuracy of 0.90, which means that it correctly classifies approximately 90% of the instances in the dataset (Fig. 15). Relatively high accuracy indicates that the logistic regression model works well on this particular dataset. One can notice that class 0 has accuracy 0.45, i.e. 45% cases are correctly classified as class 0. Class 1 has high accuracy 0.94, indicating high percentage of correct predictions. Class 2 has also accuracy 0.84. Recall values representing percentage of actual positive correctly classified cases are relatively high for all classes. For class 0 recall is of 0.25, for class 1 recall is of 0.95, and for class 2 recall is of 0.89. Estimations F1 combining accuracy and recall demonstrate the classifier's performance additionally. F1 estimation is 0.32 for class 0, F1 estimation is 0.94 for class 1, F1 estimation is 0.87 for class 2. Thus, the logistic regression classifier achieves high accuracy and works well in all classes. It has a high accuracy and recall value that is a good balance between correct positive predictions and overall coverage of positive cases. F1 estimations verified classifier's effectiveness in classifying different classes.

Random Forest classifier achieves accuracy of 0.88, hence it correctly classifies approximately 88% instances in dataset (Fig. 16). This relatively high accuracy indicates that the Random Forest model works well on this particular dataset. Examining values of accuracy, recall and F1 estimation, it can be seen that classifier works differently for each class. Class 0 has accuracy 0.40, i.e. 40% cases are correctly classified as class 0. Class 1 has high accuracy 0.90, indicating high percentage of correct predictions. Class 2 has also relatively high accuracy 0.83. Recall values representing percentage of actual positive correctly classified cases vary for different classes. Recall value is of 0. 19 for class 0, recall value is of 0.95 for class 1, and recall value is of 0.76 for class 2. F1 combining accuracy and recall into one indicator demonstrates the classifier's overall performance. F1-estimations for classes 0, 1 and 2 are 0.25, 0.93 and 0.79 respectively. Consequently, Random Forest classifier achieves high accuracy and works good for class 1 but shows relatively low effectiveness for classes 0 and 2. Accuracy and recall values vary for different classes that means dependence of classifier's performance on class. F1 scores provide the general estimation of classifier's effectiveness.

ROC AUC (Receiver Operating Characteristic Area Under the Curve) is a metrics used to evaluate the quality of binary classification model (Table 3). It measures the ability of the model to differ two classes and is defined as the area under the receiver performance curve (Receiver Operating Characteristic curve). ROC-curve shows the relationship between the proportion of correct classifications True Positive Rate (TPR) (sensitivity) and the proportion of incorrect classifications False Positive Rate (FPR). ROC AUC measures the overall difference of the model between two parameters. The ROC AUC values range from 0 to 1, where a value closer to 1 indicates a better-quality model that is able to correctly classify the data. ROC AUC in an important metric while evaluating the performance of classifiers, especially in case when the dataset has an unbalanced class distribution or the accuracy of prediction for both classes is important. Using of ROC AUC allows to compare and choose the best classification models based on their ability to distinguish between.

ROC AUC (Receiver Operating Characteristic Area Under the Curve) values ensure evaluation of performance for logistic regression classifier for each class and overall performance. KNN model is relatively good at distinguishing class 1 and class 2 from other classes, but has lower performance in distinguishing class 0. General

Table 3 Values of ROC AUC for classes 0–2

Method	Class 0	Class 1	Class 2	Overall AUC
KNN	0.71	0.87	0.89	0.82
Naive Bayes	0.56	0.58	0.70	0.61
Decision Tree	0.64	0.85	0.91	0.80
Logistic Regression	0.84	0.94	0.98	0.92
Random Forest	0.79	0.92	0.97	0.89

AUC indicates that the model has good enough performance for all classes, but there is still room for improvement (Fig. 13, Table 3). Model has ROC AUC:

- 0.71 for Class 0. It means that model has a moderate discriminative ability to distinguish class 0 from the other classes. It is better than random guessing, but there is still room for improvement.
- 0.87 for Class 1. It means that model has good discriminative ability to distinguish class 1 from the other classes. It does a relatively good job of identifying instances belonging to Class 1.
- 0.89 for Class 2. It means that model has good discriminative ability to distinguish Class 2 from the other classes. It does a relatively good job of identifying instances belonging to Class 2.

General AUC 0.82 indicates that the model has a fairly good performance, but there may be differences in performance in different classes.

Naive Bayes model poorly discriminates between class 0 and class 1, but has moderate discriminatory ability in distinguishing class 2 from other classes (Fig. 14; Table 3). General AUC 0.61 indicates that the model has limited overall performance in classifying instances in all classes. Model has ROC AUC:

- 0.56 for Class 0. It means that model has a limited discriminative ability to distinguish class 0 from the other classes. It does not work well when identifying instances belonging to class 0.
- 0.58 for Class 1. It means that model has a limited discriminative ability to distinguish Class 1 from the other classes. It does not work well when identifying instances belonging to Class 1.
- 0.70 for Class 2. In particular, model has a moderate discriminative ability to distinguish class 2 from the other classes. It is good in identifying instances belonging to Class 2.

General AUC: general AUC of 0.61 represents the average performance of the model across all classes. It provides general estimation of the model ability to classify instances across all classes correctly. The model has limited overall performance and limited ability to distinguish different classes.

Decision Tree model performs moderately when distinguishing class 0, well when distinguishing class 1, and perfectly when distinguishing class 2 from other classes (Fig. 15; Table 3). General AUC 0.80 indicates that the model has good overall performance in classification of instances across all classes. The model has ROC AUC:

- 0.64 for Class 0. It means that model has a moderate discriminative ability to distinguish class 0 from the other classes. It works quite good in identifying instances belonging to class 0.
- 0.85 for Class 1. It means that model has good discriminative ability to distinguish class 1 from the other classes. It works good in identifying instances belonging to Class 1.

- 0.91 for Class 2. It means that model has perfect discriminative ability to distinguish Class 2 from the other classes. It works very good in identifying instances belonging to Class 2.

General AUC: general AUC of 0.80 represents the average performance of the model across all classes. It provides general estimation of the model ability to classify instances across all classes correctly. AUC 0.80 indicates good model performance with a decent ability to distinguish between different classes.

Logistic Regression model has ROC AUC:

- 0.84 for Class 0. It means that for Class 0 (hate) classifier achieved value of AUC 0.84. The higher the AUC value is, the better the classifier distinguishes between positive (Class 0) and negative (other classes) samples. AUC of 0.5 indicates random performance, while AUC of 1.0 means perfect discrimination.
- 0.94 for Class 1. For Class 1 (offensive) classifier achieved AUC value of 0.94. It means that classifier distinguishes between class 1 and other classes better than class 0.
- 0.98 for Class 2. For Class 2 (no hate) classifier achieved maximum AUC value of 0.98. It means that classifier is very effective in distinguishing class 2 from other classes.

General AUC: general AUC of 0.92 represents the average performance of the model across all classes. In this case classifier achieved general AUC 0.92 that indicates good distinguish effectiveness across all classes. These AUC values determine how well a logistic regression classifier is able to split different classes in a dataset. Higher AUC values usually indicate better performance in terms of classification accuracy.

The Random Forest model is good at distinguishing class 0, very good at distinguishing class 1, and excellent at distinguishing class 2 from other classes. Model has ROC AUC:

- 0.79 for Class 0. It means that model has good discriminative ability to distinguish Class 0 from the other classes. It works good in identifying instances belonging to Class 0.
- 0.92 for Class 1. It means that model has very good discriminative ability to distinguish class 1 from the other classes. It works very well while identifying instances belonging to Class 1.
- 0.97 for Class 2. It means that model has perfect discriminative ability to distinguish Class 2 from the other classes. It works perfectly in identifying instances belonging to Class 2.

General AUC: general AUC of 0.89 represents the average performance of the model across all classes. It provides general estimation of the model ability to classify instances across all classes correctly. AUC 0.89 indicates that model has very good performance with a strong ability to distinguish between different classes.

7 Conclusions

The information technology for identifying hate speech in online communication based on machine learning methods is developed and described. The technology is based on the implementation of the following stages: data collection from reliable sources and formation of datasets, data preprocessing (noise removal, text normalization, stop word removal, tokenization), labeling and marking of data (hate or no hate), extraction of significant linguistic features (based on Bag-of-Words, TF-IDF, Word2Vec, GloVe, BERT), selection of a machine learning method, model implementation, model training and evaluation, and classifier model accuracy evaluation. The technology is based on the implementation of a pipeline consisting of the following steps: data cleaning, dataset splitting, model training, fasttext and prediction. Accuracy estimates for various classifiers such as KNN (0.832), Naive Bayes (0.315), Decision Tree (0.878), Logistic Regression (0.904), and Random Forest (0.879) were performed. Logistic Regression achieves the highest accuracy score of 0.904, followed by Random Forest with an accuracy of 0.879. Decision Tree and KNN have similar accuracy rates of about 0.878 and 0.832, respectively. Naive Bayes has the lowest accuracy of 0.315. It is important to note that accuracy alone may not provide a complete evaluation of a classifier's performance, especially when it comes to unbalanced datasets or when different classes have different importance. Therefore, it is recommended to consider other evaluation metrics, such as precision, recall, and F1 score, to have a more complete understanding of the classifier's performance in different classes. Based on the provided ROC AUC performance values for different classifiers, the following conclusions can be formulated:

- KNN (K- nearest neighbors): ROC AUC values for Class 0, Class 1 and Class 2 are moderate or high (0.71, 0.87 and 0.89, respectively). General AUC is 0.82 indicating that the model has good enough performance in classifying instances for all classes. KNN demonstrates decent distinguish ability and is good at distinguishing class 1 and class 2.
- Naive Bayes: ROC AUC values for Class 0, Class 1 and Class 2 are relatively low (0.56, 0.58 and 0.70, respectively). General AUC is 0.61 indicating low effectiveness of instances classification across all classes. Naive Bayes may not be the best classifier for this particular problem, because it is difficult to distinguish between different classes effectively.
- Decision Tree: ROC AUC values for Class 0, Class 1 and Class 2 are moderate or high (0.64, 0.85 and 0.91, respectively). General AUC is 0.80 indicating that the model has good performance in classifying instances for all classes. Decision Tree demonstrates decent discriminative ability and is good at distinguishing class 1 and class 2.
- Logistic Regression: ROC AUC values for Class 0, Class 1 and Class 2 are high (0.84, 0.94 and 0.98, respectively). General AUC is 0.92 indicating that the model has perfect performance in classifying instances for all classes. Logistic Regression demonstrates strong discriminatory ability and distinguishes well between all classes.

- Random Forest: ROC AUC values for Class 0, Class 1 and Class 2 are high (0.79, 0.92 and 0.97, respectively). General AUC is 0.89 indicating that the model has perfect performance in classifying instances for all classes. Random Forest demonstrates strong discriminatory ability and distinguishes well between all classes.

Overall, Random Forest and Logistic Regression stand out as the best performing classifiers based on their high ROC AUC values for all classes and overall AUC. Decision Tree and KNN also show decent performance, while Naive Bayes lags behind other classifiers in terms of discriminative ability and overall classification performance.

The practical value of developing a hate speech detection model lies in its subsequent potential real-world applications and social impact.

1. Protecting online spaces. The developed system can be integrated into social media platforms, online forums and other digital environments to automatically detect and flag hate cases. It helps create safer online spaces by promoting inclusive and respectful communication.
2. Improved content moderation. The system can help content moderators and platform administrators effectively identify and handle hate speech content. By automating the initial detection process, it reduces manual workload and enables timely response and intervention.
3. Support for human rights policy. Hate detection systems can provide valuable information. This information can be used to identify trends, monitor hate activity and develop policies and regulations.
4. Prevention of digital wellbeing. By reducing the spread and impact of hate, the system contributes to the creation of a healthy online environment. This has a positive impact on the mental health of individuals, the building of online communities, and the overall well-being of Internet users.
5. Raising awareness and education. An ensemble of artificial intelligence methods for identifying hate speech and hate speech can raise awareness of the prevalence and consequences of hate speech, contributing to educational initiatives aimed at increasing digital literacy, empathy and responsible behavior online.

Overall, the scientific novelty and practical value of developing a hate speech detection system lies in developing our understanding of the dynamics of hate speech, developing effective detection algorithms, and creating safer and more inclusive digital spaces for individuals and communities.

There are several prospects for further research in the area of hate speech detection and classification. Some potential areas for research may be proposed.

1. Advanced machine learning models. Exploring the use of more advanced machine learning models, such as deep learning models (e.g. recurrent neural networks or transformers), to detect hate. These models have shown promising results in natural language processing tasks and can further improve the accuracy and reliability of hate speech classification systems.

2. Multilingual hate speech detection. Extending the hate speech detection system to handle multiple languages. This process will include training the models on a variety of datasets in different languages and exploring methods to overcome language barriers and cultural nuances in identifying hate speech.

3. Contextual analysis. Considering the context and intent of the text while identifying hate speech. This involves analysing not only individual words, but also the surrounding context, sarcasm and hidden meanings to better record and accurately classify cases of hate speech.

4. Online streaming data. Adapting a hate speech detection system to process real-time steaming data from social media platforms or online forums. This will require the development of efficient algorithms to process and classify data in real time and deal with issues such as data volume, speed and noise.

5. Integration of user feedback. Incorporating user feedback systems and user-based moderation systems to continuously improve the hate detection system. User feedback can provide valuable insights and help refine models, identify new patterns, and adapt to evolving hate speech trends.

6. Transfer of learning and domain adaptation. Exploring transfer learning techniques to use pre-trained models for related tasks or domains and fine-tune them for hate detection. This can be particularly useful when tagged data is restricted to a specific domain or language.

7. Ethical considerations. Exploring the ethical implications of hate detection and classification systems, including bias, fairness and privacy concerns. Researching methods to mitigate bias and ensure fairness in the classification process is crucial to creating reliable and unbiased systems.

8. Multimodal approaches. Integrating other modalities, such as images, video or audio, alongside text to improve hate detection. Multimodal approaches can capture additional contextual signals and provide a more comprehensive understanding of hate speech incidents.

These research perspectives aim to improve the accuracy, effectiveness and ethical considerations of hate speech detection systems, enabling better identification and mitigation of harmful content on online platforms.

References

1. Sandaruwan, H.M.S.T., Lorensuhewa, S.A.S., Kalyani, M.A.L.: Sinhala hate speech detection in social media using text mining and machine learning. In: 19th International Conference on Advances in ICT for Emerging Regions, vol. 250, pp. 1–8. IEEE (2019)

2. William, P., Gade, R., Chaudhari, R., Pawar, A.B., Jawale, M.A.: Machine learning based automatic hate speech recognition system. In: International Conference on Sustainable Computing and Data Communication Systems (ICSCDS), pp. 315–318. IEEE (2022)

3. Pawar, A.B., Gawali, P., Gite, M., Jawale, M.A., William, P.: Challenges for hate speech recognition system: approach based on solution. In: International Conference on Sustainable Computing and Data Communication Systems (ICSCDS), pp. 699–704. IEEE (2022)

4. Mossie, Z., Wang, J.H.: Social network hate speech detection for Amharic language. Comput. Sci. Inf. Technol. 41–55 (2018)
5. Mykytiuk, A., Vysotska, V., Markiv, O., Chyrun, L., Pelekh, Y.: Technology of fake news recognition based on machine learning methods. In: CEUR Workshop Proceedings, vol. 3387, pp. 311–330 (2023)
6. Khanday, A.M.U.D., Rabani, S.T., Khan, Q.R., Malik, S.H.: Detecting Twitter hate speech in COVID-19 era using machine learning and ensemble learning techniques. Int. J. Inf. Manag. Data Insights 2(2), 100–120 (2022)
7. Sultan, D., et al.: Cyberbullying-related hate speech detection using shallow-to-deep learning. Comput. Mater. Continua 74(1), 2115–2131 (2023)
8. Duwairi, R., Hayajneh, A., Quwaider, M.: A deep learning framework for automatic detection of hate speech embedded in Arabic tweets. Arabian J. Sci. Eng. 46, 4001–4014 (2021)
9. Akuma, S., Lubem, T., Adom, I.T.: Comparing bag of words and TF-IDF with different models for hate speech detection from live tweets. Int. J. Inf. Technol. 1–7 (2022)
10. Velankar, A., Patil, H., Joshi, R.: A Review of Challenges in Machine Learning Based Automated Hate Speech Detection (2022). arXiv:2209.05294
11. Fernando, W.S.S., Weerasinghe, R., Bandara, E.R.A.D.: Sinhala hate speech detection in social media using machine learning and deep learning. In: 22nd International Conference on Advances in ICT for Emerging Regions (ICTer), pp. 166–171. IEEE (2022)
12. Chhabra, A., Vishwakarma, D.K.: A literature survey on multimodal and multilingual automatic hate speech identification. Multimed. Syst. 1–28 (2023)
13. Defersha, N.B., Kekeba, K., Kaliyaperumal, K.: Tuning hyperparameters of machine learning methods for Afan Oromo hate speech text detection for social media. In: 4th International Conference on Computing and Communications Technologies, pp. 596–604. IEEE (2021)
14. Mohapatra, S.K., Prasad, S., Bebarta, D.K., Das, T.K., Srinivasan, K., Hu, Y.C.: Automatic hate speech detection in English-Odia code mixed social media data using machine learning techniques. Appl. Sci. 11(18), 8575 (2021)
15. Alshalan, R., Al-Khalifa, H.: A deep learning approach for automatic hate speech detection in the Saudi Twittersphere. Appl. Sci. 10(23), 8614 (2020)
16. Lingiardi, V., Carone, N., Semeraro, G., Musto, C., D'Amico, M., Brena, S.: Mapping Twitter hate speech towards social and sexual minorities: a lexicon-based approach to semantic content analysis. Behav. Inf. Technol. 39(7), 711–721 (2020)
17. Chhikara, M., Malik, S.K.: Classification of cyber hate speech from social networks using machine learning. In: 11th International Conference on System Modeling & Advancement in Research Trends (SMART), pp. 419–423. IEEE (2022)
18. Laaksonen, S.M., Haapoja, J., Kinnunen, T., Nelimarkka, M., Pöyhtäri, R.: The datafication of hate: expectations and challenges in automated hate speech monitoring. Front. Big Data 3, 3 (2020)
19. Watanabe, H., Bouazizi, M., Ohtsuki, T.: Hate speech on Twitter: a pragmatic approach to collect hateful and offensive expressions and perform hate speech detection. IEEE Access 6, 13825–13835 (2018)
20. Bisht, A., Singh, A., Bhadauria, H.S., Virmani, J., Kriti: Detection of hate speech and offensive language in Twitter data using LSTM model. Recent Trends Image Signal Process. Comput. Vis. 243–264 (2020)
21. Al-Hassan, A., Al-Dossari, H.: Detection of hate speech in social networks: a survey on multilingual corpus. In: 6th International Conference on Computer Science and Information Technology, vol. 10, pp. 83–100 (2019)
22. Prokipchuk, O., Vysotska, V., Pukach, P., Lytvyn, V., Uhryn, D., Ushenko, Y., Hu, Z.: Intelligent analysis of Ukrainian-language tweets for public opinion research based on NLP methods and machine learning technology. Int. J. Mod. Educ. Comput. Sci. (IJMECS) 15(3), 70–93 (2023). https://doi.org/10.5815/ijmecs.2023.03.06
23. Ullmann, S., Tomalin, M.: Quarantining online hate speech: technical and ethical perspectives. Ethics Inf. Technol. 22, 69–80 (2020)

24. Thiago, D.O., Marcelo, A.D., Gomes, A.: Fighting hate speech, silencing drag queens? Artificial intelligence in content moderation and risks to LGBTQ voices online. Sex. Cult. **25**(2), 700–732 (2021)

25. Yadav, A.K., Kumar, M., Kumar, A., Shivani, Kusum, Yadav, D.: Hate speech recognition in multilingual text: Hinglish documents. Int. J. Inf. Technol. **15**(3), 1319–1331 (2023)

26. Roy, S.G., Narayan, U., Raha, T., Abid, Z., Varma, V.: Leveraging Multilingual Transformers for Hate Speech Detection (2021). arXiv:2101.03207

27. Mozafari, M., Farahbakhsh, R., Crespi, N.: A BERT-based transfer learning approach for hate speech detection in online social media. In: Complex Networks and Their Applications VIII: Volume 1 Proceedings of the Eighth International Conference on Complex Networks and Their Applications Complex Networks, vol. 8, pp. 928–940. Springer (2020)

28. Chiu, K.L., Collins, A., Alexander, R.: Detecting Hate Speech with GPT-3 (2021). arXiv:2103.12407

29. Fitria, T.N.: Artificial intelligence (AI) technology in OpenAI ChatGPT application: a review of ChatGPT in writing English essay. ELT Forum: J. Engl. Lang. Teach. **12**(1), 44–58 (2023)

30. Kwarteng, J., Perfumi, S.C., Farrell, T., Third, A., Fernandez, M.: Misogynoir: challenges in detecting intersectional hate. Soc. Netw. Anal. Min. **12**(1), 166 (2022)

31. Zannettou, S., ElSherief, M., Belding, E., Nilizadeh, S., Stringhini, G.: Measuring and characterizing hate speech on news websites. In: 12th ACM Conference on Web Science, pp. 125–134 (2020)

32. Kim, J., Wohn, D.Y., Cha, M.: Understanding and identifying the use of emotes in toxic chat on Twitch. Online Soc. Netw. Media **27**, 100180 (2022)

33. Rieder, B., Skop, Y.: The fabrics of machine moderation: studying the technical, normative, and organizational structure of perspective API. Big Data Soc. **8**(2), 20539517211046181 (2021)

34. Fortuna, P., Soler, J., Wanner, L.: Toxic, hateful, offensive or abusive? What are we really classifying? An empirical analysis of hate speech datasets. In: 12th Language Resources and Evaluation Conference, pp. 6786–6794 (2020)

35. Davidson, T.: Hate-Speech-and-Offensive-Language Dataset. https://github.com/t-davidson/hate-speech-and-offensive-language/blob/master/data/labeled_data.csv. Last accessed 21 June 2023

36. Ali, S.S.: BDA_Project_Hate_Speech_Detection Dataset. https://www.kaggle.com/code/shaikhsaadali/bda-project-hate-speech-detection. Last accessed 21 June 2023

37. Samoshyn, A.: Hate Speech and Offensive Language Dataset. https://www.kaggle.com/datasets/mrmorj/hate-speech-and-offensive-language-dataset. Last accessed 21 June 2023

38. Lees, A., Tran, V.Q., Tay, Y., Sorensen, J., Gupta, J., Metzler, D., Vasserman, L.: A New Generation of Perspective API: Efficient Multilingual Character-Level Transformers (2022). arXiv:2202.11176

39. Jahan, M.S., Oussalah, M.: A systematic review of hate speech automatic detection using natural language processing. In: Neurocomputing, p. 126232 (2023)

40. Schmidt, A., Wiegand, M.: A survey on hate speech detection using natural language processing. In: Proceedings of the Fifth International Workshop on Natural Language Processing for Social Media, pp. 1–10 (2017)

41. Biradar, S., Saumya, S., Chauhan, A.: Hate or non-hate: translation based hate speech identification in code-mixed Hinglish data set. In: IEEE International Conference on Big Data (Big Data), pp. 2470–2475. IEEE (2021)

42. What is FastText? https://fasttext.cc/. Last accessed 21 June 2023

43. Herwanto, G.B., Ningtyas, A.M., Nugraha, K.E., Trisna, I.N.P.: Hate speech and abusive language classification using FastText. In: International Seminar on Research of Information Technology and Intelligent Systems (ISRITI), pp. 69–72. IEEE (2019)

44. Sazany, E., Budi, I.: Deep learning-based implementation of hate speech identification on texts in Indonesian: preliminary study. In: International Conference on Applied Information Technology and Innovation (ICAITI), pp. 114–117. IEEE (2018)

45. Popova, I.: Top 10 Python Libraries for Machine Learning. https://light-it.net/blog/top-10-python-libraries-for-machine-learning/. Last accessed 21 June 2023

46. Luna, J.C.: Choosing Python or R for Data Analysis? An Infographic. https://www.datacamp.com/community/tutorials/r-or-python-for-data-analysis. Last accessed 21 June 2023
47. Malik, U.: Python for NLP: Working with Facebook FastText Library. https://stackabuse.com/python-for-nlp-working-with-facebook-fasttext-library/. Last accessed 21 June 2023
48. Bouzenia, I.: Train Python Code Embedding with FastText. https://medium.com/nerd-for-tech/train-python-code-embedding-with-fasttext-1e225f193cc. Last accessed 21 June 2023
49. Naïve Bayes Classifiers. https://www.ibm.com/topics/naive-bayes
50. K-Nearest Neighbors Algorithm. https://www.ibm.com/topics/knn. Last accessed 21 June 2023
51. What Is a Decision Tree? https://www.ibm.com/topics/decision-trees. Last accessed 21 June 2023
52. Sruthi, E.R.: Understand Random Forest Algorithms with Examples? https://www.analyticsvidhya.com/blog/2021/06/understanding-random-forest/. Last accessed 21 June 2023
53. Swaminathan, S.: Logistic Regression—Detailed Overview. https://towardsdatascience.com/logistic-regression-detailed-overview-46c4da4303bc. Last accessed 21 June 2023
54. Jain, P.: Basics of CountVectorizer. https://towardsdatascience.com/basics-of-countvectorizer-e26677900f9c. Last accessed 21 June 2023

Method for Counting Animals in Motion for the Milking Plant Information Systems

Pavlo Kulakov⬥, Volodymyr Kucheruk⬥, Tetiana Neskorodieva⬥,
Olena Semenova⬥, Roman Lishchuk⬥, Serhii Kontseba⬥,
Wiktoria Mankovska⬥, and Anna Kulakova⬥

Abstract If there are errors in the radio frequency identification of animals, during their movement to the group milking plant, information about the fact of the animals entry is lost. The number of the group milking plant stall strictly corresponds to the animal number in the queue, therefore, the information system server receives incorrect information about the correspondence of the animals numbers in the herd to the group milking plant stall numbers. Thus, the results of milking process measured parameters are being obtained with a false correspondence to the animals numbers in the herd. As a result, information related to all animals in the group is lost. To reduce the risk of information loss, group milking plants use means of counting animals during movement. Based on this, in order to obtain reliable information about the measured milking parameters of individual animals at group milking plants, it is necessary to ensure an accurate count of animals during their movement to the stall. Existing means of counting animals, which are based on video analysis, interruption or reflection of the optical radiation flow from animals during movement, do not always ensure their accurate counting. To detect the animals radio frequency identification errors at group milking plants, the method of counting animals is proposed, which is based on optimal linear filtering of the output signal of the animal photoelectric presence sensor. The implementation of the proposed method ensures an increase in the animal counting accuracy, which leads to the effective detection of

P. Kulakov (✉) · V. Kucheruk · T. Neskorodieva · R. Lishchuk · S. Kontseba · W. Mankovska
Uman National University of Horticulture, 1 Institutska Str., Uman 20305, Ukraine
e-mail: kulakovpi@gmail.com

S. Kontseba
e-mail: kontseba@meta.ua

O. Semenova
Vinnytsia National Technical University, 95 Khmelnytske Shosse Str., Vinnytsia 21021, Ukraine
e-mail: semenova.o.o@vntu.edu.ua

A. Kulakova
Vasyl' Stus Donetsk National University, 21 600-Richchia, Vinnytsia 21021, Ukraine

A. Semenov et al. (eds.), *Data-Centric Business and Applications*, Lecture
Notes on Data Engineering and Communications Technologies 195,
https://doi.org/10.1007/978-3-031-54012-7_16

371

radio frequency identification errors and an increase in the reliability of information about the measured milking parameters in the group milking plants information systems.

Keywords Counting animals · Information system · Milking plant · Movement of animals · Animal identification

1 Introduction

In radio frequency identification (RFID) systems, data exchange between the RFID-reader and the transponder is carried out using radio communication, the transponder has a unique code in its memory, which it transmits using the built-in radio transmitter when it enters the working area of the RFID-reader [1]. Radio frequency identification systems for animals are widely used in the agro-industrial complex [2], on modern dairy farms as part of information systems designed to determine the activity of animals [3], their localization [4], accounting, analysis and control of animal parameters and milk production technological process parameters [5, 6].

To reduce the impact of animals radio frequency identification errors during movement on the results of the livestock farms information systems [7], specialized means are used that ensure the counting of all animals, including those whose identification did not take place [8, 9]. Such means are implemented on the basis of video analysis, ultrasonic or photoelectric sensors of presence [10], the principle of operation of which is based on the interruption or reflection of the radiation flow by the animal. In most cases, when an animal passes through the working zone of the RFID-reader, its transponder code is read, after which the animal passes through the working zone of the animal presence sensor. Such a sensor consists of an emitter and a receiver of ultrasound or optical radiation. When an animal enters the working zone of the sensor, the radiation flow is interrupted or reflected, as a result, a pulse of a certain duration is formed at the output of the receiver, the presence of which is a sign of the animals passage [11]. The important factor for obtaining reliable information about the milk production technological process parameters in the group milking plants information systems is the accurate counting of animals during their movement to the stalls. Therefore, increasing the efficiency of existing methods of counting animals while moving is an actual task.

2 Setting the Task

In Fig. 1 the scheme is shown, which explains the influence of the animals radio frequency identification error on obtaining information about milking parameters when using information systems for group milking plants.

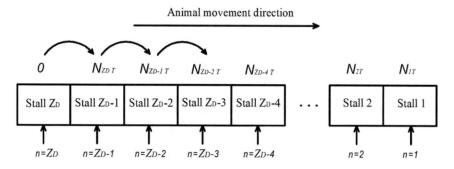

Fig. 1 The influence of the animals radio frequency identification error on obtaining information about milking parameters when using information systems for group milking plants

The group milking plant consists of two parallel parts, each part has Z_D milker workplaces, each milker workplace is equipped with the stall. After the animals enter, in the absence of the identification error, animal with number $n = 1$ in the queue, which has a transponder with code N_{1T}, will be in the stall number 1, animal with number $n = 2$, which has a transponder with code N_{2T}, will be in the stall number 2, and etc. In this case, there will be no loss of information about the milking parameters of animals in the information system. If, for example, the transponder is not identified when the animal with number $n = Z_D$[unknown template] passes through the RFID-reader working zone, the information system server will not be able to detect the presence of this animal in the stall with number $Z_D - 3$. As a result, the next animal in queue, with number $n = Z_D$[unknown template] and transponder code N_{Z_D-2T}, will be identified as being in stall with number $Z_D - 3$. Accordingly, each subsequent animal with a queue number greater than $n = Z_D - 2$ will be identified as being in stall with number $n - 1$, although in reality it is in stall with number n. The last animal of the group with number $n = Z_D$ in the queue will be in stall with number Z_D, but the value of its transponder code and number in the herd will not be determined by the server. Algorithmically, in this case, it is impossible to determine the number of the stall in which the unidentified animal is located.

Let determine the probability of losing information about milking parameters of the group of Z_D animals in the event of the identification error. Let denote by r the number of animals that were not identified in a group of Z_D animals. The probability that r unidentified animals will appear on one part of the milking plant is determined by the expression

$$p(r) = \frac{C_{d_S}^r}{C_{K_V}^{Z_D}} C_{K_V-d_S}^{Z_D-r}, \tag{1}$$

where

$$C_{d_S}^r = \frac{d_S!}{r!(d_S - r)!}; \tag{2}$$

$$C_{K_V}^{Z_D} = \frac{K_V!}{Z_D!(K_V - Z_D)!}; \tag{3}$$

$$C_{K_V - d_S}^{Z_D - r} = \frac{(K_V - d_S)!}{(Z_D - r)!(K_V - d_S - Z_D + r)!}, \tag{4}$$

where d_S is the average number of unidentified animals out of K_V animals in the herd served by one milker on one part of the milking plant.

According to the results of experiments conducted by the authors, when using a long-range RFID-reader, an average of 3% of moving animals on a group milking plant are not identified. In Fig. 2 shows the graph of the probability row of function (1) for a typical project of the group milking plant "Yalynka 2×8", which has two parts, eight stalls in each part, and which serves a herd of 600 heads, that is, for the given graph, $K_V = 300$, $d_S = 9$, $Z_D = 8$.

As follows from Fig. 2, the probability that $r = 0$ (all animals in the group are identified) is 0.74. The probability that at least one animal from the group will not be identified, and as a result all data on milking results of this animals group will be lost in the information system, is determined by the expression

$$p_G = 1 - p(0). \tag{5}$$

For the typical milking plant we are considering, the value of this probability is 0.26. Thus, the probability of losing data about a group of animals in one part of a group milking plant with existing methods of building such information systems is high.

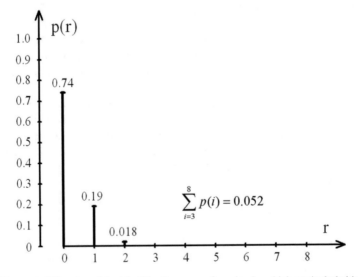

Fig. 2 The probability row of the identification error of r animals, which are included in a group of Z_D animals, for the typical group milking plant "Yalynka 2×8"

If there is one identification error in the group, the information about the milking parameters of animals on one of the parts of the milking plant is unreliable. It is impossible to identify a specific animal whose identification error led to the loss of the data of the entire group.

Based on this, in the absence of animal counting at group milking plants, there is a high risk of losing information about milking parameters due to identification errors. In the presence of animal counting, the loss of information is minimized, the further development of methods and algorithms for the implementation of animal counting is an important and actual task. The research task is to develop the method of counting animals when they enter a group milking plant, which is based on the use of optimal linear filtering of the output signal of the photoelectric animal presence sensor.

3 Research Results

In Fig. 3 shows the scheme of the animals movement along the RFID-reader antenna of the radio frequency identification system and the animal photoelectric presence sensor.

Each transponder code, received by the RFID-reader, corresponds to the pulse output signal of the animal presence sensor. If the radio frequency identification of one or more animals for certain reasons did not take place, the fact of their passage is still established. Thus, provided there is the means of detecting the animal passage, an unidentified animal is detected and its number in the queue is determined based on the counting results. In group milking plants, this makes it possible to prevent the loss of information about the entire group of animals.

The movement of animals is random, which is due to the peculiarities of their behavior. They can stop, push, move in the opposite direction, make chaotic movements, move in a direction perpendicular to the direction of the flow of animals, crouch, raise or lower their head. The typical time diagram of the animal presence sensor output signal during animals movement is shown in Fig. 4.

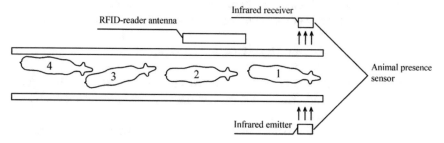

Fig. 3 Movement of animals along the RFID-reader antenna of the radio frequency identification system and the animal photoelectric presence sensor

Fig. 4 Typical time diagram of the animal presence sensor output signal during animals movement

In works [12, 13] based on experimental studies, it was established that the speed of animals movement is 0.4–0.6 m/s, and the average animal length is 2.6 m. When one animal passes through the working area of the animal presence sensor, the sensor generates a signal, the duration of which is determined by the expression

$$T_C = \frac{l_C}{v_C}.$$ (6)

where l_C is the animal length; v_C is the animal speed.

Due to the fact that the nature of animal movement depends on a large number of random factors that are weakly correlated, and there are no dominant factors among them, on the basis of the first limit theorem, it can be assumed that the duration of the impulse output signal of the animal presence sensor is a random variable that is distributed according to Gauss law with expected value T_{CM} and standard deviation σ_{CM}. Accordingly, the minimum T_{CMIN} and maximum T_{CMAX} value of the time it takes for one animal to pass the sensor working zone, determine the limits of this parameter in accordance with the three-sigma rule of thumb.

The mutual location of animals during their movement affects the reliability of their passage identification. The existing means [14, 15] of detecting the passage of an animal use the following criteria. If the movement of animals occurs evenly, with gaps between animals (animal 1 and 2 in Fig. 3), then each animal corresponds to a separate impulse output signal of the animal presence sensor. In this case, the criterion for identifying the animal passage is the fulfillment of the condition

$$T_C \in [T_{CMIN}; T_{CMAX}].$$ (7)

If the animals move one after another without a gap, or they are combined (see animals 3 and 4 in Fig. 3), then the criterion for identifying the passage of n_C animals through the working zone of the animal presence sensor is the fulfillment of the ratio

$$T_C \in [k_C n_C T_{CMIN}; k_C n_C T_{CMAX}]$$ (8)

where k_C is the constant coefficient, which takes into account the possible combination of animals during their movement and the different value of the individual animals movement speed.

Due to the fact that the coefficient k_C is not a constant value, but has a random character, in the case of combining animals during movement or in the case of their movement without a gap between them, an error in detecting the passage of the animal may occur. Based on the random nature of animal movement, the output signal of the animal presence sensor can be considered as a mixture of two signals—the useful pulse signal $s_M(t)$ and the random uncorrelated pulse signal $s_N(t)$. By detecting the signal $s_M(t)$, the passage of the animal is identified. As is known, the uncorrelated sequence $s_N(t)$ has a uniform energy spectrum, which gives reason to consider it as white noise.

To increase the reliability of detecting the passage of the animal during its movement, it is suggested to use optimal linear filtering. Let consider the principle of operation of the optimal linear filter, which ensures the maximization of the signal-to-noise ratio (SNR) between the peak value of the useful signal and the standard deviation of the noise signal. As is known, the complex energy spectrum of the signal $s_M(t)$ is determined by the expression

$$S_M(j\omega) = \int_{-\infty}^{+\infty} s_M(t)e^{-j\omega t}\,dt, \tag{9}$$

where ω is the cyclic frequency of spectral components; j—imaginary unit.

The noise signal $s_N(t)$ has the character of white noise with a uniform energy spectrum, which is determined by the expression

$$W_N(j\omega) = \int_{-\infty}^{+\infty} s_N(t)e^{-j\omega t}\,dt = W_0. \tag{10}$$

The peak value of the signal at the linear optimal filter output at the certain time t_0 is determined by the generalized expression

$$s_P(t_0) = \frac{1}{2\pi} \int_{-\infty}^{+\infty} S_M(\omega) K_{OPT}(\omega) e^{j(\omega t_0 + \varphi_M(\omega) + \varphi_K(\omega))}\,d\omega, \tag{11}$$

where $S_M(\omega)$ is the module of the complex energy spectrum of the useful signal; $K_{OPT}(\omega)$ is the module of the complex transfer characteristic of the filter; $\varphi_M(\omega)$ is the phase characteristic of the useful signal spectrum; $\varphi_K(\omega)$ is the phase-frequency characteristic of the optimal filter.

The root-mean-square (RMS) value of the noise signal at the linear optimal filter output is determined by the expression

$$\sigma_N = \sqrt{\frac{W_0}{2\pi} \int_{-\infty}^{+\infty} K_{OPT}^2(\omega)\,d\omega}, \tag{12}$$

Accordingly, the SNR between the peak value of the useful signal and the RMS value of the noise signal is determined by the expression

$$R_{SN} = \frac{|s_P(t_0)|}{\sigma_N} = \frac{\left| \frac{1}{2\pi} \int\limits_{-\infty}^{+\infty} S_M(\omega) K_{OPT}(\omega) e^{j(\omega t_0 + \varphi_M(\omega) + \varphi_K(\omega))} d\omega \right|}{\sqrt{\frac{W_0}{2\pi} \int\limits_{-\infty}^{+\infty} K_{OPT}^2(\omega) d\omega}}. \tag{13}$$

In accordance with the Cauchy-Buniakovsky inequality.

$$\left| \int\limits_{-\infty}^{+\infty} S_M(\omega) K_{OPT}(\omega) e^{j(\omega t_0 + \varphi_M(\omega) + \varphi_K(\omega))} d\omega \right|^2 \leq \int\limits_{-\infty}^{+\infty} S_M^2(\omega) d\omega \cdot \int\limits_{-\infty}^{+\infty} K_{OPT}^2(\omega) d\omega. \tag{14}$$

Then expression (13) can be represented in the form

$$R_{SN} = \frac{\left| \frac{1}{2\pi} \int\limits_{-\infty}^{+\infty} S_M(\omega) K_{OPT}(\omega) e^{j(\omega t_0 + \varphi_M(\omega) + \varphi_K(\omega))} d\omega \right|}{\sqrt{\frac{W_0}{2\pi} \int\limits_{-\infty}^{+\infty} K_{OPT}^2(\omega) d\omega}} \leq \sqrt{\frac{\int\limits_{-\infty}^{+\infty} S_M^2(\omega) d\omega}{2\pi W_0}}. \tag{15}$$

It follows from expression (15) that in order for the SNR to reach a maximum, the inequality must be transformed into equality. This will happen if the following conditions are met

$$\omega t_0 + \varphi_M(\omega) + \varphi_K(\omega) = 0; \tag{16}$$

$$K_{OPT}(\omega) = A_0 S_M(\omega), \tag{17}$$

where A_0 is the constant coefficient.

It follows that the transfer function of the optimal linear filter has the form [16]

$$K_{OPT}(j\omega) = A_0 S_M(\omega) e^{-j\omega t_0} e^{-j\varphi_M(\omega)}. \tag{18}$$

A complex-conjugate function with relatively to $S_M(\omega)$ is defined by the expression

$$S_M^*(\omega) = S_M(\omega) e^{-j\varphi_M(\omega)}. \tag{19}$$

Based on this, expression (18) can be represented in the form

$$K_{OPT}(j\omega) = A_0 S_M^*(\omega)e^{-j\omega t_0}. \tag{20}$$

The signal at the optimal linear filter output is determined by the expression

$$s_F(t) = \frac{1}{2\pi} \int\limits_{-\infty}^{+\infty} S_M(\omega) K_{OPT}(j\omega)e^{j\omega t}d\omega. \tag{21}$$

When the animal passes by, the useful output signal of the animal presence sensor is the rectangular pulse of a certain duration T_C, which depends on the movement speed and the animal length. The function that describes this signal is defined by the expression

$$s_M(t) = \begin{cases} 0, t < 0 \\ A, 0 \leq t \leq T_C, \\ 0, t > T_C \end{cases} \tag{22}$$

where A is the value of the animal presence sensor output signal.

As known, the transfer function of the linear optimal filter for the rectangular pulse is [16]

$$K_{OPT}(j\omega) = \frac{A_0}{j\omega}\left(1 - e^{-j\omega T_{OPT}}\right) \tag{23}$$

where T_{OPT} is the pulse duration which is matched to the optimal linear filter.

Consider Fig. 5, which shows the structural scheme of the linear optimal filter for a rectangular pulse of duration T_{OPT}, which has a transfer characteristic determined by expression (23).

Using the integrator, the input signal $s_M(t)$ is integrated, after which the integration result is fed to the inputs of the subtractor and the signal delay line for time T_{OPT}. At the subtractor output, the output signal $s_F(t)$ is formed, which is equal to the difference between the integral function of the input signal and the integral function of the input signal delayed for time T_{OPT}. If the linear optimal filter is matched to the input signal, i.e. $T_C = T_{OPT}$, the output signal of the filter is given by the expression

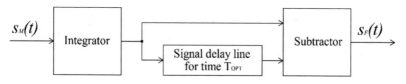

Fig. 5 Structural scheme of the linear optimal filter for rectangular pulse of duration T_{OPT}

$$s_F(t) = \begin{cases} 0, t < 0; \\ At, 0 \leq t < T_{OPT}; \\ A(T_{OPT} - t), T_{OPT} \leq t < 2T_{OPT}; \\ 0, t \geq 2T_{OPT}. \end{cases} \tag{24}$$

The peak value of the filter output signal at $T_C = T_{OPT}$ is equal to

$$s_{FOPT} = A \cdot T_{OPT}. \tag{25}$$

The SNR in the output signal of the optimal linear filter has a maximum value at $T_C = T_{OPT}$ and is determined by the expression

$$R_{SNOPT} = \frac{s_{POPT}}{\sigma_N} = \frac{A \cdot T_{OPT}}{\sigma_N}. \tag{26}$$

Let denote by T_L the duration of the optimal linear filter input signal, which is less than T_{OPT}. If $T_C = T_L < T_{OPT}$, the output signal of the optimal linear filter is determined by the expression

$$s_F(t) = \begin{cases} 0, t < 0; \\ At, 0 \leq t < T_L; \\ AT_L, T_L \leq t < T_{OPT}; \\ A(T_L - t), T_{OPT} \leq t < T_{OPT} + T_L; \\ 0, t \geq T_{OPT} + T_L. \end{cases} \tag{27}$$

The peak value of the filter output signal, in this case

$$s_{FL} = A \cdot T_L. \tag{28}$$

The SNR in the optimal linear filter output signal, if the input signal duration is less than the matched signal duration, is determined by the expression

$$R_{SNL} = \frac{s_{PL}}{\sigma_N} = \frac{A \cdot T_L}{\sigma_N}. \tag{29}$$

Let denote by T_H the duration of the optimal linear filter input signal, which is greater than T_{OPT}. If $T_C = T_H > T_{OPT}$, the output signal of the optimal linear filter is defined by the expression

$$s_F(t) = \begin{cases} 0, t < 0; \\ At, 0 \leq t < T_{OPT}; \\ AT_{OPT}, T_{OPT} \leq t < T_H; \\ A(T_H - t), T_H \leq t < T_{OPT} + T_H; \\ 0, t \geq T_{OPT} + T_H. \end{cases} \tag{30}$$

In this case, the peak value of the optimal filter output signal

$$s_{FH} = A \cdot T_{OPT}. \tag{31}$$

The SNR in the optimal linear filter output signal, if the input signal duration is more than the matched signal duration, is determined by the expression

$$R_{SN\,H} = \frac{s_{P\,H}}{\sigma_N} = \frac{A \cdot T_{OPT}}{\sigma_N}. \tag{32}$$

Operation time diagrams of the optimal linear filter for a rectangular pulse at different values of the input signal duration are shown in Fig. 6.

As follows from the analysis of expressions (24)–(32) and Fig. 6, at $T_C = T_L < T_{OPT}$ the output signal of the optimal filter is a trapezoidal function, the peak value of which is smaller than at $T_C = T_{OPT}$ and $T_C > T_{OPT}$. At $T_C = T_{OPT}$ and $T_C > T_{OPT}$, the peak value of the linear optimal filter output signal reaches its maximum value and is equal to $A \cdot T_{OPT}$. Based on this, the maximum SNR in the output signal of the optimal filter is reached at $T_C = T_{OPT}$ and remains the same at $T_C > T_{OPT}$. Thus, when detecting the passage of the animal, in order to achieve the maximum value of SNR, it is necessary to ensure the fulfillment of the expression

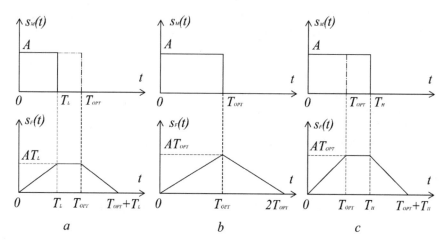

Fig. 6 Operation time diagrams of the optimal linear filter for a rectangular pulse at different values of the input signal duration: **a** $T_C = T_L < T_{OPT}$; **b** $T_C = T_{OPT}$; **c** $T_C = T_H > T_{OPT}$

$$T_{OPT} = T_{C\,MIN}. \tag{33}$$

The implementation of a linear optimal filter for the animal passage identification is provided using digital signal processing methods in real time mode. At certain moments of time t_1, t_2, t_3 ..., the instantaneous values of the animal presence sensor output signal $s_M(t_1)$, $s_M(t_2)$, $s_M(t_3)$ ... are determined, and the interpolating function $s_{AF}(t)$ is determined based on the obtained values. After that, the functions are calculated

$$s_{IAF}(t) = \int_0^t s_{AF}(t)dt, \tag{34}$$

and

$$s_{DIAF}(t) = s_{IAF}(t - T_{C\,MIN}). \tag{35}$$

The sign of the signal presence that corresponds to the animal passage is the fulfillment of equality

$$A \cdot T_{C\,MIN} = s_{IAF}(t) - s_{DIAF}(t). \tag{36}$$

As follows from the analysis of expression (30) and Fig. 6, at $T_C = T_H > T_{OPT}$, the time during which the output signal of the optimal linear filter has the maximum value $A \cdot T_{OPT} = A \cdot T_{C\,MIN}$ is determined by the expression

$$T_{CH} = T_H - T_{C\,MIN}. \tag{37}$$

Based on the results of measuring time T_{CH}, the passage of two or more animals is identified if they move with or without a gap between them. If the relation holds

$$T_{CH} \in [T_{C\,MIN}; T_{C\,MAX}], \tag{38}$$

a decision is made to identify the passage of one animal.

If T_{CH} exceeds the value of T_{CMAX}, and this occurs during the movement of animals with or without a gap between them, the decision to identify the passage of T_{CMAX} animals is made in the case of fulfilling the relation

$$T_{CH} \geq n_C T_{CM}. \tag{39}$$

In this case, the number of animals whose passage is identified is determined by the expression

$$n_C = \left[\frac{T_{CH}}{T_{CM}} \right]. \tag{40}$$

After that, the above procedures start from the beginning. The consequence of maximizing the SNR when using an optimal linear filter is an increase in the reliability of identification of the animal passage in comparison with existing methods, which ensures an increase in the accuracy of their counting.

4 Experimental Results

Experimental studies of the model sample of animal counting device, in which the developed method is implemented, were carried out on a "Yalynka 2×8" type group milking plant (ensures simultaneous milking of sixteen animals), the herd consisted of 327 animals. The PNL-4060-3 panel reader, manufactured by Allflex, was used for radio frequency identification of animals, WE-T3AD sensor, manufactured by Highly Electric, was used as an animal presence sensor. The optimal linear filter is implemented in software [17], based on the ATMega8 microcontroller produced by Microchip Technology, Inc. Acquires Atmel.

During the experiment, the BCR-01 device produced by the TDV "Bratslav" company [14], was installed on one part of the group milking plant, which includes the animal counting tool, that does not use optimal linear filtering of the animal presence sensor output signal. On the other part, the model sample of animal counting device was installed, which was implemented on the basis of optimal linear filtering of the animal presence sensor output signal.

Research was carried out for four days, milking on the farm was carried out twice a day. When using the BCR-01 device, 60–70% of unidentified animals were detected. With the help of the model sample of animal counting device, which provided optimal linear filtering of the animal presence sensor output signal, it was possible to detect 80–90% of unidentified animals.

5 Conclusions

The method for counting animals in motion is proposed, the model sample of animal counting device during their entry into a group milking plant was developed and researched. The developed method is based on the use of optimal linear filtering of the output signal of the animal presence sensor. Implementation of the developed method ensures an increase in the accuracy of animal counting. As a result, the number of detected radio frequency identification errors increases and the amount of lost information about animals milking parameters on group milking plants decreases. When using existing means of counting animals, it is possible to detect 60–70% of unidentified animals on group milking plants. As a result of the conducted experiments, it was found that when using the developed method of counting animals based on optimal linear filtering of the animal presence sensor output signal, this indicator increases to 80–90%.

References

1. RFID Journal [Electronic resource]. RFID journal LLC. http://www.rfidjournal.com
2. Saha, H.N., Chakraborty, S., Roy, R.: Integration of RFID and sensors in agriculture using IOT. In: AI, Edge and IoT-Based Smart Agriculture, pp. 361–372 (2021). https://doi.org/10.1016/B978-0-12-823694-9.00004-9
3. Ranches, J., De Oliveira, R.A., Vedovatto, M., Palmer, E.A., Moriel, P., Arthington, J.D.: Use of radio-frequency identification technology to assess the frequency of cattle visits to mineral feeders. Trop. Anim. Health Prod. **53**(3) (2021). https://doi.org/10.1007/s11250-021-02784-2
4. Achour, B., Belkadi, M., Saddaoui, R., Filali, I., Aoudjit, R., Laghrouche, M.: High-accuracy and energy-efficient wearable device for dairy cows' localization and activity detection using low-cost IMU/RFID sensors. Microsyst. Technol. **28**(5), 1241–1251 (2022). https://doi.org/10.1007/s00542-022-05288-7
5. Rudyk, A.V., Semenov, A.O., Kryvinska, N., et al.: Measuring quality factors of the radio-frequency system components using equivalent circuits. J. Comput. Electron. **20**, 1977–1991 (2021). https://doi.org/10.1007/s10825-021-01770-z
6. Noinan, K., Wicha, S., Chaisricharoen, R.: (2022). The IoT-based weighing system for growth monitoring and evaluation of fattening process in beef cattle farm. Paper presented at the 7th International Conference on Digital Arts, Media and Technology, DAMT 2022 and 5th ECTI Northern Section Conference on Electrical, Electronics, Computer and Telecommunications Engineering, NCON 2022, pp. 384–388. https://doi.org/10.1109/ECTIDAMTNCON53731.2022.9720346
7. Ding, X., Chen, L., Gong, Y.: An application of information collection method based on RFID and WSN technology in cow breeding. Paper presented at the Proceedings of 2019 IEEE 4th Advanced Information Technology, Electronic and Automation Control Conference, IAEAC 2019, pp. 2663–2666 (2019). https://doi.org/10.1109/IAEAC47372.2019.8998074
8. Kucheruk, V.Y., Palamarchuk, E.A., Kulakov, P.I.: The statistical models of machinery milking duration by group milking machines. East.-Eur. J. Enterp. Technol. **4**(4(70)), 13–17 (2014). https://doi.org/10.15587/1729-4061.2014.26287
9. Kucheruk, V.Y., Palamarchuk, E.A., Kulakov, P.I., Gnes, T.V.: The statistical model of mechanical milking duration of farmyard milking installation. East.-Eur. J. Enterp. Technol. **2**(4(68)), 31–37 (2014). https://doi.org/10.15587/1729-4061.2014.23120
10. Lancaster, P., Gyawali, P., Mavrogiannis, C., Srinivasa, S.S., & Smith, J.R.: Optical proximity sensing for pose estimation during in-hand manipulation. Paper presented at the IEEE International Conference on Intelligent Robots and Systems, 2022-October, pp. 11818–11825 (2022). doi:https://doi.org/10.1109/IROS47612.2022.9981692
11. Polikarpus, A., Grasso, F., Pacelli, C., Napolitano, F., De Rosa, G.: Milking behaviour of buffalo cows: entrance order and side preference in the milking parlour. J. Dairy Res. **81**(1), 24–29 (2014). https://doi.org/10.1017/S0022029913000587
12. Shepley, E., Lensink, J., Vasseur, E.: Cow in motion: a review of the impact of housing systems on movement opportunity of dairy cows and implications on locomotor activity. Appl. Anim. Behav. Sci. **230**, 105026 (2020). https://doi.org/10.1016/j.applanim.2020.105026
13. Su, L., Zhang, Y., Wang, J., Yin, Y., Zong, Z., Gong, C.: Segmentation method of dairy cattle gait based on improved dynamic time warping algorithm. Nongye Jixie Xuebao/Trans. Chin. Soc. Agric. Mach. **51**(7), 52–59 (2020). https://doi.org/10.6041/j.issn.1000-1298.2020.07.007
14. TDV "Bratslav" [Electronic resource]. TDV "Bratslav". https://www.bratslav.com/. Accessed 19 Aug 2023
15. Pallar LTD Co. & Musson Co. [Electronic resource]. Corporate website of companies Pallar LTD Co. & Musson Co. www.pallar.com.ua. Accessed 19 Aug 2023

16. Semenov, A., Voznyak, O., Osadchuk, O., Baraban, S., Semenova, O., Rudyk, A., Klimek, J., Orazalieva, S.: Development of a non-standard system of microwave quadripoles parameters. In: Photonics Applications in Astronomy, Communications, Industry, and High-Energy Physics Experiments 2019 (2019). https://doi.org/10.1117/12.2536704
17. Tsmots, I., Rabyk, V., Kryvinska, N., Yatsymirskyy, M., Teslyuk, V.: Design of the processors for fast cosine and sine Fourier transforms. Circuits Syst. Signal Process. **41**, 4928–4951 (2022). https://doi.org/10.1007/s00034-022-02012-8

Cryptocurrency as a Tool for Attracting Investment and Ensuring the Strategic Development of the Bioenergy Potential of Processing Enterprises in Ukraine

Vladyslav Malinov⬤, Viktoriia Zhebka⬤, Ivan Kokhan⬤, Kamila Storchak⬤, and Tymur Dovzhenko⬤

Abstract The article explores how cryptocurrency could revolutionize financing within Ukraine's agricultural sector, specifically for processing enterprises. This sector, pivotal for numerous economies, grapples with financial challenges stemming from conventional funding methods—high interest rates, demanding collateral prerequisites, and prolonged approval procedures. Blockchain-based cryptocurrencies offer distinct advantages: transaction transparency, reduced costs, and the capability to execute global payments sans intermediaries. These benefits could significantly aid agricultural businesses seeking operational optimization and investment influx. The article delves into cryptocurrencies' role in agriculture, discussing fundamental principles like Bitcoin, Ethereum, and their potential applicability in the sector. It particularly underscores the prospect of employing "smart contracts" to automate payments and risk management. Additionally, the authors address probable hurdles and constraints associated with implementing cryptocurrencies in agriculture, encompassing regulatory, educational needs for new technologies, and the risks tied to cryptocurrency price fluctuations. The study's primary findings underscore cryptocurrencies' vast potential in reshaping agriculture, opening new economic avenues for farmers and processing entities. Notably, cryptocurrencies could serve as a pivotal financing tool for innovative agricultural technology ventures, ultimately contributing to the sector's sustainable development. Overall, this article significantly contributes to understanding cryptocurrencies' role in agriculture, offering valuable

---

V. Malinov · V. Zhebka · I. Kokhan (✉) · K. Storchak · T. Dovzhenko
State University of Information and Communication Technologies, 7 Solomenskaya Str., Kyiv 03110, Ukraine
e-mail: ivan.v.kokhan@gmail.com

V. Zhebka
e-mail: viktoria_zhebka@ukr.net

K. Storchak
e-mail: kpstorchak@ukr.net

T. Dovzhenko
e-mail: timurdov@ukr.net

© The Author(s), under exclusive license to Springer Nature Switzerland AG 2024
A. Semenov et al. (eds.), *Data-Centric Business and Applications*, Lecture Notes on Data Engineering and Communications Technologies 195, https://doi.org/10.1007/978-3-031-54012-7_17

387

insights for researchers, practitioners, and policymakers intent on optimizing and modernizing the agricultural landscape.

Keywords Cryptocurrency · Bioenergy · Processing companies · Tokenisation · Efficient mechanism · Bioenergy potential · Economic situation · Processing industry · Agriculture · Alternative energy sources · Biogas

1 Introduction

The agricultural sector, including processing enterprises, has significant potential for bioenergy production. However, effective use of this potential requires significant investment. Traditional methods of financing may be insufficient or unavailable to provide the necessary resources.

One of the innovative strategies for attracting investment is the use of cryptocurrency. In the context of the strategic development of agricultural processing companies, cryptocurrency can act as a new financing mechanism. This will allow processing companies to receive investments for the development of bioenergy potential, bypassing traditional financial constraints.

Nevertheless, incorporating cryptocurrencies into agriculture presents its own set of hurdles. These encompass regulatory concerns, the necessity for education and familiarity with emerging technologies, and the exposure to risks linked with cryptocurrency fluctuations. Consequently, thorough investigation becomes imperative to guarantee the secure and efficient application of cryptocurrencies within the agricultural domain.

The issues of studying the attraction of investments and the prospects for the use of bioenergy are the basis of academic works of many scientists, in particular, thorough research in this area is carried out by Geletukha [1], Denysenko [2], Dankevych [3], Denysiuk [4], Khorak [5], Kudria [6, 7], Chetveryk and Stoian [8], Andrusevych [9] and others [10–16]. However, it is necessary to consider the possibility of using cryptocurrencies to finance the bioenergy potential of processing enterprises, creating a token as an investment tool based on practical experience on the example of Agri Coin and Solar Token.

The article aims to explore how cryptocurrency can be utilized as a means to draw investments and assess the potential for strategic growth concerning bioenergy within agro-industrial processing firms in Ukraine.

2 Presentation of the Main Material

The agricultural processing industry holds substantial potential for generating bioenergy. Yet, harnessing this potential necessitates considerable investments, often challenging to obtain through traditional funding avenues [17]. The article delves into an innovative strategy—leveraging cryptocurrency—to attract investments for developing the bioenergy capacity of processing enterprises.

As a novel approach within the strategic advancement of agricultural processing firms, employing cryptocurrency for financing emerges. Hence, we'll explore the intricacies of using cryptocurrency to solicit investments in the bioenergy potential of these enterprises.

This study aims to introduce the concept of cryptocurrency and elucidate its relevance in the agricultural domain. By highlighting cryptocurrencies' potential in agriculture, stakeholders can make informed decisions regarding the integration of this new financial instrument into their operational frameworks. Overall, this study aims to identify cryptocurrencies as a viable means of financing bioenergy projects in the agro-processing sector. We will analyse the advantages and limitations of cryptocurrency investments, optimal implementation strategies, and the broader implications of adopting this financing model. The gained knowledge can help agricultural enterprises to use cryptocurrencies strategically, in order to realise untapped bioenergy opportunities.

Often referred to as the future of finance, cryptocurrencies have made significant inroads into various sectors of the economy. Agriculture, the backbone of many countries, has been left out of this financial evolution. In this article, we will look at the basics of cryptocurrency and its growing importance in the agricultural sector.

A cryptocurrency, as defined by its nature, is a form of digital or virtual currency employing cryptography to safeguard against counterfeit operations [10]. The renowned Bitcoin, established in 2008 by the unidentified entity Satoshi Nakamoto, stands as one of the most prominent examples of cryptocurrencies [10].

These digital currencies operate on blockchain technology—a decentralized ledger system that encompasses all network transactions. This technology permits every participant to access a comprehensive database containing transaction histories. Transactions are logged as "blocks" and interconnected in a chain, establishing an exceedingly intricate system where altering any singular transaction record would necessitate modifying all successive blocks [11].

While Bitcoin retains its status as the most widely recognised cryptocurrency, thousands of alternative cryptocurrencies have emerged, including Ethereum, Ripple and Litecoin, each with different functionalities and applications.

Thus, cryptocurrencies are transforming finance through innovative encryption and decentralised elements. Understanding their growing role in agriculture is imperative for stakeholders who want to navigate this financial evolution.

Cryptocurrencies offer numerous benefits to agricultural sector participants, primarily by bypassing intermediaries and speeding up more affordable transactions. For example, a farmer in Letychiv can directly sell his produce to a European buyer via bitcoin, bypassing banks and currency exchange, thereby saving transaction costs [11]. The blockchain technology underpinning cryptocurrencies provides unprecedented transparency, as all transactions are recorded and available to network participants. This makes it possible to trace the origin of agricultural products, ensuring their ethicality and compliance with quality standards [12].

Many domestic farmers do not have access to traditional banking services. Cryptocurrencies offer these farmers the opportunity to participate in the global economy by providing a means to save, invest and expand their operations.

Ethereum and related cryptocurrencies allow for the creation of "smart contracts"—autonomous codes in which the terms of transactions are programmed. For agriculture, this can automatically make payments upon delivery confirmation, ensuring timely compensation and mitigating disputes [10].

Cryptocurrencies can serve as a risk management tool. Farmers can hedge against volatility or adverse weather conditions by using cryptocurrency-based financial products. In addition, an initial coin offering is one way for agricultural start-ups to raise capital, where investors buy tokens that are jointly owned [18].

The integration of cryptocurrencies into agriculture has huge potential, with benefits that include simplifying direct transactions, increasing transparency and improving traceability. However, as with any innovation, challenges remain, including regulation and promoting widespread understanding. As agriculture continues to evolve, cryptocurrencies are likely to play an increasingly prominent role [19].

Agriculture remains heavily dependent on traditional financing and payments, which contributes to systemic inefficiencies and hinders sustainable development. However, the innovative properties of cryptocurrencies have enormous potential to transform agricultural systems and unlock new economic opportunities. Cryptocurrencies can provide access to capital through initial public offerings, increase transparency in supply chain transactions, reduce remittance costs, facilitate seamless micropayments, and provide incentives through smart contracts. Overall, cryptocurrencies can accelerate growth, promote financial inclusion, and support the development of agricultural and food systems [20].

The objective of this research is to evaluate the existing utilization, advantages, obstacles, and forthcoming trajectories of cryptocurrencies within the agricultural domain. The primary focal points of investigation encompass decentralized finance (DeFi) applications in agro-technology, tokenized frameworks for cooperative farming, and the integration of blockchain into supply chain management within this sector. Our approach involves a comprehensive analysis—both from a technological and economic standpoint—to ascertain the potential transformative effects of cryptocurrency innovations on the agricultural industry, aiming to assess their capacity for expansion and evolution.

The use of cryptocurrencies in the agro-processing sector, which is critical to global food systems, has significant potential. This sector of the economy often faces difficulties in securing traditional financing, including the following issues:

- High interest rates, as financial institutions perceive agricultural enterprises as risky due to their dependence on weather conditions and market volatility.
- Collateral requirements that small businesses may not be able to meet.
- Lengthy loan approval processes that can impede timely financing.
- Limited financial products that may not meet the unique needs of processors.

- Geographical and infrastructural limitations that prevent access to services in rural areas.

Cryptocurrencies have advantages over traditional finance [21]:

- Global accessibility, enabling decentralised transactions without intermediaries.
- Reduced transaction costs by offsetting traditional banking systems.
- Instant processing of transactions, providing fast access to finance.
- Flexible financial products, such as decentralised finance platforms, tailored to processors. No collateral requirements, instead relying on project potential.
- Transparency and immutability of transactions through blockchain, which increases reliability.

To sum up, while conventional financing is still fraught with many challenges, cryptocurrencies are a promising solution for the agro-industrial processing sector. Their global reach, cost savings and innovative products can contribute to growth, but risks require careful implementation [22].

The main advantages of cryptocurrency financing include [23]:

- Access to global capital outside of regional banks.
- No collateral requirements.
- Lower barriers to participation.
- Partial ownership through tokenisation.
- Transparent accounting based on blockchain.
- Potential for higher profitability compared to traditional lending.

With improved access to capital, agro-processing companies will be able to invest in critical equipment, infrastructure, R&D work, and more to ensure growth, efficiency, and sustainability. Further research should quantify these benefits in developing optimal implementation strategies [24, 25].

The structured stages of using cryptocurrencies for financial innovation in the processing industry sector of the agro-industrial complex are offered to consideration.

Before a token can drive the digital finance industry, it must be conceptualised. This involves defining its raison d'être, technological basis, and maximum amount of issuance. The attractiveness lies in the simplicity brought by platforms such as the Ethereum ERC-20 [26, 27].

Advantages of creating tokens:

- Direct search for funds without intermediaries.
- Access to a global pool of potential investors.
- Attracting more commitment from stakeholders by giving them the opportunity to receive a part of the product through token ownership.

Initial Coin Offering (ICO)—this stage is characterised by the offering of freshly minted tokens to potential investors. A comprehensive technical document, similar to a project plan, describing the project's features, team skills and the anticipated roadmap is indispensable here [11].

Establishing a strong digital presence on social media, forming alliances with cryptocurrency leaders, and attending cryptocurrency-specific conferences can be beneficial.

The importance of allocating funds:

- Additional funding for project growth and daily operating costs.
- Branding and outreach activities.
- Ensuring all legal conditions.
- Create a fund for emergencies[12].
- Consistent updates, periodic audits and transparent dialogue is a key to building investor confidence.

Once the ICO has passed, the tokens can be listed on cryptocurrency exchanges, but this requires a maze of rigorous technical assessments and legal hurdles. At this stage, trading platforms such as Binance, Coinbase and Kraken come into play. Liquidity is paramount, not only to ensure ease of trading, but also to add a level of stability to the value of the token, deterring extreme price fluctuations.

The accumulated profits can either be returned to token holders, similar to dividends, or invested in the project. Striking the right balance is crucial for the longevity of the project.

Cryptocurrency financing could well be the boost the agro-processing industry needs. But, like all powerful tools, it requires careful and careful deployment, with an emphasis on legal compliance, transparent transactions and building stakeholder trust. With structured steps such as token creation and pre-sale, setting specific token rights, developing smart contracts and assessing pre-demand [11], the path to successful cryptocurrency-based financing in agriculture is well laid out.

Bioenergy, derived from organic materials, is a renewable and green source of energy. However, the establishment and operation of bioenergy projects are associated with significant costs. Let's consider a comprehensive financial analysis of such a project.

Potential return on investment during the payback period: Assume that a bioenergy project sells its energy at a price of $0.10 per kWh and produces 100,000 kWh annually:

- Annual revenue = $0,10/kWh * 10,000,000 kWh = $1,000,000.
- Net annual profit (from year 1) = Annual revenue − Total annual operating costs = $1,000,000-$650 000 = $350,000.
- Return on investment = (Net annual profit/Total initial investment) * 100% = ($350,000/$2,650,000) * 100% = 13.2%.
- Payback period = Total initial investment/Net annual profit = $2,650,000/ $350,000 = 7.57 years.

Although bioenergy projects require a significant upfront investment, they offer an acceptable return on investment of 13.2%. A payback period of approximately 7.57 years indicates that investors can recover their initial investment within this timeframe, making them a viable long-term investment.

The capital requirements for a bioenergy project can be broken down into key stages:

- R&D works and engineering—$5 million. Detailed feasibility study and design.
- Permitting and preparation for construction—$8 m, environmental and regulatory approvals.
- Construction—$12 million. Procurement, site development and construction of processing facilities.
- Working capital—$5 mn. Inventory, operation and testing.

Total upfront investment is $30 million over 2–3 years of development to full operation. The ICO, which aims to raise $35 million, provides flexibility and reserves. Annual operating costs are projected at $15 million, with potential revenue of $25 million once the plant reaches full capacity of 50,000 MWh per annum.

Potential barriers to adoption among investors and stakeholders:

- Lack of understanding means that many potential investors do not have a full understanding of how cryptocurrencies work, which leads to hesitation.
- Security concerns—high-profile hacking attacks and security breaches make investors wary.
- Lack of tangibility—the digital nature of cryptocurrencies, with no physical representation, can be a barrier for some traditional investors.
- Integration challenges—businesses and merchants face similar challenges in integrating cryptocurrency payment systems.

While cryptocurrencies offer many opportunities, it is important for investors and stakeholders to be aware of the risks and challenges involved. By understanding and navigating these aspects, it is possible to make informed decisions in the crypto environment.

The combination of cryptocurrencies and traditional sectors, especially agriculture, offers a glimpse into the future of finance. By looking at real-world examples of the introduction of tokenisation in the renewable energy sector, we will examine the examples and challenges faced by businesses that have ventured into this area.

AgriCoin is a process of financing crop diversification. AgriCoin was ingeniously conceived as a multi-faceted solution that uses blockchain technology and digital tokens to stimulate and catalyse crop diversification among Thai farmers. The main goal was to reduce the vulnerability of the country's agricultural sector to external shocks while improving the economic well-being of farming communities. It is in this context that we explore the profound impact of AgriCoin on agricultural diversification.

The main mechanism of AgriCoin was to provide farmers with a new form of currency, AgriCoin. These digital tokens had an intrinsic value in the entire agricultural ecosystem, as they could be exchanged for fiat currency or used to purchase a wide range of agricultural goods. By using this financial incentive, AgriCoin sought to stimulate behavioural change in the traditionally conservative agricultural sector.

In its first year of operation, AgriCoin managed to attract more than 10,000 farmers to the program. This significant growth reflects the ongoing enthusiasm of the farming community for diversification, driven by the attractiveness of AgriCoin. It has effectively bridged the gap between financial capability and agricultural diversification, thereby breaking the cycle of over-reliance on rice cultivation.

The empirical results have been impressive: crop diversification has increased by 15% in participating regions. This transformation in a sector deeply rooted in centuries-old practices and traditions highlights the potential of direct incentives to accelerate change. The success of AgriCoin is a clear testament to the effectiveness of innovative financial instruments in guiding traditional sectors to more sustainable and resilient development options.

Moreover, AgriCoin's triumph highlights the paramount importance of community engagement in the success of such innovative initiatives. The partnership between the cryptocurrency project and farming communities was built on trust, cooperation and shared goals. The active involvement of local stakeholders, agricultural cooperatives and government agencies has been crucial in facilitating the smooth introduction and adoption of AgriCoin.

SolarToken is a renewable energy financing platform launched in Germany to finance the installation of solar power plants in communities. Investors purchased SolarTokens during an ICO. The funds were used to install solar panels, and token holders received dividends from the sale of the electricity generated.

These examples highlight the potential of cryptocurrency models to unlock new capital, streamline operations and engage a wider range of stakeholders.

The methodology for attracting investments for Ukrainian processing enterprises through tokenisation and cryptocurrencies is presented as follows:

1. **Selection of an enterprise**. Select a processing company that needs investment and is ready to implement blockchain and cryptocurrency technologies.
2. **Analysis of assets**. Evaluate assets and determine which assets can be tokenised. These can be stocks, bonds or other securities that will be represented in the form of tokens on the blockchain.
3. **Select a blockchain platform**. Choose a suitable blockchain platform for asset tokenisation. Ethereum and BinanceSmartChain are popular choices for this purpose.
4. **Create tokens**. Create tokens that will represent the assets of the enterprise on the selected blockchain platform.
5. **Regulatory aspects**. Taking into account the legislation of Ukraine, comply with all necessary rules and regulations related to tokenisation and cryptocurrencies.
6. **Marketing and promotion**. Develop a marketing campaign to attract investors. Include information about the benefits of investing in tokenised assets.
7. **Conducting an STO** (Security Token Offering). Conduct an STO where investors can purchase company tokens in exchange for investments in cryptocurrency or fiat money.
8. **Token management**. Ensure effective token management, including the issuance, circulation and redemption of tokens.

9. **Enterprise development**. Use the attracted investments to develop the company and implement blockchain technologies in its activities.
10. **Regular reporting**. Provide investors with regular financial statements and information on the development of the enterprise.

This algorithm can help Ukrainian processing enterprises attract investment through tokenisation and cryptocurrencies accelerate development and ensure greater transparency in financial activities.

It is appropriate to describe in detail the fourth stage of the proposed methodology, which consists in the creation of tokens on the selected blockchain platform.

The algorithm for creating tokens on the blockchain can be described in detail as follows:

1. **Selecting a blockchain platform**. Determining the blockchain platform on which tokens will be created. The most common platforms are Ethereum, BinanceSmartChain, and others. The choice depends on the needs and circumstances of your project.
2. **Determining the type of token**. Choose the type of token you want to create. There are three main types of tokens: ERC-20, ERC-721 (e.g., useless tokens), and others, each of which has its own properties and purpose.
3. **Development of a smart contract**. Creation of a smart contract that will be responsible for the creation and management of tokens. It defines rules, such as the volume of issuance, token distribution, token transfer functions, etc.
4. **Programming the token functionality**. Defining the functionality that will be associated with your tokens. This may include the ability to transfer, store, exchange, or redeem tokens, depending on the purpose.
5. **Smart contract audit**. Conducting an audit of the smart contract to ensure that it is secure and does not contain any errors that could lead to the loss of tokens or other problems.
6. **Deploying the smart contract**. Placing the smart contract on the chosen blockchain platform. This includes gas costs (gas commissions) associated with this process.
7. **Token issuance**. Using a smart contract to issue tokens. Setting the volume and distribution of tokens according to your needs.
8. **Providing liquidity**. Ensuring the availability of places to exchange tokens for currency or other tokens on various crypto exchanges or through decentralized exchanges.
9. **Advertising and marketing**. Conducting an advertising campaign to attract users and investors to your token.
10. **Token management**. Development and implementation of a token management system, including redemption, circulation, and additional development.
11. **Reporting and compliance**. Ensure reporting and compliance with all relevant laws and regulations regarding cryptocurrencies and tokens.
12. **Support and development**. Providing ongoing support and development of your token to ensure its sustainability and value in the future.

The proposed algorithm can be adapted depending on the specific project and user needs.

The mathematical model for attracting investment in processing companies using cryptocurrencies can be complex and include various parameters, such as the amount of investment, the dynamics of cryptocurrency rates, market demand, etc.

Let's introduce the following notation:

- I is investment in dollars or other fiat currency.
- C is the total amount of tokens that will be issued to investors.
- T is the price of one token in dollars or other fiat currency at the beginning.
- I is the number of tokens issued at the beginning of the process (Initial Token Offering, ITO).
- R—the dynamics of token price growth (for example, daily percentage growth).

Then the following recursive calculation for the model can be built:

1. At the beginning of the process:

   - The number of tokens for sale (K) and their price (T) are announced.
   - The amount of investment (I) in fiat currency (for example, dollars) is determined.
   - Calculating the total number of tokens to be issued ($C = K + I/T$).

2. After each period (for example, a day):

   - A new token price is calculated ($T = T * (1 + R)$).
   - Investors can buy additional tokens (I), which are added to the total number of tokens (C).
   - The new investment volume (I) is calculated.

3. The process continues until a certain final investment amount is reached or until a certain number of issued tokens is reached.

This simplified model lacks consideration for various complexities, such as fluctuations in cryptocurrency rates, investor engagement, marketing expenses, commissions, and additional variables. Real-world projects necessitate more intricate models that encompass a multitude of factors. Moreover, acknowledging the substantial risks associated with investing in cryptocurrencies, it's crucial to conduct comprehensive analyses and seek guidance from financial and legal experts before employing such a model.

If we consider the model in the form of differential equations, it will look like this:

1. The differential equation for the price of tokens T:

$$dt/dT = R \cdot T$$

where R is the coefficient of growth of the token price.

2. The differential equation for the volume of investment I:

$$dt/dI = D - F$$

where D is the market demand for tokens and F is the fees and costs associated with the investment.

3. The differential equation for the total volume of tokens C:

$$dt/dC = TI$$

These equations determine how the price of tokens, the amount of investment, and the total amount of tokens change over time. The calculations can start with the initial values of the token price (T), the volume of investment (I), and the volume of tokens issued at the beginning (C).

This model allows us to analyse the dynamics of attracting investments in processing enterprises through cryptocurrency, taking into account market factors, commissions, and the distribution of tokens over time.

Initial coin offerings allow projects to raise significant seed capital in the early stages of development, while tokenised revenue sharing agreements provide ongoing incentives for project participants. Platforms such as DeFi and smart contracts cut out intermediaries, simplifying financing workflows. The transparent recording of transactions in blockchain ledgers also helps to build trust and engage stakeholders (Table 1).

After exploring bioenergy and mining, we introduced a novel concept termed "biomining." This process involves utilizing enterprise outputs to generate biogas, which, upon combustion, produces electricity utilized to power specialized mining equipment.

Biomining specifically involves creating bitcoins by solving intricate mathematical puzzles using hardware fueled by renewable energy derived from biomass—organic matter generated through photosynthesis (Fig. 1).

The primary conclusion drawn is the pioneering introduction and description of the "biomining" concept. While most researchers previously focused on utilizing state subsidies via the green tariff paradigm, aiming to reduce CO_2 emissions from waste processing and sell enhanced fertilizers, our study showcases an alternative perspective. It demonstrates that with modern technologies and optimal resource utilization, becoming a hub for cryptocurrency extraction via biomining is feasible.

Table 1 Factors affecting the liquidity of tokens

Factor	Description
Trade volume	Higher trading volumes generally indicate higher liquidity
Stock exchange lists	Tokens listed on multiple, reputable exchanges tend to have better liquidity
Marker utility	Lexemes with real-world applications or specific functionalities may have higher liquidity
Market sentiment	Positive news or project developments can increase trading activity and liquidity
Regulatory environment	Supportive rules can facilitate trade, while strict rules can discourage it
Token	Limited or restricted supply can affect demand and liquidity
Market depth	A balanced number of purchase and sales orders in the order book can indicate good liquidity

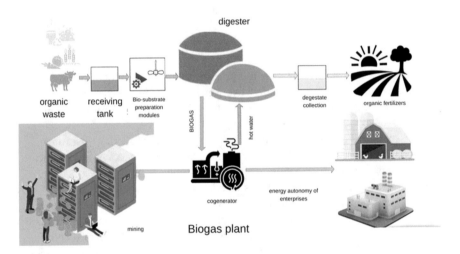

Fig. 1 Biomining is the process of generating bitcoins by solving a complex mathematical puzzle

This approach has the potential to elevate the country's GDP by trading crypto coins on exchanges.

The investigation, involving 20 companies utilizing biogas derived from waste processing for electricity generation, resulted in the production of 303.6 million kW of electricity in 2021. This electricity, sold to the grid, attracted approximately €36 million in state subsidies at a rate of €0.12 per kWh. Leveraging this electricity, Bitcoin cryptocurrency mining became feasible, with over 10 thousand devices potentially being utilized across these enterprises.

The outcome of mining in 2021 relies on several factors—mining complexity, Bitcoin's price, and electricity expenses. However, assuming consistent mining difficulty, an estimate can be made of the potential number of bitcoins mined based on generated electricity. Considering current bitcoin prices and electricity costs, the projected revenue from mining stands at approximately 93 million euros.

This significant income potential from electricity generated via biogas usage can offset the implementation costs of a biomining system—a substantial investment for businesses. This generates added economic advantages for enterprises, rendering biomining an appealing technology for harnessing bioenergy potential.

Conclusively, this study underscores the efficacy of biomining as a mechanism for optimizing the bioenergy potential within agricultural processing enterprises. Utilizing electricity from biogas for bitcoin mining presents an innovative application with considerable income potential for enterprises. Although challenges exist, the promising benefits warrant further exploration of biomining.

From an analysis of 20 enterprises utilizing biogas derived from waste processing to generate electricity, a substantial output of 303.6 million kW of electricity was achieved in 2021. This electricity was successfully sold to the grid, garnering approximately €36 million in state subsidies at a rate of €0.12 per kWh. Utilizing this electricity, Bitcoin mining became feasible. Based on the enterprises' capacities, over 10 thousand bitcoin miners were required for the mining operation in 2021. The outcomes of this mining, depending on Bitcoin's price, project significant revenues:

With Bitcoin at a price of 40.2 thousand euros in January 2022, the mining yielded over 119 million euros.

At a Bitcoin price of 20.1 thousand euros in May 2022, the mining garnered more than 59 million euros.

Considering taxation rates of 5 and 1.5%, the respective revenues to the state budget would amount to 7.7 million euros and 3.8 million euros. Thus, our dissertation research has established an effective mechanism for leveraging the bioenergy potential of processing enterprises. This mechanism not only diminishes the reliance on green tariff revenues but also generates additional income for the state budget. Such a mechanism can be extended to solar and wind energy producers, as well as the construction of new biogas plants, ultimately channeling produced electricity into cryptocurrency mining.

Exploring a correlation model between general grid sales under the green tariff and cryptocurrency mining, considering the number of mining nodes and Bitcoin value, could yield valuable insights. Developing this model could offer a comprehensive understanding of the potential advantages of biomining technology for businesses. Specifically, this study delves into correlating investment with five variables relevant to biomining: the green tariff, cryptocurrency production at various Bitcoin price points, electricity generation, and ASIC quantities (Table 2).

The initial variable under examination is the green tariff, signifying the rate at which companies sell biogas-generated electricity to the grid. The subsequent variables encompass cryptocurrency mining at distinct Bitcoin prices—20.1 thousand euros in May 2022 and 40.2 thousand euros in January 2022. The third variable measures electricity production in millions of kW during 2021, reflecting the output of electricity generated through biomining technology by enterprises. The fourth variable accounts for the number of ASICs in millions, symbolizing the computational power utilized for cryptocurrency mining.

Table 2 Summary data for correlation and regression analysis of the influence of factor variables X1, X2, X3, X4, X5

	Investments (Y)	Green tariff (X1)	Cryptocurrency mining at BTC price = 20.1 thousand euros May 2022 (X2)	Cryptocurrency mining at BTC price = 40.2 thousand euros January 2022 (X3)	Generated electricity million kW 2021 (X4)	Number of asics million (X5)
Teofipol Energy Company LLC	40.00	10.02	16.40	32.79	83.5	0.00297365
Vinnytsia Poultry Farm LLC	27.00	6.85	11.21	22.43	57.1	0.00203348
Korsun Eco Energy LLC	18.00	4.55	7.44	14.88	37.9	0.00134972
PJSC Oril leader	15.00	3.86	6.32	12.65	32.2	0.00114672
Gorodische Pustovarivske LLC	12.90	3.25	5.32	10.64	27.1	0.0009651
LLC Agrofirma im Chkalova	12.30	3.10	5.07	10.13	25.8	0.0009188
LLC Józefo-Nikolaivska bioenergy	11.00	2.77	4.54	9.07	23.1	0.00082265
Clear Energy Odesa LLC	8.00	2.03	3.32	6.64	16.9	0.00060185
LNK LLC	7.80	1.98	3.24	6.48	16.5	0.00058761
Clear Energy LLC	6.30	1.58	2.59	5.18	13.2	0.00047009
Clear Energy Kherson LLC	5.10	1.28	2.10	4.20	10.7	0.00038105
Biogas Ukraine LLC	4.00	1.02	1.67	3.34	8.5	0.00030271

(continued)

Table 2 (continued)

	Investments (Y)	Green tariff (X1)	Cryptocurrency mining at BTC price = 20.1 thousand euros May 2022 (X2)	Cryptocurrency mining at BTC price = 40.2 thousand euros January 2022 (X3)	Generated electricity million kW 2021 (X4)	Number of asics million (X5)
Komertsbudplast LLC	3.70	0.95	1.55	3.10	7.9	0.00028134
Clear Energy Kremenchuk LLC	2.80	0.72	1.18	2.36	6	0.00021368
Lancast LLC	2.20	0.58	0.94	1.89	4.8	0.00017094
Energo Sich LLC	2.05	0.52	0.84	1.69	4.3	0.00015313
Clear Energy Chernihiv LLC	2.00	0.50	0.82	1.65	4.2	0.00014957
Tis Eco LLC	3.50	0.50	0.82	1.65	4.2	0.00014957
AEO Energy LLC	2.90	0.43	0.71	1.41	3.6	0.00012821
PJSC Ecoprod	5.40	0.43	0.71	1.41	3.6	0.00012821
Total	144.2	36.4	59.6	119.2	303.6	0.01081197

To construct the correlation model, data were gathered from various businesses employing biomining technology, capturing information on investment, the green tariff, cryptocurrency mining specifics, electricity production, and ASIC quantities. A correlation analysis was conducted to gauge the strength and direction of the relationships between these variables.

The outcomes of the correlation analysis demonstrated positive correlations between investment and each variable. Firstly, a positive correlation emerged between investment and the green tariff, suggesting that higher feed-in tariffs drive greater investment in biomining technology due to increased revenue potential for enterprises.

Likewise, positive correlations were observed between investment and cryptocurrency production at different Bitcoin values, as well as investment and electricity production. Elevated Bitcoin values and increased electricity production both contribute to greater investment in biomining technology, reflecting heightened revenue prospects for businesses.

Additionally, a positive correlation was identified between investment and the number of ASICs utilized. Greater computational power for cryptocurrency mining via increased ASIC quantities corresponds to amplified investment in biomining technology, offering augmented revenue potential for enterprises.

A regression model serves as a crucial statistical tool, elucidating the relationship between a dependent variable and one or more independent variables. In this model, the dependent variable signifies the outcome of a function of the independent variables (Table 3).

Within this context, the multiple coefficients of determination, known as R-squared, serve to gauge the model's fitting accuracy to the dataset. Ranging between 0 and 1, a value of 1 signifies a flawless alignment of the model with the data. In this instance, the R-squared value of 0.9916499 underscores the model's high suitability.

Moreover, the normalized R-squared, particularly significant when multiple independent variables are present, attests to the model's alignment with actual data. The value of 0.991186005 further emphasizes the model's strong alignment with the real dataset, reliant on the correlation of the independent variables.

The standard error portrays the model's precision, derived from the regression's standard errors. This value, 0.90495824, aids in establishing confidence intervals and comparing diverse regression models.

Table 3 Regression statistics indicators

Regression statistics	Value
Multiple R	0.995816198
R-square	0.9916499
Normalized R-squared	0.991186005
Standard error	0.90495824
Observations	20

The count of observations, numbering 20 in this case, holds significance in determining the model's accuracy and alignment with the dataset. In summary, a regression model proves invaluable for analyzing variable relationships and foreseeing dependent variable values based on known independent variables. The findings derived from a regression model offer insights instrumental in shaping management and decision-making strategies.

A typical regression model involves two key types of variables: dependent and independent. The dependent variable is the quantity slated for prediction or estimation, while independent variables are factors capable of influencing the dependent variable.

Linear regression stands as the most common form of the regression model, presuming the dependent variable to be a linear function of independent variables. In this scenario, least squares methodology determines the regression model's parameter values by minimizing the sum of squared deviations between observed and predicted dependent variable values.

Variance analysis, a statistical method, enables the assessment of factor influence on a variable by comparing mean values. Presented below is a table showcasing the outcomes of variance analysis for a model encompassing both regression and residuals.

The variance analysis results reveal that the regression model comprises one degree of freedom, yielding a sum of squares (SS) value of 1750.637611. Conversely, the residual model consists of 18 degrees of freedom, resulting in an SS value of 14.74108948. The cumulative SS for the entire model stands at 1765.3787, accounting for the summation of regression and residual SS.

The F-statistic serves as a tool to gauge the statistical significance of the regression. It's calculated by dividing the regression mean square (MS) by the residuals mean square (MS). In this scenario, the F-value amounts to 2137.66269. Despite the remarkably small F-statistic value (3.67399E-20), it signifies the statistical significance of the regression model, implying a substantial effect of the regression factor on the dependent variable.

3 Conclusions

The article delves into cryptocurrencies' role in attracting investments and fostering the strategic advancement of bioenergy potential within agro-industrial processing firms in Ukraine. Specifically focusing on processing enterprises, the agricultural sector exhibits substantial promise in bioenergy production. Introducing cryptocurrencies as a means of financing stands as a fresh and innovative strategy to draw investment into the agricultural sector's bioenergy potential. Cryptocurrencies bring forth several advantages for agriculture, such as transaction transparency, reduced costs, and direct access to global markets. Blockchain technology and cryptocurrencies wield significant potential in revolutionizing agricultural systems, advancing sustainability, and fostering financial inclusivity. The exploration encompasses the utilization of blockchain-driven smart contracts to streamline and enhance agricultural transactions. Moreover, the analysis scrutinizes cryptocurrencies' impact on the economic stability and development of agro-industrial processing enterprises. The instances provided underscore the relevance of employing cryptocurrencies as an effective mechanism for harnessing the bioenergy potential inherent in agro-industrial processing enterprises.

References

1. Geletukha, G., Dragniev, S., Kucheruk, P., Matvieev, Y.: Practical guide on the use of biomass as fuel in the municipal sector of Ukraine (for representatives of the agricultural sector). https://www.ua.undp.org/content/ukraine/en/home/ (2017). Accessed 01.05.2023
2. Denysenko, V.O.: Assessment of the Potential of Biomass in Ukraine. Agrosvit, №24, pp. 84–89 (2019)
3. Dankevych, A., Perevozova, I., Nitsenko, V., Lozinska, L., Nemish, Y.: Effectiveness of bioenergy management and investment potential in agriculture: the case of Ukraine. In: Koval, V., Olczak, P. (eds.) Circular Economy for Renewable Energy. Green Energy and Technology. Springer, Cham (2023). https://doi.org/10.1007/978-3-031-30800-0_6. Accessed: 07.05.2023
4. Horák, J., Bilan, Y., Dankevych, A., Nitsenko, V., Kucher, A., Streimikiene, D.: Bioenergy production from sunflower husk in Ukraine: potential and necessary investments. J. Bus. Econ. Manag. **24**(1), 1–19 (2023). https://doi.org/10.3846/jbem.2023.17756. Accessed 10.05.2023
5. Kudria, S.O.: Atlas of Energy Potential of Renewable Energy Sources of Ukraine, 55 p. ViolPrint LLC, Kyiv (2008)
6. Kudria, S.O.: Electricity potential of renewable energy sources in Ukraine. In: Renewable Energy and Energy Efficiency of the XXI Century: Materials of the XXI International Conference, Kyiv, 14–15 May 2020, pp. 26–33
7. Chetveryk, G.O.: Energy-efficient conversion of liquid waste of biomass gasification in a biogas plant, 160 p. Ph.D. in Technical Sciences: Speciality: 05.14.08. National Academy of Sciences of Ukraine, Kyiv (2018)
8. Andrusevych, A., Andrusevych, N., Kozak, Z.: European Green Deal: Opportunities and Threats for Ukraine. https://dixigroup.org/storage/files/2020-05-26/european-green-dealwebfinal.pdf (2020). Accessed 01.05.2023
9. Malinov, V., Zhebka, V., Zolotukhina, O., Franchuk, T., Chubaievskyi, V.: Biomining as an effective mechanism for utilizing the bioenergy potential of processing enterprises in the agricultural sector. In: CEUR Workshop Proceedings this link is disabled, vol. 3421, pp. 223–230. https://ceur-ws.org/Vol-3421/short8.pdf (2023). Accessed 06.07.2023
10. Hu, Z., et al.: High-speed and secure PRNG for cryptographic applications. Int. J. Comput. Net. Inf. Secur. **12**(3), 1–10 (2020). https://doi.org/10.5815/ijcnis.2020.03.01(accessed:07.07. 2023)
11. Obushnyi, S., et al.: Autonomy of Economic Agents in Peer-to-PeerSystems, Cybersecurity Providing in Information and Telecommunication Systems, vol. 3288, pp. 125–133 (2022)
12. Official website of the State Statistics Service of Ukraine. http://www.ukrstat.gov.ua/. Accessed 01.05.2023
13. The official website of the NEURC "Information on alternative energy facilities that have been granted a feed-in tariff as of 01.01.2020". http://www.nerc.gov.ua/data/filearch/elektro/energo_pidpryem/stva/stat_zelenyi-taryf.01-2020.pdf. Accessed 01.05.2023
14. Regulation 2019/1009 on the regulation of the EU market for fertilisers. Regulation (EU) 2019/1009 of the European Parliament and of the Council of 5 June 2019. https://eur-lex.europa.eu/eli/reg/2019/1009/oj. Accessed 01.05.2023

15. Statistical report of the European Biogas Association: European overview for 2019. European Biogas Association Statistical Report: 2019 European Overview. https://www.europeanbiogas.eu/eba-statistical-report-2019european-overview/. Accessed 01.05.2023
16. Kokhan, I., Kudryavtsev, S., Merzhyievskyi, D.: Preparation of aurum chloride complex and its use for industrial acetylene hydration catalyst process. East.-Eur. J. Enterp. Technol. **3**(6 (117)), 14–22 (2022). https://doi.org/10.15587/1729-4061.2022.260279
17. Stoian, O.Yu.: State Regulation of Renewable Energy Development in Ukraine: Theory, Practice, Mechanisms: Monograph, 387 p. Mykolaiv (2014)
18. Ahmed, W., Rasool, A., Javed, A.R., Kumar, N., Gadekallu, T.R., Jalil, Z., Kryvinska, N.: Security in next generation mobile payment systems: a comprehensive survey. IEEE Access **9**, 115932–115950 (2021). https://doi.org/10.1109/ACCESS.2021.3105450
19. Kryvinska, N., Kaczor, S., Strauss, C., Greguš, M.: Servitization—Its Raise through Information and Communication Technologies. Lecture Notes in Business Information Processing, pp. 72–81 (2014). https://doi.org/10.1007/978-3-319-04810-9_6
20. Kryvinska, N., Zinterhof, P., van Thanh, D.: An analytical approach to the efficient real-time events/services handling in converged network environment. In: Network-Based Information Systems, pp. 308–316 (2007). https://doi.org/10.1007/978-3-540-74573-0_32
21. Ivanukh, R.A., Dusanovsky, S.L., Bilan, E.M.: Agrarian Economy and Market, 305 p. "Zbruch", Ternopil (2003)
22. Gustavsson, J., Cederberg, C., Sonesson, U., van Otterdijk, R., Meybec, A.: Global Food Losses and Food Wastes—Extent, Causes and Prevention. FAO, Rome (2011)
23. REN21: Renewables 2013 Global Status Report, p. 177. REN21 Secretariat, Paris (2013)
24. Smeets, E.M.V., Faaij, A.P.C., Lewandowski, I.M., Turkenburg, W.C.: A bottom-up assessment and review of global bioenergy potentials to 2050. Energy Combust. Sci. **33**, 56–106 (2007)
25. IEA Bioenergy: A Sustainable and Reliable Energy Source. Main Report. International Energy Agency, Paris (2009)
26. Kampman, B., Bergsma, G., Schepers, B., Croezen, H., Fritsche, U.E., Henneberg, K., et al.: BUBE: Better Use of Biomass for Energy. Background Report to the Position Paper of IEA RETD and IEA Bioenergy, p. 151. CE Delft/Öko-Institut, Darmstadt (2010)
27. Lamers, P., Hamelinck, C., Junginger, M., Faaij, A.: International bioenergy, trade—a review of past developments in the liquid biofuels market. Renew. Sustain. Energy Rev. **15**(6), 265–267 (2011)

Printed in the United States
by Baker & Taylor Publisher Services